DINOSAUR ODYSSEY

DINOSAUR ODYSSEY

Fossil Threads in the Web of Life

Scott D. Sampson

UNIVERSITY OF CALIFORNIA PRESS

Berkeley   Los Angeles   London

University of California Press, one of the most distinguished university presses in the United States, enriches lives around the world by advancing scholarship in the humanities, social sciences, and natural sciences. Its activities are supported by the UC Press Foundation and by philanthropic contributions from individuals and institutions. For more information, visit *www.ucpress.edu*.

University of California Press
Berkeley and Los Angeles, California

University of California Press, Ltd.
London, England

© 2009 by Scott D. Sampson
First Paperback Printing 2011

Library of Congress Cataloging-in-Publication Data

Sampson, Scott D.
    Dinosaur odyssey : fossil threads in the web of life / Scott D. Sampson.
        p. cm.
    Includes bibliographical references and index.
    ISBN 978-0-520-26989-7 (pbk.: alk paper)
1. Dinosaurs. 2. Paleontology. I. Title.

    QE 861.4.S26 2009
    567.9—dc22   2009006150

Manufactured in Canada
16   15   14   13   12
10   9   8   7   6   5   4   3   2

The paper used in this publication meets the minimum requirements of ANSI/NISO Z39.48-1992 (R 1997) (*Permanence of Paper*).

Cover illustration: The horned dinosaur *Achelousaurus horneri*. Painting by Michael Skrepnick.

For my mother, Catherine June Sampson,
who ignited and fueled a young boy's passion for dinosaurs

CONTENTS

Plates follow page 148

FOREWORD

Philip J. Currie

*Professor and Canada Research Chair
of Dinosaur Paleobiology, University of Alberta*

I was only 6 years old when I "dug up" my first dinosaur from the inside of a cereal box. The plastic model inspired my imagination in a powerful way that led to regular visits to the dinosaur galleries at the Royal Ontario Museum in Toronto, Canada. Several times a week, I would go to Sixteen Mile Creek near my home to scramble up and down the cliffs of Ordovician sediments, collecting marine invertebrate fossils while I fantasized about discovering dinosaurs. I read (and reread) every book that was available to me about any fossils from anywhere. After reading *All about Dinosaurs* by Roy Chapman Andrews when I was 11 years old, I knew that I wanted to be a dinosaur hunter. Such is the power of the written word. Unfortunately, in the 1950s and 1960s, there were relatively few printed words to meet my insatiable appetite for detailed information on dinosaurs. Dinosaurs were popular with the public, but little scientific research was being done on them. As a substitute, I filled in the void by reading comic books, science fiction stories, and anything else that even remotely mentioned a dinosaur.

Scientific research on dinosaurs reached a watershed in 1962 when a young scientist named John Ostrom discovered specimens of a small meat-eating dinosaur in Montana. The discovery went virtually unnoticed at first, but when his influential publication naming *Deinonychus antirrhopus* appeared in 1969, things really started to happen. An article by Robert T. Bakker in *Scientific American* in 1975 made it official with its title: "Dinosaur Renaissance."

Throughout the 1970s, publication of scientific papers increased every year, and popular books kept in step with the new dinosaur discoveries. In 1976, I took my first job as a dinosaur collector and researcher for the Provincial Museum of Alberta in Edmonton. My influences were many, but one book published in 1977 in particular characterized my approach. *A Vanished World: The Dinosaurs of Western Canada* by Dale A. Russell, illustrated with magnificent paintings by Eli Kish and spectacular photos by Susanne Swibold, integrated a broad spectrum of sciences to bring the dinosaurian world to life. A curator at the National Museum of Natural Sciences (now the Canadian Museum of Nature) in Ottawa, Russell was ahead of his time in developing multidisciplinary research programs. I followed his lead in the creation of the curatorial and research programs and the displays of the Royal Tyrrell Museum of Palaeontology in Drumheller.

It is difficult to assess the importance of any single publication. There are books that I consider poor in terms of scientific accuracy and presentation. Yet, if any of these books captures the imagination of even one person, who am I to complain? On the other hand, many good books have been written about dinosaurs in recent years. Some that I have considered excellent have gone out of print in record time and have probably had little influence on the public perception of our profession. Others are narrowly focused but fill a valuable niche. But I have never seen another volume that is as broad-based, inclusive, and multidisciplinary as Russell's book. At least, not until I read Scott Sampson's *Dinosaur Odyssey*.

Dinosaur Odyssey looks not just at dinosaurs but also at the myriad life-forms that shared their ecosystems, from bacteria to birds. This is done deliberately to show how life-forms interact to form complex, interdependent systems. But *Dinosaur Odyssey* goes beyond Russell's book by exploring the science of paleontology alongside the dynamic world of dinosaurs.

I first met the author of this book more than 20 years ago when he was an undergraduate student. Even then, it was evident that Scott Sampson would become a noteworthy dinosaur paleontologist. His interests were broad, and like a kid in a candy shop, he found it difficult to focus on a single direction. Luckily, Jack Horner of the Museum of the Rockies had an ideal project that encompassed the description of two new species of horned dinosaurs. Not content to just name the animals and work out their relationships, Scott pushed the boundaries to try to unravel the aspects of their biology that made them so unique. The resulting doctoral thesis was exemplary and led to the publication of several influential papers.

One of the most successful and fascinating groups of animals that ever evolved, dinosaurs are the heart and soul of this book. Scott has succeeded in translating their history into an epic tale. They appeared late in the Triassic and dominated the planet for all of the Jurassic and Cretaceous periods. In fact, their living descendants—birds—are still more diverse than mammals. Residing on the same branch of the family tree as dinosaurs, birds number more than 10,000 species, compared with 6,000 species of reptiles and amphibians, and 4,000 species of mammals. Nevertheless, these living dinosaurs are numerically overwhelmed by fish, insects, and other groups of organisms.

Although dinosaurs dominated the Mesozoic world by virtue of their size and diversity, they were likewise overwhelmed by smaller denizens. Like us, they were pestered by mosquitoes, flies, and ticks. And while many people assume that mammals did not exist or were rare as long as dinosaurs ruled the Earth, there always would have been many more furballs around than giant dinosaurs. Certain ecosystems were likely home to a greater diversity of mammals than of dinosaurs. Mammals were common enough for small dinosaurs like *Sinosauropteryx* to hunt and eat (as we know because one specimen of this remarkable little feathered dinosaur from northeastern China has two mammals in its stomach). Was it always the mammals that got picked on? Apparently not, because from the same part of the world a dog-sized mammal known as *Repenomamus* was discovered with baby dinosaurs in its stomach!

The interactions between dinosaurs, mammals, and all other forms of life comprise a major theme of this book. Evolution and extinction require that the actors in the theater of life are always changing, but fundamentally the play has remained constant. Dinosaurs put on some of the most spectacular versions of "the play of life" during Mesozoic times, and if we hope to understand and learn from their performances, we clearly need to be good understudies.

This book succeeds admirably in showing how the study of dinosaurs is approached by paleontologists today and in documenting the state of this popular science early in the twenty-first century. Exciting aspects of dinosaurian biology are opening up through the use of new tools and techniques by highly creative people. *Dinosaur Odyssey* will not be the final word on dinosaurs, nor does it aspire to be. As much as we know about these animals, there is so much more that we need to learn. Thousands of species remain to be discovered, among which will undoubtedly be forms bizarre enough to surprise even the most experienced dinosaur hunters.

Teachers from kindergarten to graduate school have long realized that our fascination with dinosaurs makes them an effective vehicle for communicating the nature of deep time, the epic of evolution, and the inevitability of extinction. Few general books on dinosaurs incorporate a more readable, integrated approach than that employed by Scott Sampson. He writes engagingly, telling personal stories of relevance and interest and adding a dash of humor throughout. Given its marvelous breadth of content, I think that *Dinosaur Odyssey* is the best general-audience dinosaur book since the Dinosaur Renaissance began in the 1970s. It is exactly the kind of book that I would have liked to write, covering much of the same material that I use in one of my university courses. The approach is perfect: teach biological (and to a lesser extent geological) ideas by using dinosaurs as a hook. This is not an introduction to dinosaurs as much as it is a book to take the reader deeper into the subject, exploring some of the most important questions that paleontologists address, and describing implications that reach far beyond simply naming the biggest or most bizarre species. Indeed, *Dinosaur Odyssey* places dinosaurs into context with an evolving planet. It will entice many into thinking about the importance of nature (particularly evolution and ecology, the origin of life, and the inevitability of extinction). In his juxtaposition of past and present, Scott Sampson has made it clear that understanding the ancient world of the dinosaurs is more relevant than ever as we struggle to cope with climate change and extinction in our fast-moving, constantly evolving world.

PREFACE

Dinosaur Odyssey explores two worlds: the ancient world of dinosaurs and the present-day world of dinosaur paleontology. Combining current science with personal stories from the field, the book conveys the entire prehistoric odyssey of dinosaurs, from humble origins on the supercontinent Pangaea, to the largest land animals the planet has ever known, and, finally, to an abrupt and catastrophic demise. Rather than focus solely on the "charismatic megafauna" of the time, the scope extends to encompass the shifting web of life on Earth and the roles that dinosaurs played in this evolutionary drama. Woven into the story are the latest discoveries and controversies. Why did gargantuan body sizes evolve so many times within dinosaurs? Were these famed prehistoric animals warm- or cold-blooded? In what ways was the hothouse Mesozoic world similar to and different from our own? How did dinosaurs go extinct and what meaning do they have for our lives today? *Dinosaur Odyssey* tackles these questions and many more.

Arguably, the most recent previous attempt by a paleontologist to synthesize the cutting edge of dinosaur paleontology was Robert Bakker's 1986 book, *The Dinosaur Heresies*. Bakker, a leading figure in the 1970s "Dinosaur Renaissance," portrayed these animals as highly dynamic—hot-blooded, active, and intelligent—much more similar to modern-day mammals and birds than to "dim-witted" reptiles. Bakker's efforts to popularize this revitalized view of dinosaurs, combined with those of other paleontologists such as Jack Horner, reinvigorated popular interest in dinosaurs, eventually resulting in books and movies like *Jurassic Park*.

Dinosaur paleontology also underwent a dramatic surge in the wake of the Dinosaur Renaissance. Now rescued from their swamp-dwelling, tail-dragging torpor, these super-charged giants became a major research focus for enthusiastic young scientists. Indeed, the quarter century since the mid-1980s represents the most active period ever in dinosaur science. As a result, more dinosaurs have been named during the past 25 years than in all prior history. Entire groups of dinosaurs such as the plant-eating therizinosaurs and the meat-eating abelisaurs were virtually unknown prior to the 1990s. Many of these creatures were discovered on Southern Hemisphere landmasses, opening an entirely new window into Mesozoic Earth. Fieldwork in Asia has revealed a surprising and won-drous menagerie of feathered dinosaurs, documenting an intimate evolutionary rela-tionship between *T. rex* and your Thanksgiving turkey. Meanwhile, in North America, the pace of research greatly accelerated, yielding amazing insights into dinosaur repro-duction, growth, behavior, and ecology. This period also witnessed the first comprehen-sive use of high technology to study dinosaurs, with techniques like electron microscopy, histology, and computed tomographic (CT) scanning providing stunning answers to previously unapproachable questions.

To date, no book for a general audience has addressed this exciting time in the most popular of sciences. *Dinosaur Odyssey* fills that gap. To my mind, all science writing should follow Albert Einstein's dictum: "Make everything as simple as possible, but not simpler." Throughout the chapters that follow, I have done my best to achieve this goal. Use of some technical terminology is, of course, unavoidable, but recurring terms that may be unfamiliar to many readers (noted with capital letters in the text) are defined in a glossary section at the end of the book. As with any general interest science book, this one represents my own unique perspective. Along with simplicity, I have aspired toward a balanced synthesis of the various ideas currently debated by dinosaur experts, yet the end result inevitably reflects my own professional experiences and biases.

The story builds in a stepwise fashion, progressing through three major parts, each more integrative than its predecessor. Following an introduction in chapter 1, the first of these parts (chapters 2–6) presents some of the raw materials necessary for reconstruct-ing the Mesozoic web of life. First up is the amazing evolutionary history that gave rise to Mesozoic Earth, from its stardust beginnings 14 billion years ago to the Triassic origin of dinosaurs about 230 million years ago. Chapter 3 shifts focus to the physical aspects of the Mesozoic world, including the restless dynamics of Earth's interior. Chapter 4 then looks at how solar energy captured by plants and microbes fuels the workings of ecosystems. Finally, chapters 5 and 6 explore ecology and evolution, respectively—the dual themes of this book and the twin forces that guided the building (and rebuilding) of the dinosaurs' world.

The next section of the book (chapters 7–11) delves deeper, further integrating dinosaurs into their prehistoric web of life and making use of the previously described raw materials to weave a number of critical threads. Successive chapters discuss the flow of energy and the cycling of nutrients, first from plants to herbivores, then from herbivores to carnivores, and, finally, from all of these life-forms to decomposers and back to plants. Dinosaurs take their rightful place in this ancient web of relationships as the dominant, large-bodied, land-dwelling consumers of plants and meat. I then turn to the spectacular array of horns, crests, spikes, and other bony "bells and whistles" in an attempt to figure out why so many dinosaurs were not only big but truly bizarre. Also discussed here is dinosaur physiology—the famous warm- versus cold-blooded debate—in which I present a new hypothesis linking the evolution of metabolism with the recurring appearance of giants.

The final section of the book (chapters 12–15) is the most synthetic. The ecoevolutionary threads of earlier chapters are now used to spin some Mesozoic webs and present an up-to-date summary of the dinosaur odyssey. We begin with the origin of dinosaurs in the Triassic Period, investigating whether their marvelous success on land was the result of good fortune or evolutionary superiority. Moving forward in time to the Jurassic Period, I tackle a profound paleontological conundrum: how did so many varieties of dinosaurs—many of them larger than modern elephants—manage to live side by side? In the succeeding Cretaceous Period, the story shifts to the best-known case study in dinosaur evolution—the parade of forms that existed on the island landmass of western North America for tens of millions of years. Eventually, of course, the dinosaurs went extinct. Or did they? The saga now turns to the disappearance of the dinosaurs 65.5 million years ago, and to their continued legacy to the present day. Finally, the epilogue explores the meaning of dinosaurs for humans, arguing that they still have much to teach us; indeed, dinosaurs might even play a significant role in the persistence of our species.

In the end, *Dinosaur Odyssey* is more than a book about dinosaurs. It's a book about how the world works, with dinosaurs serving as the primary protagonists. Highlighting the twin themes of ecology and evolution, I attempt to demonstrate how any ecosystem, past or present, is the culmination of moment-to-moment energy flow (ecology) married to the flow of information over millions of years (evolution). Many of the topics addressed—such as extinction and global warming—have direct relevance to our lives today. Although the book is intended for anyone with an interest in dinosaurs and science, it is my hope that science educators in particular will embrace some of the approaches presented here, using dinosaurs as a vehicle to address a broad range of topics. Ultimately, it is my hope that readers will come to see the odyssey of dinosaurs not merely as a tale of prehistoric monsters in some distant land, but rather a major chapter in our own ongoing story.

ACKNOWLEDGMENTS

Despite the fact that dinosaurs figure prominently in popular culture, relatively few people make a career out of studying them. Many readers will be surprised to learn that only about one hundred professional dinosaur paleontologists are currently active worldwide. We are a small community, and most of us know one another. Conferences like the annual meeting of the Society of Vertebrate Paleontology offer valuable opportunities to exchange ideas and hear about the latest research. I graciously acknowledge the entire community of vertebrate paleontologists, past and present, particularly the dinosaur workers, whose research forms the backbone, flesh, and blood of this volume. Science is a cumulative process, and the conclusions presented here are founded as much on the insights of earlier workers like Sir Richard Owen, Charles Sternberg, and Edwin Colbert as they are on the current generation of dinosaur paleontologists. I also acknowledge the Society of Vertebrate Paleontology for its critical role in furthering this vibrant corner of the historical sciences.

Many will also be surprised to learn that paleontology is one of the few sciences where volunteers make a profound research contribution. At the Utah Museum of Natural History, where I am affiliated, an amazing team of more than fifty volunteers assists in everything from finding and excavating the fossils to preparing and curating these ancient remains. Some of these folks—like Joe Gentry, Jerry Golden, Jay Green, and Sharon Walkington—put in hundreds of hours each year walking rugged badlands in search of fossils or meticulously removing rocky matrix from around the fragile bones. Much of the science presented in this book is the direct result of the commitment, patience, and diligence of volunteers, for which I am deeply thankful.

Grand Staircase–Escalante National Monument and the Bureau of Land Management have provided steadfast support of the Kaiparowits Basin Project in Utah, the core of my present research and a focus of one of the chapters. In particular, Alan Titus (GSENM) has been absolutely essential to the project's success. For their generous assistance with the Mahajanga Basin Project in Madagascar, I also gratefully acknowledge Armand Rasoamiaramanana (Université d'Antananarivo), Benjamin Andriamihaja (Institut pour la Conservation des Ecosystèmes Tropicaux), the villagers of Berivotra, and the many students and volunteers who have participated in field expeditions.

As sciences go, paleontology is thrifty (and thus a solid investment). Someone (presumably with too much time on their hands) allegedly calculated that a single space shuttle launch costs more than has been spent in the entire history of dinosaur paleontology. Regardless, adequate funding is critical to paleontological research, and I acknowledge the following organizations for their generous support of my work: the United States Bureau of Land Management, the National Science Foundation, the National Geographic Society, Discovery Channel, The Dinosaur Society, and the University of Utah.

This book was conceived, developed, and partially written while I was living in Salt Lake City and serving in a dual position at the University of Utah. I am grateful to the staff of the Utah Museum of Natural History (UMNH) and to the faculty of the Department of Geology and Geophysics for their support. Special thanks go to Mike Getty for working with me to develop the UMNH vertebrate paleontology program. Many thanks also to my graduate students—Bucky Gates, Mark Loewen, Eric Lund, Joe Sertich, Josh Smith, and Lindsay Zanno—who, over the years, conversed with me about many of the ideas presented herein. Mark Loewen in particular frequently darkened my office doorway, and participated in the development of many of the ideas presented in the following chapters. I must also express gratitude to the many hundreds of undergraduate students who took my World of Dinosaurs class over a period of seven years; they unknowingly provided the testing ground for many of the topics and approaches in this book.

A career in science is directed in large part by interactions with colleagues. I am indebted to Jack Horner, first for providing an amazing fossil sample for my doctoral dissertation, and second for his continued support. David Krause changed the course of my career (and, ultimately, of this book) by offering me a marvelous, early-career opportunity to work in Madagascar; I thank him for his friendship and for his efforts to help the children of Madagascar. Peter Dodson, Jim Farlow, and Dale Russell all inspired me to take a broader, ecoevolutionary approach to understanding dinosaurs. Larry Witmer, Matt Carrano, Cathy Forster, and Michael Ryan were key research collaborators at different stages in my career, and the resulting projects have been pivotal in guiding my thinking.

In writing a book of such broad scope, feedback from colleagues is an essential element. The following individuals kindly read and commented on one or more chapter drafts: Michael Benton, Matt Carrano, Karen Chin, Doug Emlen, Andrew Farke, Jim Farlow, David Fastovsky, Nick Fraser, Mike Getty, Thomas Holtz, Jack Horner, Randy Irmis,

Kirk Johnson, David Krause, Mark Loewen, Dale Penner, Eric Rickart, Judy Scotchmoor, Alan Titus, and Lindsay Zanno. Philip Currie and Greg Erickson went a major step further, reviewing the entire manuscript and providing thoughtful input that significantly improved the final product. I also thank Phil for encouraging my dinosaur research aspirations in the early days and for being a stalwart supporter ever since. My friend Kirk Johnson is to be heartily commended for his generous assistance in all matters paleobotanical.

I am grateful to Michael Skrepnick, paleoartist extraordinaire, for his exquisite chapter frontispieces and for many other pieces of artwork included herein. The artistic talents of Marjorie Leggitt, Mark Loewen, Lukas Panzarin, and Lindsay Zanno are also featured liberally in the book's numerous original figures. Mark Loewen's relentless shepherding of the figures through multiple iterations was an additional critical element in the book's completion.

Many thanks to my literary agents, Katinka Matson and John Brockman, for their business acumen and sage advice. Thanks also to the University of California Press, in particular my patient editor, Blake Edgar, and the book's production director, Scott Norton.

Tim (T.J.) Moore and Dale Penner offered countenance at critical times in the process, and my walking buddy, John Gillette, enabled me to maintain both focus and a sense of humor in the latter stages of writing. Warmest thanks are reserved for my wife, Toni, who supplied the unwavering encouragement, confidence, and love that made it possible to tackle (and complete) this project. Finally, I am ever-grateful to my mother, June Sampson, for getting the whole thing started.

Traffic jam on a Madagascar floodplain. An apex predator, the abelisaur theropod *Majungasaurus*, feeds on a fallen sauropod, *Rapetosaurus*, while an opportunistic *Masiakasaurus* waits patiently for scraps. A pair of *Simosuchus*—odd, blunt-snouted, herbivorous crocodilians—observe the spectacle briefly before moving on.

1

TREASURE ISLAND

A SPONTANEOUS BARRAGE OF EXPLETIVES rang through the air, bringing my coworkers scrambling over the hill. Madagascar's sweltering midday heat no longer mattered. There before me, beneath a clod of freshly dislodged sediment, were four shining teeth, exposed to the light of day for the first time in more than 65 million years. Most kids could have confirmed that these sharp, recurved, chocolate brown objects, each topped with fine serrations, once lined the mouth of a meat-eating dinosaur, a theropod. Best of all, these teeth were still attached to a jawbone. Further digging revealed a complete and undistorted jaw, with every tooth in place. Over the next couple of days we found more bones of the same, exceptionally preserved skull—part of the eye socket, another jaw with teeth, a gnarly bone from the nose region. Soon it became clear that most of the skull was buried here, although the individual bones had fallen apart and now lay strewn over several square meters. We could barely contain our excitement. Field paleontology relies as much on serendipity as on know-how and hard work, and the fates had smiled down upon us. Yet, as more and more bones of the ancient predator were unearthed, we began to get nervous. A key portion of the skull remained missing, leaving a mystery unsolved.

In Lewis Carroll's classic tale, *Through the Looking-Glass and What Alice Found There*, Alice gazes into a mirror to find a world similar to her own yet distinctly different. Her view of this reflected world varies dramatically depending on where she stands and how she holds the mirror. And Alice dreams of actually stepping through the looking glass to experience firsthand the wonders beyond. Like Alice, paleontologists attempt to gaze through the looking glass of time in order to catch glimpses of other, distant worlds. We, too, find that our perspective is always limited, changing considerably depending on how we hold the mirror and indeed which mirror we choose. And we, too, dream of witnessing these worlds firsthand. The ongoing efforts to open windows into ancient landscapes and their inhabitants comprise the science of PALEONTOLOGY, the study of ancient life.

Dinosaur paleontology, my particular specialty, is a peculiar profession. After all, how many people can claim to have a job that is the envy of most 6-year-olds? Telling others that you're a dinosaur paleontologist often results in the usual questions. "How do you know where to dig for them? What was the biggest dinosaur? Why do *you* think dinosaurs went extinct?" Of the most common queries, the one that I find most amazing and dismaying is "Don't we already know everything about dinosaurs?"

People tend to think of science as the gradual, steady accumulation of facts that has been ongoing for centuries. So it's often imagined that today we scientists are merely adding insignificant grains to an enormous, established mountain of knowledge. This view could hardly be further from the truth. The vast majority of nature's secrets have yet to be revealed. In the evocative words of biologist and environmentalist David Suzuki: "It is as if we are standing in a cave holding a candle; the flame barely penetrates the darkness, and we have no idea where the cave walls are, let alone how many caves there are beyond. Standing in the dark, cut off from time, and place, and from the rest of the universe, we struggle to understand what we are doing here alone."[1] Rather than being daunted by our overwhelming ignorance, I am inspired by the multitude of new discoveries that patiently await us, entombed within the earth, carefully preserved in museum drawers, and tucked away in the corners of our imaginations. It's an exciting time to be a paleontologist.

If the overriding aim of science is to understand and describe as accurately as possible the workings of nature, certainty turns out to be a scarce commodity. To speak of "scientific facts" is to border on using an oxymoron. Most scientists would agree that there is a single, physical reality to comprehend. To borrow the slogan of a recent popular television show, "the truth is out there." Yet the best we can offer are successive approximations of that truth, formulated as alternative explanations, or HYPOTHESES. The scientific method involves sorting among these various alternatives. Consequently, testability is an integral part of the process, and only the strongest THEORIES, like gravity and EVOLUTION, withstand decades of testing and become accepted as fact.

But how can paleontologists test ideas? Like geology, paleontology is a historical science, concerned predominantly with understanding and interpreting past events. Historical sciences differ in at least one fundamental way from nonhistorical fields such as physics and chemistry. Paleontologists cannot test a hypothesis through direct experimentation for the simple reason that it is impossible to reproduce past events. For example, barring the highly unlikely cloning of a dinosaur from its DNA or the invention of a time machine (even less likely), we clearly can't investigate the metabolism of *Tyrannosaurus rex* directly. Similarly, geologists cannot observe the rifting and collisions of ancient continents. Given the strong emphasis on reproducibility—the ability to run the same experiment multiple times in order to test for similar results—some have even argued that the inability of historical sciences to reproduce results should disqualify them as scientific disciplines.

Yet the historical sciences are able to circumvent the conundrum of time's arrow, at least to some degree, through an elegant loophole. Although the inexorable march of time prohibits actual reproduction of past events, it's possible to observe multiple examples of such events. If these examples are consistent with a stated hypothesis, it gains support. If not, the hypothesis is falsified or at least brought under closer scrutiny.

Take evolution, for example. Darwin's theory states that all organisms past and present share common ancestry and that life evolved from simple, single-celled beginnings. Thus, we predict that the order of appearance of particular groups of organisms should mimic the branching pattern of evolution, with a trend toward increasing complexity through time. Convincing evidence against evolution would be the discovery of any animal that lived long before its supposed time of its origin—say, for example, the fossilized remains of a rabbit (or human or dinosaur, for that matter) dating to 400 million years ago. With hundreds of paleontologists working around the globe in rocks that span most of Earth history, this amounts to hundreds of thousands of opportunities annually to discredit evolutionary theory. Yet, invariably, we continue to find groups of organisms restricted to rocks of a specific age range. In all the years I have been hunting for dinosaurs in Mesozoic-aged deposits, I have never found any indication of advanced mammals such as cats, whales, or aardvarks, let alone humans. And the same is true for all of my paleontological colleagues, because such a find would be headlined in the media worldwide and bring with it the potential for all forms of academic accolades, as well as research funding. In short, through study of multiple examples of past phenomena, paleontology and geology are anchored on testability.

Science grows in fits and starts. Research occurs within a particular theoretical framework, or paradigm, that guides scientific thinking. Occasionally, a new overarching theory, sometimes triggered by a dramatic discovery, causes an entire scientific field to reassess its assumptions and ask new kinds of questions. A fundamental breakthrough of such magnitude is called a PARADIGM SHIFT, because it requires restructuring or even wholesale replacement of an old theoretical framework. Prime examples of paradigm shifts in the history of science include the Copernican and Darwinian revolutions. The first of these devastated the then-current worldview of a fixed and finite universe with Earth presiding at the core, forcing humans to regard their planetary home as being far removed from the celestial center stage. The second knocked us further off the pedestal of centrality, relegating *Homo sapiens* as one of millions of species that together represent merely the latest wave in an unfathomably deep ocean of evolutionary change. Importantly, with rare exceptions like the Copernican revolution, paradigm shifts do not entail the wholesale tossing out of previous ideas. Science proceeds by building on what has come before, and many ideas within science are known with great confidence, unlikely ever to change. As the architecture of the building is modified, however, occasional large-scale renovations are necessary.

Beginning in the late 1960s, dinosaur paleontology experienced its own, humbler paradigm shift. As a child of the baby boom generation, my first exposure to paleontology occurred just prior to this shift, when dinosaurs were regarded as sluggish, dim-witted behemoths. I fondly remember flipping the pages of large dinosaur books with awe-inspiring illustrations of long-necked sauropods (aka "brontosaurs") fully submerged in lakes except for the tops of their heads. Prevailing thinking viewed these animals as simply too gargantuan to support themselves on land. Those dinosaurs that did walk on terra firma were generally depicted as slow and awkward. Giant bipedal CARNIVORES such as *Tyrannosaurus* were reconstructed as Godzilla-like, with upright bodies and massive tails dragging behind. Four-footed plant eaters like *Stegosaurus* were portrayed with sprawled, lizardlike front limbs, low-slung bodies, and dragging tails. The overall impression was one of awkward giants lumbering across ancient landscapes, with brains barely capable of carrying them from day to day.

Then, virtually overnight it seemed, dinosaurs received a stunning makeover. Sauropods emerged from the water to strut on land with elephantine limbs, and scientists came to argue that an aquatic lifestyle would have been impossible for these behemoths because of water pressure compressing the chest cavity. Meanwhile, *T. rex* pivoted to a sleeker, horizontal body posture. No longer trailing uselessly behind, its rigid tail projected rearward to counterbalance the head and trunk. *Stegosaurus* and its armored, four-footed kin were also transformed, bestowed with upright limbs and nimble, airborne, potentially lethal tails. Reconstructions like these signaled much more active, agile animals. In addition to their redesigned bodies, these freshly envisioned dinosaurs were considered more intelligent, engaging in such behaviors as pack hunting, herding, and parental care.

What caused this fundamental change in our conception of dinosaurs? The answer is a paradigm shift triggered by a combination of discovery and insight. The pivotal FOSSIL discovery was a sickle-clawed "raptor" theropod, recovered in Montana in 1964 by an expedition from Yale University. The revolutionary insights came from Yale paleontologist John Ostrom. In his 1969 description of this extraordinary predator, which he called *Deinonychus* ("terrible claw"), Ostrom argued that at least some dinosaurs were considerably more active than previously assumed. Shortly thereafter, noting a large number of birdlike features on the skeleton of *Deinonychus* and related predators, Ostrom revived the nineteenth-century idea that birds evolved from dinosaurs. He catalogued numerous bony characteristics linking theropod dinosaurs with birds, and subsequent workers have added many more, bringing the total number of shared, specialized features to greater than one hundred. Today, most experts agree that birds are the direct descendants of dinosaurs and thus are, in a very real sense, dinosaurs themselves.

Faced with this new evidence and a fresh perspective, paleontologists quickly began to view all dinosaurs as more like birds than lizards. Investigators returned to existing museum collections and began to reassess long-held views and biases. Soon, the mounted skeletons of bipedal carnivores and HERBIVORES were reengineered to assume a more horizontal posture. Further support for the revised stance came from the discovery of dinosaur trackways that showed no signs of dragging tails.

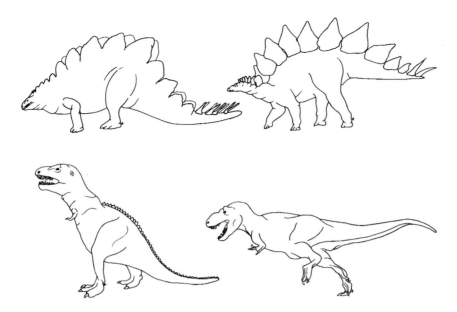

FIGURE 1.1

Reconstructions of *Stegosaurus* (top) and *Tyrannosaurus* (bottom), depicting earlier, "prerenaissance" postures (left) and more recent, "postrenaissance" postures (right).

This paradigm shift spawned novel research programs and heated debates. Were dinosaurs warm-blooded? Did some forms exhibit parental care? What were the intellectual and behavioral capacities of the different dinosaur groups? In an attempt to address these questions, paleontologists have applied a range of analytical tools old and new, from detailed anatomical comparisons with living animals to computed tomographic (CT) scanning of fossils. Several decades of extremely active research have led to numerous insights, many of which are discussed in this book.

As is often true of cultural and scientific trends, once in motion, the pendulum of a paradigm shift tends to swing in a wide arc. This has certainly been the case with dinosaurs. If John Ostrom ignited the paradigm shift, the fuel for the subsequent explosion was Robert Bakker, Ostrom's flamboyant, iconoclastic student and avid champion of the dinosaur renaissance. Not long after Ostrom's original argument for more active, potentially warm-blooded dinosaurs, Bakker rescued sauropods from their aquatic torpor and even had them rearing up on their hind legs to battle marauding theropods. Similarly, *Tyrannosaurus* and its large carnivorous kin, no longer awkward and lumbering behemoths, were depicted as agile predatory machines capable of running speeds in excess of 65 kilometers (40 miles) per hour. Then there were the small raptorlike, sickle-clawed theropods such as *Deinonychus* and *Velociraptor*, traveling in packs and utilizing a combination of cunning and cooperative behavior to take down prey of much greater body sizes. The *Jurassic Park* movie series brought these new ideas to popular audiences via the big screen and stretched science to the breaking point.

Today, the pendulum is swinging back toward the middle, as paleontologists generate more rigorous, tempered, yet undoubtedly richer reconstructions of dinosaurs and their worlds. One example is the recent work indicating that *Tyrannosaurus* and other large theropods could not attain the remarkable, jeep-pursuing speeds previously reported and that they were likely incapable of true running. It has even been suggested that *T. rex* was not the predatory tyrant-king long depicted but rather a lowly scavenger, eking out a living from remains of the dead. The same type of scrutiny is being applied to plant eaters as well. Not only have investigators questioned whether sauropods could rear up on their hind legs. Some have also argued that these successful giants could not elevate their elongate necks much above the horizontal because of, among other things, the difficulty of pumping blood up to the head.

On July 16, 2005, John Ostrom passed away at the age of 77. A few years before, Ostrom confessed to me that he would give just about anything to be back near the beginning of his career. He talked about the new age of discovery in dinosaur paleontology and the exciting work yet to be done. Yet, unlike the great majority of us, John leaves behind a deep and lasting legacy. His discoveries and vision triggered a revolution in our perception of dinosaurs, a true paradigm shift has enabled us to see these long-dead animals with new eyes, and to explore questions never previously conceived. Many of Ostrom's students went on to become leaders in the field, and the dinosaur renaissance he initiated has ushered in new generations of scientists eager to conduct research on these long-dead yet suddenly more fascinating beasts. Today there are about 100 dinosaur paleontologists worldwide, more than ever before. And the number of new dinosaur species named in the past 25 years exceeds that found in all prior history, with no end in sight.

It's tempting to romanticize the hunt for dinosaurs, even for those of us engaged in it. Yet the majority of fieldwork is anything but romantic—heat, insects, and tedious labor in remote, lunarlike landscapes typically comprise the bulk of the daily routine. Add to this list the lack of running water, dearth of culinary options, and living outdoors with a small group of people, and most folks would choose to bow out. So why do we venture around the globe in search of places referred to as "badlands" and endure often harsh conditions for weeks or even months on end?

Well, for one thing, fossil hunting is a form of time travel. As I walk up a steep slope of sandstones and mudstones, a piece of fossilized skull found in one rock layer may come from an animal that lived thousands, hundreds of thousands, or even millions of years apart from an animal whose toe bone I spot a few steps farther on. On a good day, as I stoop to pick up a 75-million-year-old jaw fragment, the chasm of time will suddenly open, instantly transporting me back to the Cretaceous. Now crouching unseen in the shadows, I gaze on the giant duck-billed dinosaur cropping conifer needles with its broad beak and slicing up this green energy with formidable batteries of teeth. I listen as this gargantuan animal takes deep draughts of the crisp morning air, and I inhale its musty odor. For a moment, I even become this dinosaur and *feel* its world. As the feeling subsides and I return to the present day, the experience invites a new, broader perspective of myself as part of the single, unbroken flow of life and energy through deep time. Who would have thought that a chunk of old bone could wield such power?

There's another, perhaps more common reason why people are drawn to the search for fossils: discovery. Few feelings compare to being the first person ever to set eyes on a previously unknown ancient animal or to be part of a crew that unearths a well-preserved, fossilized skeleton. Walking across the rocky terrain with eyes trained on the surface, you never know if, perhaps just around the next corner, the remains of some magnificent prehistoric creature await you. During the heyday for fossil hunting early in the twentieth century, it was relatively easy for any intrepid paleontologist to find pristine badlands when in search of a fossil "grail." Today, even the most fortunate fieldworker has only a handful of opportunities to be the first paleontologist in an unexplored region.

Such places still exist in Madagascar.

Madagascar, situated off the southeast coast of Africa, is the fourth largest island in the world, bigger than the state of California. A mountainous spine runs for most of the 1,000-mile length of the island from north to south. Along the eastern side lie tropical rain forests, though these have been decimated by human activity. The western and southern sides of the island, in the rain shadow of the highlands, are home to tropical dry forests, thorn forests, deserts, and shrublands. Because of its great size, diverse geography, and lengthy isolation from other landmasses (for about 85 million years), Madagascar is home to a highly unusual biota, with approximately 80 percent of its plant and animal species known nowhere else. By far the most famous of these are the lemurs, a group of primates found only on Madagascar that today numbers more than seventy species. Humans settled on the island about 2,000 years ago, likely from Southeast Asia, and soon drove virtually all of the largest animals to extinction. Recently extinguished species include gorilla-sized lemurs, pygmy hippos, and giant, flightless elephant birds. Thanks to its fine weather, abundant food, and numerous secluded coves, Madagascar served as a pirate hideout and stronghold for several decades during the late fifteenth and early sixteenth centuries. Such infamous figures as William Kidd, John Bowen, and Thomas Tew plundered merchant ships in the Red Sea, the Persian Gulf, and the Indian Ocean, stealing silks, cloth, spices, jewels, gold, silver, and coins.

In search of a different kind of treasure, David Krause, a mammal paleontologist from Stony Brook University, launched an expedition to Madagascar in the early 1990s. He wondered where Madagascar's wondrous marooned trove of unique plants and animals had come from. Were the ancestors of lemurs and other modern forms present when Madagascar became an island 85 million years ago? Or did they arrive much later, crossing major water gaps separating the island from other landmasses, such as mainland Africa? The first place you might think to look for answers is the fossil record preserved on the island. Unfortunately, however, other than the very recent past (about 26,000 years) there is virtually no fossil record of land animals on Madagascar since the major extinction of the dinosaurs 65 million years ago. One has to go all the way back to fossil-rich deposits of Late Cretaceous age, at the end of the Age of Dinosaurs, to search for the origins of today's Malagasy FAUNA.

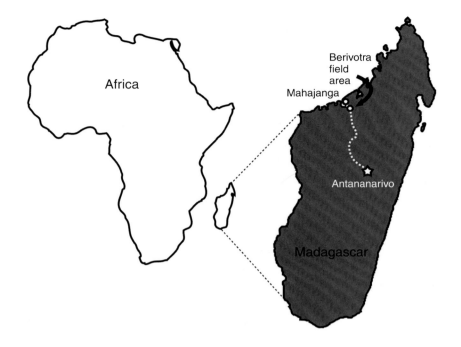

FIGURE 1.2

Map of Madagascar in relation to Africa. Close-up view of Madagascar (shaded, on right) depicts the country's capital city (Antananarivo), the city of Mahajanga, and the Berivotra field area that has produced abundant fossils of Late Cretaceous dinosaurs and other vertebrates.

Krause decided to investigate a small outcropping of rocks on the northwest part of the island, surrounding a tiny village called Berivotra, an hour's drive from the bustling port city of Mahajanga. Like most of Madagascar, the land around Berivotra was long ago deforested by human activity. Annual mass burnings of grasslands, both voluntary and involuntary, together with agriculture and grazing by livestock, now ensure that the forest ecosystem cannot regain a foothold. This widespread habitat loss has devastated ecosystems all over the island. Many regions, including the northwest where we work, are dominated by spear grass, an exotic invader species entirely deserving of its name. The removal of forest has also led to rampant soil erosion. Every year, monsoonal rains scour the outermost skin of red soil, which then bleeds into rivers like the Betsiboka and ultimately into the Indian Ocean. From the air, it looks as if the entire island is hemorrhaging, a description not far from the truth. Ironically, this ongoing decimation of the landscape has benefited us paleontologists, who search out areas with minimal vegetation and depend on erosion to bring ancient bones to light.

Starting in the Berivotra field area was not a shot in the dark. The French invaded Madagascar and declared it a colony in 1896. The year prior, a French military physician named Félix Salètes was charged with constructing a temporary hospital about 45 kilometers from Mahajanga (known to the French as Majunga). Salètes recognized the paleontological potential of the region. Lacking time to check out the area himself, he dispatched his regimental staff officer Landillon to carry out a survey for fossils. Landillon excelled in this task, collecting a diverse range of fossils that were shipped back to Paris, where they came under the study of renowned French paleontologist Charles Depéret. Depéret identified two large dinosaurs from the collection: one herbivore, a long-necked sauropod; and one carnivorous theropod, which he dubbed *Megalosaurus crenatissimus.* *Megalosaurus* was previously known from several sites in Europe, and the Malagasy example was established solely on the basis of two teeth, three VERTEBRAE (elements of the "backbone"), and a single toe bone. In the middle of the twentieth century, following construction of a major new road through the region, another Frenchman, René Lavocat, surveyed the deposits around Berivotra. Lavocat apparently conducted few excavations but nevertheless collected many fossils eroding from the surface. One of these was a fragmentary lower jawbone (dentary) that Lavocat regarded as sufficiently distinct from *Megalosaurus* to erect a new name, *Majungasaurus.*

Our story then jumps to 1976, when French paleontologist Philippe Taquet received a shipment of Malagasy fossils at the Natural History Museum in Paris that were being held in storage elsewhere in the city. The collection included a chunk of bone that Taquet recognized to be part of the skull roof above the brain compartment. The specimen was strange, however, topped by a roughened, domelike structure. Notes accompanying the fossil documented that the specimen had been collected in the "Majunga District" of northwestern Madagascar early in the twentieth century, but little else was known. Together with Hans-Dieter Sues (now of the Smithsonian Institution's National Museum of Natural History), Taquet published a paper in 1979 describing the specimen and concluding that it belonged to a group of dome-headed herbivorous dinosaurs known as pachycephalosaurs. They gave it a new name, *Majungatholus atopus.* Identifying *Majungatholus* as a pachycephalosaur seemed reasonable based on its conspicuous bony dome. Yet the assertion was also remarkable because, other than this single fragmentary specimen, pachycephalosaurs were (and still are) otherwise known only from Northern Hemisphere continents. So *Majungatholus* seemed a long way from home.

Enter David Krause and the joint Stony Brook University/University of Antananarivo expedition of 1993, hoping to find the first evidence of Mesozoic mammals in Madagascar. On first arriving in Berivotra, the crew jumped out of the vehicles and headed to a nearby rock outcrop in search of fossils. (It's a very long trip from North America, and folks are typically anxious to find fossils as soon as they arrive.) Within minutes, keen eyes led to the discovery of a partial mammal tooth—a tremendous start to the project. Although this would be the only scrap of mammal fossil recovered in the 6-week field season, the crew did find plenty of dinosaur remains, as well as those of many other backboned animals. The dinosaur finds prompted David to enlist the help of dinosaur paleontologists, first Peter Dodson (University of Pennsylvania) and then Cathy Forster (now at George Washington University) and me.

During the following field season, in 1995, we found plenty more fossils, including lots of dinosaur bones. Curiously, however, there was no further sign of the putative dome-headed pachycephalosaur. Close examination of the original specimen housed in the National Museum of Natural History in Paris made us suspect that *Majungatholus* might not be a plant-eating pachycephalosaur at all. We noted that the dome itself is covered with a roughened, wrinkly texture unlike that known for any other member of the group. Moreover, a few anatomical features suggested that the partial skull might instead belong to a carnivorous dinosaur. We were pretty convinced that *Majungatholus* and *Majungasaurus* would turn out to be one and the same animal, but we needed more fossils to prove it.

Early in the 1996 season, we checked out a known locality that appeared to have promise. Walking around the low hill, I spotted a small vertebra eroding from the surface and was excited to recognize it as a part of a theropod tail. Using rock hammer and pick, it took only a few minutes to uncover another tail vertebra, and then another. With eager anticipation, we literally chased the tail into the hill. A few hours later, after exposing several more vertebrae, we removed a chunk of rock to reveal a large limb bone, a clue that much of the skeleton might be interred within the site. But it soon became clear that the limb bone belonged to a giant herbivorous sauropod, the most common kind of dinosaur in these deposits. Our hearts sank, because it appeared that the locality consisted merely of a jumble of bones from various kinds of animals.

Then my awl lifted a piece of sediment to expose those four shiny teeth mentioned at the outset of this chapter. The team continued to excavate this small area, uncovering additional skull bones of the Cretaceous predator while watching closely for the skull roof. Peter Dodson exposed a spectacular, tooth-filled lower jawbone. Cathy Forster found large portions of the side of the face. Finally, the rear portion of the skull roof appeared, attached to the rest of the bony compartment that long ago housed the animal's brain. And there, as predicted, was a rounded, roughened, domelike structure. We now had conclusive evidence that *Majungatholus*, the supposed thick-headed herbivore, and *Majungasaurus*, the top predator, were the same animal. A hypothesis had been confirmed and a mystery solved. Pachycephalosaurs did not inhabit Madagascar. Rather, the island was home to a strange, dome-head carnivore. Because *Majungasaurus* was named first, and ultimately we could not tell it apart from *Majungatholus*, the former name has taken precedence and is the one we use today.

After all the fragile skull elements were excavated, shipped back to New York, cleaned of the rocky MATRIX, and copied using a process known as molding and casting, the duplicates were put together to reconstruct the original appearance of the skull. Amazingly, the bones fit back together almost perfectly, definitive evidence that the skull elements had undergone minimal postburial distortion. We joked that the entire set of casts could be marketed as a kids' dino-skull kit. Uncovering this exquisite specimen ranks as the highlight discovery of my career—not one I expect to top.

Near the end of the twentieth century, the foundations of science once more began to rumble, the tremors triggered by what appears to be another large-scale paradigm shift.

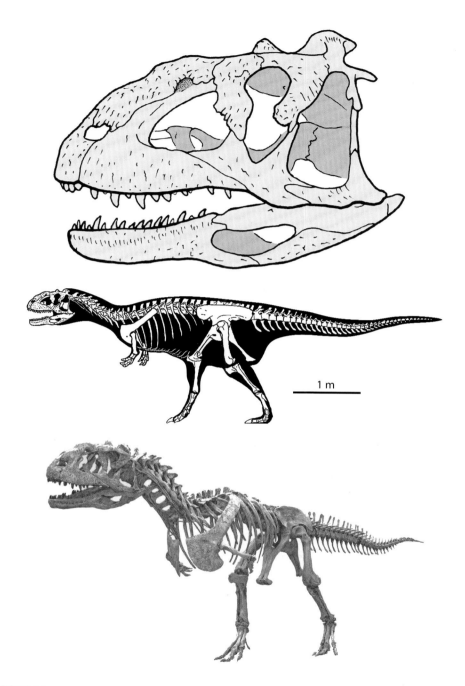

FIGURE 1.3

The abelisaur theropod *Majungasaurus*, from the Late Cretaceous of Madagascar (about 70 million years ago), showing side view of skull (top), illustration of skeleton (middle), and photograph of mounted skeleton (bottom).

For more than 400 years, the scientific conception of nature has been dominated by a mechanistic worldview inspired by sixteenth- and seventeenth-century scientists like Francis Bacon, René Descartes, and Isaac Newton. The underlying assumption of this perspective is that the Universe is a machine, and its secrets can be revealed only by dissecting and compartmentalizing nature into ever smaller parts. This atomistic approach, often called REDUCTIONISM, has yielded remarkable insights into the structure and function of the universe, from cosmic to subatomic scale. Accompanying these insights have been numerous technological innovations such as satellites, computers, and artificial hearts. In the academic realm, entirely new disciplines have continually emerged as scientists focused their efforts on successively smaller units of nature. Within biology, fields of inquiry today include ecology, systematics, developmental biology, physiology, cytology, genetics, and molecular biology.

So entrenched have we become in the minutiae of nature that we have forgotten that the separate disciplines recognized throughout science are human inventions—categories that relate to how scientific research is practiced and not how the universe itself is structured. Moreover, despite its grand successes, the tendency to fragment nature has severely limited our ability to comprehend interconnectedness.

In recent decades, it has become increasingly evident that natural phenomena tend to be complex and highly sensitive to initial conditions. That is, small-scale changes in initial conditions can be amplified into large-scale effects. The commonly invoked metaphor is the "butterfly effect"—a butterfly flaps its wings, say, in the Kalahari savannah of Africa, which causes a tornado in Wichita, Kansas. In other words, small, seemingly insignificant events can have disproportionately large consequences. It turns out that the key to understanding the behavior of such dynamic systems—from weather to finance to ecosystems—is not to dissect all the parts and examine them with greater and greater scrutiny, but to investigate how the various components interact with one another. Consequently, investigators have begun focusing less on parts and more on wholes.

This radical new view is all about connections. Today, numerous disciplines are feeling the effects, becoming increasingly integrative and holistic. As a result, we're witnessing a reunification of once-separate fields into new disciplines with names like "geobiology" and "biocomplexity." Much effort is now devoted toward unveiling the complex, weblike links among all living and nonliving systems. Scientific think tanks aimed specifically at such problems, such as the Santa Fe Institute in New Mexico, bring together evolutionary biologists, economists, anthropologists, meteorologists, and mathematicians, as well as representatives from many other fields. It turns out that, once these diverse experts agree on a common language (no easy task, as one might imagine), they are able to see the world differently and develop exciting new insights. These integrative thinkers emphasize such topics as networks, dynamic systems, and feedback loops. They develop novel models addressing everything from climate change to stock market fluctuations. Under the reductionist worldview, physics was viewed as the "purest" of sciences, providing the deepest insights into nature. With a new, broader perspective, scientists now look to more integrative disciplines such as ECOLOGY that investigate connections.

Traditionally, paleontology has followed the reductionist lead of other sciences, fragmenting the history of life into smallest identifiable units—discovering and naming new species, and assessing their evolutionary and temporal relationships. This approach has enabled the recovery of patterns—answers to the who, what, where, and when questions—but at the expense of processes—the how and why questions. I cast no dispersions here; most of my own work falls under this same descriptive umbrella. In part, a focus on patterns makes perfect sense, because dinosaur species are often represented by only one or two fragmentary specimens. And every field must go through a period of collecting basic data before undertaking more integrative studies. Nevertheless, in recent years, there has been a subtle shift toward more holistic, connections-based studies. Dinosaur specialists interact more with experts from other disciplines, such as geochemistry, PALEOBOTANY, geophysics, and paleoecology. Instead of an unwavering spotlight on dinosaurs, much more effort now goes into reconstructing the rest of Mesozoic ecosystems. These collaborations have just begun to bear fruit, and I anticipate an exciting new era for dinosaur paleontology.

After the successful 1996 field season, I ventured with three other crew members to the southern part of Madagascar to visit my first tropical rain forest. Our destination was Ranomafana National Park, situated on the edge of the island's high plateau, where steep, forested slopes shelter a tremendous diversity of wildlife. The rugged terrain here limited deforestation, and the area was formally protected in 1986. Having pitched our tents beside a clear flowing river, we set out in search of lemurs; twelve species are known from the park. We followed a well-trodden trail into the dense forest and began looking high in the canopy for any movement. Lush lichens and mosses festooned the trees, and giant bamboo and orchids lined the waterways. Having grown up camping and hiking in the wilds of British Columbia, I had spent plenty of time spotting wild game. I expected to put this experience to good use, and my compatriots felt equally confident. Yet hours later, we returned to the camp exhausted, without so much as a glimpse of a lemur. We soon learned the necessity of a detailed knowledge of place.

The next morning, we set out with two Malagasy park guides. Within minutes, they located the first lemurs of the day. These were Milne-Edwards sifakas, astounding primate athletes that launch themselves from tree trunks. While airborne, they rotate their bodies until their limbs are posed to alight, seemingly nonchalantly, on another tree. Sifakas can leap impressive distances (up to about 7 meters, or 20 feet) between trees, and we felt fortunate to be the exclusive audience for an acrobatic display. In the next three hours, we saw three more lemur species and added a fourth during a nighttime forest walk.

In addition to lemurs and a few birds, we spotted several reptiles, including chameleons and a large leaf-tailed gecko, all of them masters of camouflage. We also saw numerous insects—caterpillars, butterflies, ants, preying mantids, and phasmids, or stick insects. Phasmids are wondrous creatures that look like something out of a Tolkien dream, appearing to be half plant and half animal. Also inconspicuous, though because of their small size, are the innumerable tiny beasties—mites, nematode worms, tardigrades, and the like—amassed in the forest's soil. Ultimately, the real action in any ecosystem takes place in the microscopic realm. It is here that most of the habitat's diversity resides in seemingly boundless forms of bacteria. A single handful of forest soil may contain thousands of different bacterial varieties and literally billions of bacterial individuals.

All of this bewildering diversity is intertwined in intricate relationships. Bacteria grab nitrogen from the soil and convert it into a form useable by plants. Lemurs, birds, bats, and various insects participate in plant reproduction by pollinating flowers and dispersing seeds. Beetles and flies, among others, feed on dung and recycle the remains of all forest dwellers as they die. Concentrating only on the myriad details may reveal volumes about those details but little about the functioning of the entire rain forest. How can we even begin to unravel such complexity, to tell the story of this place, or any natural habitat? A good place to start is with two E-words: *ecology* and *evolution*.

Ecology and evolution are flip sides of the same coin. Ecology provides a description of the complex relationships connecting all organisms and their environments. Evolution is organic change through time, encompassing the full range of biological processes that have generated the wondrous living world. These two themes are inseparable. Without evolution, ecology becomes largely a description of relationships, limited by its lack of time depth. Attempting to comprehend an ecosystem without the perspective of evolution is like trying to understand a person without knowing anything of his or her life experiences. Similarly, without ecology the processes and effects of evolution cannot truly be envisioned, because evolution occurs on the stage of the ecological theater.

From the narrow time-slice perspective of ecology, the Ranomafana rain forest (indeed, any ecosystem) is an interwoven collection of relationships, with all the players necessary to keep energy flowing and nutrients cycling. At the base of the food web are producers— from ferns to giant tropical hardwoods—that create their own food by harnessing the sun's energy. Much of this energy escapes into the environment as latent heat, keeping the forest cool and causing the abundant formation of clouds. Yet enough energy is passed on to multiple levels of consumers to keep the system in motion. First are the primary consumers—including various insects, birds, and lemurs—that feed on the plants. The primary consumers in turn are consumed by a range of carnivores, some of which rise to still higher levels by feeding on other carnivores. In the Ranomafana rain forest, these secondary consumers include spiders, snakes, chameleons, and birds. Largest of the carnivores here is a strange, lemur-eating mongoose relative called a fossa (pronounced FOO-sa), that looks like an unfortunate cross between a dachshund and a bobcat. Finally, the energy cycle is completed by a diversity of decomposers, or detritus feeders—mostly insects, bacteria, and other tiny dwellers of the forest floor—that convert wastes and dead organisms into chemicals useable by the next generation of producers. Together this multilevel web of interactions is self-regulated by various feedback loops.

Such an intricate, interwoven biological fabric did not simply pop into existence tens, hundreds, or even thousands of years ago. From an evolutionary perspective, Ranomafana resulted from millions upon millions of years of unique historical events driven in large part by coevolution—the mutual influence of species on one another. Even a modest comprehension of this amazing place requires that one consider the biological dramas unfolding through time. Many bacteria are little changed from those that have inhabited this planet for billions of years. Other, more recent microbial forms are highly specialized, forming intimate, mutualistic relationships with one or more species. The hardwood trees and varied forms of insects trace their heritage as major groups to life-forms that struggled onto land about 400 million years ago. The leaf-tailed gecko is descended from a long line of egg-laying lizards, which branched off from other vertebrates on the order of 300 million years ago. The birds owe their existence to a small, feathered dinosaur that first took flight perhaps 160 million years ago. Much of the rain forest vegetation shares common ancestry within the ANGIOSPERMS (flowering plants), which first appeared alongside the dinosaurs about 125 million years ago and subsequently exploded into global floral dominance. The lemurs, in contrast, only spawned from primitive primate stock sometime around 60 million years ago. Finally, the true newcomers are humans, primate descendants of hominid forebears that stood up on their hind legs and split from the great ape lineage 7–9 million years ago in nearby Africa, though these descendants did not arrive on the island until about 2,000 years ago. Importantly, all of these organisms share a common ancestor in the even more remote past.

So let's return to the question of the origins of Madagascar's unique biota, existing in splendid isolation in the southern Indian Ocean. David Krause's project, focused on the Late Cretaceous, has yielded important clues regarding how animal lineages first arrived on the island. Nine successful field seasons to date have uncovered a treasure trove of vertebrate fossils. The ancient booty includes not only dinosaurs but fishes, frogs, lizards, snakes, turtles, crocodiles, birds, and mammals. In addition to the dome-headed top theropod *Majungasaurus*, we have unearthed remains of a small buck-toothed predator called *Masiakasaurus* that I will introduce in chapter 3. *Rahonavis* was a raven-sized, sickle-clawed predator with feathers that was either a bird or a small nonavian dinosaur akin to *Velociraptor*. Other than fragments of turtle shell, the most common finds around Berivotra are those of long-necked sauropod giants like *Rapetosaurus*. The crocodiles are unusually diverse here, with seven distinct species discovered to date. Strangest of the crocs is *Simosuchus*; with a short, rounded nose and simple, blunted, leaf-shaped teeth, this pug-nosed animal looks more like an aberrant vegetarian bulldog than any meat-loving crocodilian. This wealth of fossil evidence demonstrates that the bizarre Late Cretaceous inhabitants of Madagascar were worthy predecessors of their modern-day and recently extinct counterparts. Remarkably, all of the animals that can be confidently identified represent species that occur only on the Great Red Island, with the great majority of these being new to science.

But what of the many other ecosystem components, such as plants, bacteria, and insects? Paleontologists are forever limited by the fact that the fossil record is restricted largely to the remains of hard parts such as bones, teeth, and shells. We have no fossilized insects, though their presence is confirmed by numerous bored holes and gouges present on many of the dinosaur bones. Nor is there any direct evidence of bacteria, though we can be certain they were ubiquitous. Even our knowledge of the vertebrates from this habitat remains grossly incomplete, with many currently represented only by isolated bones or teeth. Yet, despite this lack of evidence for most of the diversity present in this ancient ecosystem, all the primary ecological roles were undoubtedly filled. Chunks of fossilized wood and occasional carbonized plant remains provide meager evidence of plant producers. Vertebrates like the gargantuan sauropods and pug-nosed crocodiles were among the plant eaters, whereas the carnivores included predatory dinosaurs, snakes, and crocodiles. Finally, the abundant insect borings on dinosaur bones provide evidence of the detritus feeder category, those that made a living on the remains of others.

The key point here is that Madagascar has been home to a diverse assortment of communities during its lengthy tenure in the Indian Ocean (and for eons prior to becoming an island). Viewed from the perspective of deep time, the cast of characters in this subtropical paradise has changed almost continuously. Thus far, our paleontological expeditions have recovered little evidence of close evolutionary connections between the recent biota and those of the Late Cretaceous. The dinosaur-bearing rocks, on the order of 70 million years old, have yielded no bones of lemurs or chameleons, for example. Among groups like frogs, fishes, turtles, snakes, crocodiles, and mammals that are represented in both the Cretaceous and modern ecosystems, the fossil record has provided few indications of close relatives or basal stocks. It could be that we simply have not found the fossil representatives of most of these groups, but this seems highly unlikely given the great diversity of fossil forms that have been found. Rather, this pattern suggests that the ancestors of the highly specialized modern flora and fauna of Madagascar arrived sometime in the Cenozoic, long after the disappearance of the dinosaurs. To do so, ancestral members of many groups must have crossed the broad and deep Mozambique Channel, which separates the island from Africa and is 400 kilometers (250 miles) wide at its narrowest point. While some of these aquatic travelers might have swam, others most likely rafted across on chunks of floating plant debris. (In support of this seemingly improbable scenario, Caribbean iguanas have been documented traversing distances roughly equivalent to the width of the Mozambique Channel on floating logs.)

Despite such uncertainties, we can be sure that solar energy has flowed and nutrients have cycled continuously through ecosystems on the Great Red Island, sustaining a continual parade of diverse and often spectacular life-forms. Although the fossil and modern life-forms on Madagascar differ dramatically, the island's inhabitants share intimate bonds that stretch through the intervening millions of years. In a few cases, these bonds were likely forged by direct evolutionary descent; most, however, are ecological in nature. While experiencing continual changes, the island's ecosystems have persisted without hiatus. Energized by the sun, these complex communities of producers, consumers, and decomposers formed dense networks of relationships that morphed through time. Through the millennia, the carcasses of each generation served as fodder for the next, the constant cycling of nutrients blurring the distinction between eater and eaten. In this way, Madagascar's life-forms (and those elsewhere) can be linked not only through space but through time as well. The fates of all subsequent generations are inextricably tied to all those that came before.

Most popular portrayals of dinosaurs focus on dinosaur-dinosaur interactions—in particular, life-and-death struggles between predator and prey. Yet these ancient beasts did not exist in a biological vacuum. Without doubt, they were integrally embedded into their worlds, participating in expansive webs of relationships with other organisms. Many of these relationships are not preserved in the fossil record and thus are lost to us forever. Many others, however, have been uncovered through meticulous sleuthing, often involving collaboration of experts from a variety of disciplines. Others still can be speculated on with some confidence, informed by multiple lines of evidence and subject to future testing.

The remainder of *Dinosaur Odyssey* explores the dynamic world of dinosaurs, adopting a connections-based approach supported by the twin E-pillars of ecology and evolution. The web of life, it turns out, is composed of two distinctly different kinds of threads—those that link organisms at any given moment in time through the flow of energy (ecology), and those that link all life-forms through deep time via genetic information and shared common ancestry (evolution). Whereas ecology is concerned mostly with interconnections and interdependence, evolution is all about change and interrelationship. United, they provide a powerful and underutilized lens through which to observe the natural world.

So rather than merely asking who ate whom in the world of dinosaurs, the following chapters probe deeper, taking on more complex questions. Why did gargantuan body sizes evolve independently so many times within dinosaurs? What roles did dinosaurs play in Mesozoic ecosystems? How did drifting continents and shifting climates affect dinosaur diversity and distribution? What impacts did the interweaving of dinosaurian fossil threads have on the web of life that we are so fortunate to participate in today? Considerable time will be spent contemplating the living realm, because the fossil record is mute, or nearly so, with regard to several key topics. Our chief strategy, like that of scientists generally, will be to search for patterns and use these to infer underlying processes. As we embark on this journey, it's important to keep in mind that dinosaurs were simply the "charismatic megafauna" of their time. In order to tell their story, we must understand something of the other, less flashy players in this ancient drama, as well as the world they inhabited. The objective will be to construct a new looking glass with which to gaze on dinosaurs.

The resulting story comprises no less than a prehistoric odyssey, an epic journey through time that we have only begun to understand. In Homer's *Odyssey*, the hero Odysseus takes ten years to travel back to his native land after the fall of Troy, encountering numerous adventures along the way. Finally arriving home, Odysseus finds himself transformed. The Mesozoic odyssey of dinosaurs, although taking considerably longer (about 160 million years), was no less adventure filled and transformational. More and more we realize that dinosaurs were neither oversized lizards nor exactly like birds or mammals. They were something unique in the history of life, a blossoming of land-living animals that forever changed Earth's biosphere. Yet dinosaurs did not appear suddenly on Earth, as if from a vacuum. They were the culmination of a long, unbroken chain of evolving beings, a chain that has persisted to the present day. We now turn to that Great Story, which encompasses not only the dinosaurs but all life on Earth, including us humans.

The ascent of life from single-celled bacteria to dinosaurs. From bottom to top: a blue-green algae, or cyanobacteria; *Pikaia*, a wormlike animal close to the ancestry of vertebrates; *Cephalaspis*, an armored jawless fish; *Eusthenopteron*, a lobe-finned fish close to the ancestor of tetrapods; *Ichthyostega*, a fishlike amphibian; *Eryops*, a land-dwelling amphibian; *Chasmatosaurus*, a crocodile-like archosaur; *Lagosuchus*, a close dinosaur relative among archosaurs; *Apatosaurus*, a sauropod dinosaur; *Tyrannosaurus*, a theropod dinosaur; *Buteo*, a red-tailed hawk and an avian theropod dinosaur. The sequence does not depict a direct line of ancestors and descendants, but, rather, examples of representative life forms that branched off the dinosaurian thread during the epic of evolution.

2

STARDUST SAURIANS

BEING BOTH FAMOUS AND EXTINCT, dinosaurs tend to be portrayed as poster children for failure. Yet they are one of Nature's great success stories, having persisted for about 160 million years. (In contrast, primates have been around about 55 million years, our hominid cousins 7 million years, and humans a mere 200,000 years or so.) During that time, they evolved into a wondrous DIVERSITY of forms and occupied every continent, from pole to pole, becoming the dominant large-bodied land animals of the Mesozoic Era. Plus, they aren't completely extinct; abundant evidence confirms that birds are the direct descendants of dinosaurs, and so, in a very real sense, are dinosaurs themselves. With about ten thousand living species (and thousands of extinct forms), birds are merely the latest chapter in the lengthy dinosaurian story. And finally, extinction is not the shameful exception but the ultimate fate of all species. Over 99 percent of species that have ever existed on Earth are now extinct. Evolution depends as much on the deaths of species as it does on their births. Disappearance of the old makes way for the new.

It takes a blend of hubris and myopia to regard dinosaurs as unsuccessful. We like to view ourselves as the destiny of evolution, with the previous thousands of millions of years of evolution relegated to a warm-up act. Yet such anthropocentric hubris would not be possible (or at least palatable) without a severe case of temporal myopia. Most of our day-to-day concerns, and those of our hominid forebears, span mere seconds, minutes, hours, and days. Our lifetimes are measured in decades, whereas the age of the planet spans billions (thousands of millions) of years! Like cosmic distances, Earth history is so vast as to be virtually beyond human conception, and the tenure of our species amounts to a minute fraction of 1 percent of this duration.

When fathoming the immensity of geologic time, it helps to turn to metaphors. For example, if Earth history is a 24-hour clock, humans show up a few seconds before midnight. If depicted as the length of a football field, we appear just as time covers the distance and breaks the plane of the goal line. If compared to the old measure of an English yard—the distance from the tip of the king's nose to the tip of his outstretched hand—a single stroke of the royal nail file on his middle finger eliminates human history. My favorite comes from Mark Twain, who noted that if the height of the Eiffel Tower were to represent Earth history, the duration of humans would amount to the thickness of paint on the ball at the tower's pinnacle. Mocking human hubris, Twain went on to add that, of course, anybody could perceive that the tower was created for that thin layer of paint!

Along with Copernicus's sun-centered solar system and Darwin's common ancestry of life, recognition of DEEP TIME ranks as one of the most important discoveries in all human thought, forever altering our view of ourselves and the universe. Most profoundly, deep time reduces the duration of human existence to a veritable blink.

Of the hundreds of millions of species that have lived on Earth, we are the first to contemplate the vastness of geologic time. But imagine a world in which there was no linear time—no sense of events incessantly shifting from past to present. What if, instead, we viewed life as a continual sequence of repeating cycles, with no true sense of history? As difficult as this concept may be to grasp, it represents reality for peoples throughout most of human history and continues to be the norm for many indigenous cultures. As expressed by cultural ecologist David Abram in *The Spell of the Sensuous*:

> To indigenous, oral cultures, the ceaseless flux that we call "time" is overwhelmingly cyclical in character. The senses of oral people are attuned to the land around them, still conversant with the expressive speech of the winds and the forest birds, still participant with the sensuous cosmos. Time, in such a world, is not separable from the circular life of the sun and the moon, from the cycling of the seasons, the death and rebirth of animals— from the eternal return of the greening earth.[1]

For such preindustrial peoples, even unique and extraordinary events become fully assimilated into the concept of circular time and recurrent myths, morphing as the story passes through generations. Abram argues that the concept of linear time emerged alongside alphabetic writing, perhaps with the ancient Hebrews. Only with the written word can we permanently record details of historical or mythic events, creating a nonrepeating time line.

In the same way that scientists think of light both as particles and waves, life and time can be regarded as both an arrow and a cycle—a unique sequence of historical events and a repetitive set of cyclic processes. If viewed only in arrow fashion, as a set of nonrepeating events, the history of life lacks any processes. Yet our perspective is also incomplete without time's arrow, because evolution is a contingent phenomenon constantly buffeted by history. The scientific "discovery" of deep time greatly lengthened the arrow, bestowing on Earth and the universe an almost unfathomable prehuman history.

Deep time is a recent concept, percolating in the collective human consciousness only since the late eighteenth century. Prior to this time, experts generally agreed that Earth was about 6,000 years old. Several prominent individuals, including physicist Isaac Newton, attempted more precise calculations. Perhaps the most famous estimate came from Bishop James Ussher, who, after lengthy study of the Bible and other sources, determined that Earth was created on the eve prior to October 23, 4004 BC.

Science historians generally attribute the discovery of deep time to the eighteenth-century Scottish polymath James Hutton, often referred to as the father of modern GEOLOGY. In contrast to the "Neptunist" theory of the time, in which all rocks were thought to have precipitated from a single, planetwide flood, Hutton correctly inferred that heat within Earth's interior is responsible for creating new rock. This so-called Plutonist theory, detailed in Hutton's *The Theory of the Earth* (1775), also argued that Earth history is best understood as cycles of repeated events. These events, he claimed, required vast amounts of time, pushing temporal roots far deeper than previously suspected. Hutton's ideas spawned the principle of uniformitarianism, the concept that past geologic events can be explained by present natural processes, such as volcanism, uplift of Earth's crust, and erosion by wind and water. In short, the present is the key to the past. With sufficiently great time spans, recurring events literally recycle Earth's surface—wind and water erase mountain ranges and geologic uplift creates new ones.

Many of Hutton's key insights did not arise from empirical observations made in the field. Somewhat ironically, his worldview was heavily influenced by the mechanistic perspective of Newton described in the previous chapter. In an effort to forge Newtonian-type links between cosmology and geology, and thereby increase the scientific rigor of his chosen discipline, Hutton supported the idea of the cycle of time. So the man responsible for greatly extending time's arrow was also a major proponent of time's cycles. Only after publication of his major treatise did Hutton make several key field observations supporting this idea.

Chief among these observations was recognition of structures known today as unconformities. Within the borders of Scotland, Hutton found layers, or strata, of rocks that differed markedly in their orientations. The underlying rock layers were oriented almost vertically, whereas the overlying layers were horizontal. If a steady accumulation of sediment had occurred, the layers would all be horizontal. Hutton realized that a dynamic series of events must have generated this pattern. At a minimum, four successive steps were required: (1) an initial period in which sediments were deposited horizontally; (2) a subsequent period of deformation or tilting, followed by uplift; (3) erosion of the now-tilted sequence of rocks; and (4) a period of renewed sedimentation on top of the more ancient rocks. Here, then, was a repeating cycle requiring prodigious amounts of time— exactly what the Scotsman was looking for.

If Hutton is the father of modern geology, the nineteenth-century Englishman Charles Lyell is its first son. Lyell, in his famed three-volume *Principles of Geology* (1830–1833), followed Hutton in claiming that Earth's history is properly viewed in terms of everyday, slow-acting processes summed over geologic time. By documenting numerous examples in the field, Lyell became the champion of uniformitarianism, arguing that the vastness of planetary time was ample to account for the diverse array of geologic structures seen today, from the depths of the Grand Canyon to the dizzying heights of the Himalayas. His contemporary and friend Charles Darwin embraced the uniformitarian perspective for his theory of evolution. For Darwin, evolution occurred gradually, imperceptibly from the human perspective, with the abyss of deep time more than sufficient to accumulate the diversity of life we see around us. Darwin's adherence to gradualism and uniformitarianism is beautifully expressed in what is perhaps the best-known passage from his 1859 *On the Origin of Species*:

> It may be said that natural selection is daily and hourly scrutinizing, throughout the world, every variation, even the slightest; rejecting that which is bad, preserving and adding up all that is good; silently and insensibly working, whenever and wherever opportunity offers, at the improvement of each organic being in relation to its organic and inorganic conditions of life. We see nothing of these slow changes in progress, until the hand of time has marked the long lapse of ages.

Yet the uniformitarian view has not been without challengers.

Humans have a curious predilection to define problems in terms of dichotomies, and scientists are no exception. The principal challenge to uniformitarianism has been its logical opposite—catastrophism. The most brilliant catastrophist was Baron Georges Cuvier, an outstanding anatomist of the early nineteenth century and the first person to document extinction in the fossil record. Cuvier noted that certain kinds of fossils were restricted to specific rock strata, and he argued for a series of cataclysmic events that wiped out one group of animals after another. Like Lyell, Cuvier acknowledged that the brevity of human life spans blinds us to the effects of deep time. Rather than invoking slow, ongoing geologic processes, however, Cuvier saw signs in the rocks of tumultuous events such as flooding and volcanism. He noted that if these events were rare, occurring once in millions of years, they would be wholly outside the experience of humans. Yet such periodic cataclysms could, given the immensity of time, conceivably result in the world we see around us.

How are we to reconcile these polar opposites? Of the two ideas, uniformitarianism has fared better since the mid–nineteenth century, becoming the guiding principle of modern geology and paleontology. Many of the geologic processes seen today can indeed be identified in the rock record. Yet, as with many dichotomies, this one is false, because the two alternatives are not mutually exclusive. That is, catastrophic, onetime events could have happened within recurring cycles to shape Earth history. In recent decades, catastrophism has made a roaring comeback, due in part to the idea that an asteroid impact killed off the dinosaurs. This hypothesis has garnered widespread support, and other mass extinctions have been plausibly linked to collisions with extraterrestrial bodies. Whatever the cause(s), these widely separated events were incontrovertibly catastrophic, vindicating Cuvier. Today, both repeated cycles and unique historical events are regarded as integral to Earth's deep history.

While Lyell, a barrister by training, was successfully convincing the scientific world of the ubiquity of uniformitarianism, geologists in Europe were piecing together the geologic time scale. The largest and most encompassing units of time were referred to as EONS. All plants and most animals are restricted to the Phanerozoic Eon, which began about 540 million years ago. Eons are composed of ERAS, such as the Mesozoic, which in turn comprise PERIODS like the Jurassic. Smaller still are EPOCHS—for example, the recent Pleistocene or the present Holocene. These geologic units of varying magnitude are distinguished by the composition of their rocks and fossils, as well as by the presence of unconformities. Many subdivisions within the Phanerozoic coincide with the simultaneous disappearance of multiple groups of animals, now recognized as mass extinction events. Geologists also discovered that the same general sequence of rocks occurs in other parts of the world, allowing investigators to correlate geologic units in, say, North America, with those in Europe.

Although cognizant of deep time, nineteenth-century geologists were unable to determine the actual age of the rocks they studied. For example, they could not state with confidence whether dinosaurs lived tens of thousands of years ago or millions of years ago. Put simply, they lacked an appropriate clock. It was only after the discovery of radioactivity at the end of the nineteenth century that scientists devised an effective tool for this purpose.

Variants of a given chemical element possessing the same number of PROTONS in the nucleus but differing numbers of NEUTRONS are called ISOTOPES. Geologists make use of the steady decay of UNSTABLE ISOTOPES into more stable forms as clocks to measure the ticking of deep time. Specifically, they compare the abundance of naturally occurring radioactive isotopes (parent isotopes) with their decay products (daughter isotopes). Once the ratio of parent-to-daughter isotopes is assessed, the rate of decay is then used to calculate the time elapsed since decay began—that is, the age of the organic matter (in the case of radiocarbon dating) or rocks (most other methods). Fortunately, radioactive isotopes decay at different rates. For example, radiocarbon dating takes advantage of the relatively rapid decay of carbon-14 into carbon-12, estimating the age of organic remains less than about 60,000 years old. In contrast, potassium-40 decays to argon-40 more than two hundred thousand times slower, allowing age estimates in the realm of tens to hundreds of millions of years. Potassium-argon dating is a commonly used method for determining the age of dinosaur fossils and other Mesozoic remains. Based on radioactive isotope dating, Earth's age is now estimated at 4.6 billion years. Astronomers have used an entirely different set of techniques to assess the age of the universe at 14,000,000,000 (14 billion) years, so Earth has been around just over one-third of that duration.

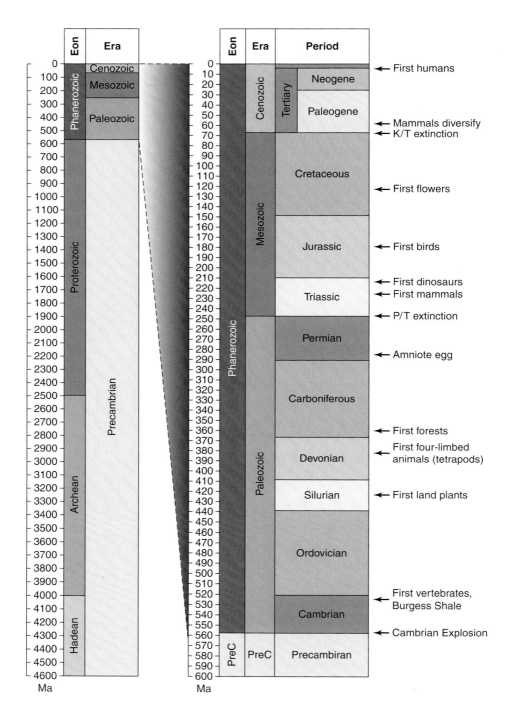

FIGURE 2.1

The geologic time scale. The left column depicts the entire 4.6 billion years of Earth history, noting eons and eras discussed in the text. The right column zooms in on the Phanerozoic Eon (the last 560 million years), showing eras and periods along with key evolutionary events. Both columns include a time scale (denoted in millions of years ago, abbreviated as "Ma") on the left side.

This temporal immensity offers a new perspective on the web of life. The web of life is vast, at any given moment encompassing the entire BIOSPHERE. Yet it has a historical aspect as well, encompassing most of Earth's duration. So life's web is both vast *and* deep. Viewed from a deep time outlook, the web is never fixed or rigid. Rather, its innumerable, interlinked parts are constantly morphing, as ecological threads spun from the sun's energy continually replace one another. Meanwhile, evolutionary strands, spun from genetic information, morph at a slower pace, transforming together with the lifeforms that temporarily house them.

Given that all organisms are descended from single-celled ancestors, the web of life forms an uninterrupted network extending back to the first BACTERIA, with inestimable numbers of transformations in the intervening eons. All lineages can be thought of as individual genetic threads, with the origin of new forms marked by the branching of distinct strands, some of which evolve into substantial threads themselves. Over time, the web became increasingly complex, with greater and greater numbers of interwoven threads. It's a fascinating thought experiment to select one particular strand and attempt to trace its convoluted evolutionary history through the abyss of time. With a solid deep time framework now in hand, I will attempt to tease out the thread leading to dinosaurs and take you on a whirlwind, "dinocentric" tour of the epic of evolution, from the big bang to the biggest animals ever to walk on Earth.

About 14 billion years ago, the universe flared forth in the big bang, unfurling space, time, light, and matter in a singular, immense explosion. Less than one minute after this "Great Radiance," protons and NEUTRONS emerged and combined to form nuclei of the simplest elements, mostly hydrogen and helium. About 300,000 years later, electrons combined with these nuclei for the first time, transforming matter into hydrogen and helium atoms. An infinitesimally small amount of these atoms would one day form the bodies of all dinosaurs. About 11 billion years ago, matter condensed into billions of swirling spiral galaxies. One of these would eventually become known as the Milky Way. Within each immense galactic pinwheel, billions of stars condensed from gas and dust, igniting nuclear fusion that converted vast amounts of hydrogen into helium. After consuming most of their hydrogen fuel, a large portion of these stars—those at least eight times the size of our sun—generated even greater extremes of heat and temperature, forging heavier elements such as carbon, nitrogen, oxygen, silver, magnesium, copper, and iron within their interior furnaces. Ultimately, the giant stars exploded, disgorging their heavy elements into surrounding galactic space. All life on Earth would one day be formed from this wandering stardust. The shock waves from these supernovae also triggered the formation of second-generation stars, resulting in additional cycles of stellar births and deaths.

About 4.6 billion years ago, our sun, an average-sized second-generation star, formed in an outlying spiral arm of the Milky Way galaxy, about 26,000 light-years from the nucleus. The remaining debris disk of gas and dust orbiting the primordial sun condensed into eight planets, together with a bunch of smaller planetoids and moons. The heavier elements—including the atoms that would one day walk the Earth as *Stegosaurus* and *Velociraptor*—were concentrated closest to the central star, forming the four rocky worlds we know as Mercury, Earth, Venus, and Mars. Abundant debris remained in the early solar system, however, and impacts with the planets were commonplace. Earth, the third rock from the sun, experienced a particularly violent impact from a Mars-sized world about 4.5 billion years ago. This fortuitous collision carved off a huge chunk of our globe, ejecting countless molten moonlets into Earth's orbit. In less than a century, these moonlets coalesced to form our moon. The moon stabilized Earth's axis of rotation and helped set up conditions for life, including the cyclical rise and fall of oceanic tides.

The extraterrestrial bombardment persisted through most of Earth's first eon, the Hadean, finally coming to an end about 3.9 billion years ago. The effects of this meteoric barrage remain visible on the heavily cratered surface of the moon, which has undergone minimal change during the succeeding billions of years. Earth, in contrast, has always been a dynamic world, receiving regular face lifts from above and below. The Hadean Eon—named after Hades, the mythological hell of the Greeks—is an apt name for Earth's primordial phase. As the planet was being pummeled from space, geologic turmoil boiled within, generating great bouts of volcanism on the heaving, molten surface. About 4 billion years ago, Earth's crust formed. Soon after (in the deep time sense), torrential rains began to fall, eventually forming expansive oceans.

Perhaps 3.8 billion years ago (the fossil evidence is equivocal), the dinosaurian thread came to life, literally, animated by single-celled bacteria. Referred to as PROKARYOTES, these microscopic bits of stardust—Earth's first life-forms—appeared shortly after the planet was cool enough to sustain water. The origin of life remains a mystery, but its rapid emergence suggests to some that life may not have been a lottery-like stroke of luck but a virtual inevitability, given the right conditions. Despite their diminutive sizes, and lacking even a nucleus, these early bacteria were complex. Like us, they had active metabolisms, consuming energy and carrying out hundreds of chemical transformations involving vitamins, proteins, sugars, nucleic acids, carbohydrates, and fats. The environment in which life first evolved remains in question; candidates include the planetary crust miles below the surface and hydrothermal fissures on the ocean floor.

Over the course of the Archean Eon, from 3.8 to 2.5 billion years ago, bacteria blossomed into an array of colors (green, purple, and red), shapes (oval, eel, and rod shaped), and metabolisms (fermenting, photosynthetic, sulfide producing, and oxygen producing). Remarkably, these millions of bacterial forms were not distinct species, at least not in the same way we differentiate lions and tigers, or tyrannosaurs and allosaurs. Like all life, they were able to reproduce, with the vital information of inheritance stored in specialized macromolecules of deoxyribonucleic acid (DNA). Lacking immune systems and rigid barriers, bacteria possess the capacity to share their DNA, enabling them to evolve quickly in response to changing environments. In a sense, then, the bacteria alive at any single moment in time, from the Archean to the present, can be thought of as a single, astoundingly variable species.

Those earliest Archean bacteria grabbed energy directly from the environment through chemical reactions, synthesizing organic compounds with the help of carbon dioxide. Then, sometime between 3.0 and 2.7 billion years ago, a subset of Earth's bacterial world "learned" to convert light energy from the sun into chemical energy, tapping into a virtually limitless energy source. These blue-green CYANOBACTERIA had discovered PHOTOSYNTHESIS, a process that uses carbon dioxide and water as raw materials and the sun as an energy supply, with oxygen generated as a by-product. This evolutionary event was critical on the road to dinosaurs, because virtually all subsequent life-forms on Earth have depended on the energy of sunlight captured by photosynthesis.

Initially, oxygen generated by the activities of water-living, photosynthetic bacteria was transformed into dissolved iron in the oceans and deposited as iron ore. Eventually, after saturating the seas and painting much of the land red, increasing oxygen levels in the Archean atmosphere triggered an environmental crisis. The biosphere turned poisonous for most bacteria, unable to metabolize this rogue element. Evolution responded to the so-called Oxygen Catastrophe of 2.4 billion years ago by evolving bacterial forms capable of respiring oxygen. Suddenly, using a form of controlled combustion, the dinosaur thread could breathe oxygen. The same substance that had been killing life now enabled it to flourish and diversify in new directions. Oxygen levels in the atmosphere stabilized at a concentration of about 21 percent.

The world had now entered the Proterozoic Eon, which lasted from about 2.5 billion years ago until 545 million years ago. The next major event in the history of the dinosaurian thread occurred about 2 billion years ago, when life became nested in cells with nuclei—the EUKARYOTES. Our present understanding is that certain bacterial cells invaded others as parasites. For some reason, the host did not digest the parasite, and the parasite did not kill its host. Over time, the intruders became permanent, cooperative residents, providing energy to the host cell in return for protection. The newly incorporated bacteria-turned-ORGANELLES were MITOCHONDRIA, destined to become the miniature powerhouses within the cells of dinosaurs (and all other complex life). Another key merger involved photosynthesizing bacteria, relatives of those ancient cyanobacteria, entering some cells containing mitochondria. The merging photosynthesizers were transformed into CHLOROPLASTS, energy-capturing organelles later passed on to plants and algae. The dinosaurian thread, however, had already veered off on a different evolutionary course, tracking with life-forms called PROTISTS.

All of the early eukaryotes were protists—slimy, nucleated organisms whose ranks would eventually include algae, amoebas, water molds, and slime molds.[2] After almost 2 billion years in bacterial guise, the dinosaur thread entered a lengthy protist phase that would last another 1.5 billion years. Throughout most of Earth history, all life, including the thread that would one day lead to dinosaurs, remained microscopic. Only in the last 700 million years has the living world exploded into the macro realm, with those slimy protists giving rise to all other complex organisms—plants, animals, and FUNGI. Life's early history of bacterial mergings also means that every cell of every complex organism that has ever lived on Earth has housed multiple collections of DNA acquired from single-celled ancestors. Dinosaurs were products of ancient evolutionary mergers, as we humans are today. So, contrary to the standard Darwinian outlook of "nature red in tooth and claw," there is much more to evolution than ruthless, dog-eat-dog-style competition. Cooperation, too, is a prevalent theme in the history of life, one that has been given short shrift by evolutionary biologists.

The first multicellular life-forms appeared about 1.2 billion years ago, consisting of simple cell colonies. Sexual reproduction originated around the same time, accelerating the rate of evolution. A bout of global glaciation embraced the world from 850 to 630 million years ago. At the end of this chilly interval, multicellular life experienced its first major blooming. This was the Ediacara Period (635–542 million years ago), the final days of the Proterozoic Eon. The "Garden of Ediacara," as it's been called, included an amazing diversity of animals, both predators and prey, ranging in size from millimeters to meters. Most had soft bodies, but a small portion developed calcified hard parts. Some resembled palm

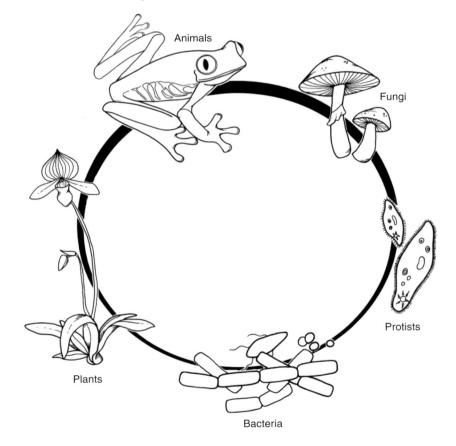

FIGURE 2.2
The five kingdoms of life: bacteria, protists, fungi, plants, and animals. This is one of several schemes used in recent years to organize the diversity of life on Earth.

fronds, others were jellyfish-like disks, and still others were reminiscent of mud-filled bags. Most of the strange creatures would soon be gone, but the dinosaur thread was there, having adopted a wormlike guise that would persist for millions of years.

The remaining 545 million years of Earth history comprise the present eon of "visible life," the PHANEROZOIC, which includes three periods: the PALEOZOIC ("ancient life"; 545–225 million years ago), MESOZOIC ("middle life"; 248–65.5 million years ago), and CENOZOIC ("new life"; 65.5 million years ago to the present). The Paleozoic, in turn, is subdivided into five periods: Cambrian, Ordovician, Silurian, Devonian, Carboniferous, and Permian.[3]

The Paleozoic opened with a bang. Often referred to as the CAMBRIAN EXPLOSION, this blast of evolutionary change about 530 million years ago resulted in an astonishing menagerie, including ancestors of most major animal groups alive today. Hard-shelled trilobites were here, as were clams, sponges, snail-like mollusks, and echinoderms—the radially structured precursors of starfish, urchins, and sand dollars. Interactions between predators and prey intensified, with teeth, drills, and claws evolving in response to protective shells. Of course, the thread that would one day connect to dinosaurs and other backboned animals (VERTEBRATES) was present as well. The closest known representative, *Pikaia*, still looked very wormy but now possessed multiple body segments. Segmented bodies were an important innovation, enabling complex interactions among parts. The notochord, a flexible rod running the length of *Pikaia*'s body, would one day become the multipart backbone that gives vertebrates their name. As in trilobites and some other Cambrian INVERTEBRATES, eyes and other sensory organs were now concentrated at one end of the body. The dinosaur thread had added sight to its sensory repertoire.

Committed to the vertebrate path, our thread of interest morphed multiple times and spawned numerous strands along the way. First came small jawless fish, circa 450 million years ago, which later developed elaborate bony armor within the skin. Gill arches were modified to form jaws, resulting in a succession of jawed fishes. By the Ordovician Period (about 460 million years ago), an interior skeleton had developed in fish, initially made of cartilage (as is still the case in sharks and rays) and later of bone. A mass extinction—the first of six in the Phanerozoic and the second-most devastating on record—hammered the biosphere at the close of the Ordovician (about 440 million years ago). Many families of trilobites and clamlike brachiopods vanished, along with entire reef-building communities. The dinosaur thread, however, managed to persist in fish form.

In the wake of the Ordovician extinction, bony fishes evolved alongside the highly successful nautilus-like ammonoids. Sometime during the Devonian, no later than 375 million years ago, one line of bony fishes carried the dinosaur thread from water to land. A long series of transitional fish-amphibians, propelled by stout, muscular fins, now endured the effects of gravity. Vertebrates were latecomers to the terrestrial realm, which had already been occupied for millions of years by all the other kingdoms (bacteria, protists, fungi, and plants) and even other animals (insects). Fungi and land plants had made the jump during the Ordovician, about 100 million years earlier, forming symbiotic partnerships that have persisted to the present day.

FIGURE 2.3

The dinosaur thread, showing the evolutionary pathway from stardust to dinosaurs, with major branches of life noted. From the center of the spiral to the outside, the branches off the main thread of dinosaur evolution are as follows: bacteria, protists, plants, fungi, sponges, jellyfish, flatworms, round-worms, mollusks, annelid worms, arthropods (the previous three forming a distinct group), echino-derms, lampreys, sharks, bony fishes, lungfish, amphibians, mammals, turtles, snakes and lizards, crocodiles, dinosaurs, and birds. The spiral galaxy at the core of the image symbolizes the fact that all life-forms are made of stardust. Life-forms are not to scale.

The second major mass extinction took place at the end of the Devonian, about 370 million years ago. Once again, the reef builders were decimated and, along with them, brachiopod clams, trilobites, and armored fishes (placoderms). Seemingly undaunted, the dinosaur thread tracked along with the amphibian niche for a substantial chunk of deep time, now borne by sprawled limbs and adapted to breathing air. These early amphibians also developed ears capable of detecting airborne sound waves, radically improving the lineage's ability to hear (previously limited to sensing low frequency sounds moving through the ground). Initially, members of this line were unable to venture far from their aquatic heritage. But that all changed during the Carboniferous (about 320 million years ago) with the appearance of the amniote egg. Featuring a semipermeable shell that allowed gas exchange without losing water, the egg acted like a portable sea, enabling AMNIOTES to move permanently onto dry land, no longer tethered by the need to reproduce in or around water. Assuming the guise of reptiles, the thread spread across Earth's landmasses, unaware of the profound changes that lay ahead.

Later in the Carboniferous, a new strand emerged that would ultimately generate countless generations of mammals and mammal relatives. It is here, in the latter half of the Paleozoic, that the evolutionary paths of humans and dinosaurs diverged. Our last common ancestor with the dinosaur thread was a lightly built lizardlike critter perhaps 20 centimeters (8 inches) long, with splay-toed feet and sprawled limbs transporting its long body and tail. This animal, akin to Carboniferous creatures like *Hylonomus*, offered few hints of the two grand lineages it was about to spawn. During the following Permian Period, another important parting of the ways occurred when the dinosaur thread split from that of lizards and snakes. Not long afterward, animals within the dinosaur thread first began to look distinctly dinosaurian. These so-called ARCHOSAURS and their near relatives—which would give rise to crocodiles and pterosaurs, as well as to dinosaurs—grew to larger sizes while retaining a predilection for carnivory. As the Permian drew to a close, all of Earth's continents docked together to form the supercontinent Pangaea, with profound implications for the subsequent course of evolution.

The boundary between the Permian and Triassic periods (and thus between the Paleozoic and the Mesozoic eras) was punctuated by another global paroxysm, the third Phanerozoic mass extinction and the worst overall. This time greater than 90 percent of Earth's species were exterminated. Somehow, for reasons that remain obscure, the dinosaur thread survived. And, with the ecological slate all but wiped clean, there was now plenty of space for evolution to explore new avenues. The archosaur-dinosaur line took full advantage, blossoming into a great variety of mostly midsized carnivorous reptiles that became interwoven into the Triassic web of life. (The evolutionary story of archosaurs will be explored in more detail in chapter 12.) Finally, late in the Triassic, the evolutionary thread that began in stardust more than 10 billion years earlier produced the first true dinosaurs.

We recognize three periods in the Mesozoic: the TRIASSIC (248–206 million years ago), JURASSIC (206–144 million years ago), and CRETACEOUS (144–65.5 million years ago). Dinosaurs underwent an initial evolutionary flourish during the Triassic. Among the earliest forms were meat eaters like *Eoraptor* and *Coelophysis*. Although these carnivores perpetuated the bloodthirsty legacy of their ancestors, they were forced to compete with many other, often larger archosaurs. The two major lineages of plant-eating dinosaurs, sauropodomorphs and ornithischians, also appeared at this time. *Plateosaurus* and other long-necked sauropodomorphs became the largest animals to have existed on Earth, as well as the most common big herbivores on many parts of Pangaea.

Following another mass extinction at the end of the Triassic, Jurassic dinosaurs seized most of the big-bodied niches, herbivore and carnivore alike. Sauropods pushed body size to record-breaking extremes that have never been topped. Also appearing in this period were ornithischian plant munchers like the four-footed stegosaurs and ankylosaurs, as well as the bipedal ornithopods. This was also a period of expansion among the meat-eating THEROPODS, with predators like *Allosaurus* and *Torvosaurus* exceeding 1,000 kilograms (2,200 pounds) for the first time in the long history of carnivory. Beginning in the latter part of the Jurassic, evolution was accelerated by the breakup of Pangaea, leaving dinosaur assemblages isolated on smaller landmasses to pursue independent evolutionary paths.

Dinosaurs reached their acme during the Cretaceous, after the dinosaur thread had spun off several new strands. Dome-headed pachycephalosaurs, horned ceratopsians, and crested hadrosaurs were added to the mix, at least in much of the Northern Hemisphere. Down south, a group of long-necks called titanosaurs reigned supreme. Abelisaur carnivores became top predators on many southern landmasses, while tyrannosaurs terrorized the north. At the very end of this period, *Tyrannosaurus* shared its evolutionary moment with *Triceratops* and *Ankylosaurus*.

The fifth and most famous of the major mass extinctions occurred about 65.5 million years ago, wiping out most of the remaining dinosaurs and virtually all other large animals alive at the time. Yet the dinosaur thread was not completely severed by this cataclysm. Sometime during the Jurassic, one group of small carnivorous dinosaurs found a way to become airborne, recruiting scales modified into filamentous, air-trapping structures. Together with the leathery-winged pterosaurs, these feathered wonders we know as birds graced the skies above Cretaceous dinosaurs. Unlike their larger, earth-bound brethren, birds managed to eke through the extinction bottleneck that brought an end to the Mesozoic. For the past 65 million years, they have diversified into a bewildering array of forms, from menacing ground-dwelling predators to tiny, hovering nectar sippers. The sheer tenacity of the dinosaur thread is underscored by the thousands of bird species alive today.

Within the last million years, barely a sneeze in deep time, the mammal thread attained self-awareness within a strand of gangly, two-legged primates, the most recent of which is *Homo sapiens*. Gifted with this newfound level of consciousness, humans became the first of Earth's organisms to contemplate the vast scope of the universe and, ultimately, the place of dinosaurs within the epic of evolution.

Most of us subdivide the living world into three major groups: plants, animals, and microbes or germs. Germs, of course, are regarded as something to be avoided or, better yet, vanquished. Biologists take a broader view based on formal rules of classification. Traditionally, their primary subdivisions of life have been referred to as KINGDOMS. The number and composition of kingdoms has changed over time, largely in response to changing conceptions about evolutionary relationships. One recent scheme, the one followed here, recognizes five kingdoms: bacteria, protists, fungi, plants, and animals. During the epic of evolution, the dinosaur thread was interwoven with three of these kingdoms: in temporal sequence, bacteria, protists, and animals.

Stepping back, several patterns emerge from this stunning, multibillion-year saga. One of these is an overall increase in complexity. The concept of increasing complexity, or progress, has a contentious history among biologists. Yet when viewed in a limited sense—as "production through time of increasingly complex and controlling organisms and societies, in at least some lines of descent"[4]—there can be little doubt this pattern applies to the history of life. Complexity clearly increases within the evolutionary sequence that includes hydrogen atoms, molecules, bacteria, protists, fishes, and elephants. The most complex structure known in the universe is also one of the most recent—the human brain. Of course, as noted in the previously cited quotation, not all evolutionary lineages increased in complexity. Indeed, most major groups, from bacteria through mammals, include descendants that differ only in minor ways from primitive forms. Nevertheless, a directional trend toward advancing complexity is apparent.

The increase in complexity was achieved through integration, yielding another pattern: smaller wholes becoming parts of larger wholes. Over and over we see the progressive development of multiple-part individuals from simpler forms. Thus, atoms become integrated into molecules, molecules into cells, and cells into organisms. At each new emergent stage, older forms are enveloped and incorporated into newer forms, with the end result being a nested, multilevel hierarchy.

So the epic of evolution has apparently been guided by two counterbalancing trends that we can refer to as complexification and unification. Complexity increased through unifying simpler forms into ever-more complex wholes. In this way, the wholes at each level fed back on one another, changing as necessary, such that all levels persisted. This journey was not an inevitable, orchestrated march but a creative, quixotic unfolding in which future biospheres could not be predicted. Perhaps most remarkable of all is the sheer scale of the transformations. A short version of our dinocentric story might go something like this: take a bunch of hydrogen atoms, wait several billion years, and you get *T. rex*.

So there you have it—a glimpse of the entire dinosaur odyssey in a single chapter. The rest, as they say, is details. But, oh, what discoveries await those who are patient enough to rummage through the details. The rest of the book explores various aspects of the world of dinosaurs, progressively building an ever-more detailed picture before attempting a step-by-step walk through the Mesozoic. First things first, however. With the dramatic preamble established, let's narrow our focus and concentrate on the diversity of dinosaur players that participated in the ever-changing Mesozoic biosphere.

A representative assemblage of dinosaurs from around the world. From left to right: *Kentrosaurus* (a stegosaur from the Late Jurassic of Africa), *Amargasaurus* (a sauropod from the Early Cretaceous of South America.), *Altirhinus* (an ornithopod from the Early Cretaceous of Asia), *Cryolophosaurus* (a theropod from the Early Jurassic of Antarctica), *Minmi* (an ankylosaur from the Early Cretaceous of Australia), and *Diabloceratops* (a ceratopsian from the Late Cretaceous of North America).

3

DRAMATIS DINOSAURAE

DINOSAURS AND ROCK STARS MAKE unlikely bedfellows. The serendipitous intersection of these organisms takes us back to the island of Madagascar. Our crew of Americans, Canadians, and Malagasy was excavating in a quarry known to us as "93-18," being the eighteenth site found during the 1993 field season. Despite its lackluster name, 93-18 is the most important fossil locality ever found on the Red Island. This remote hole in the ground near the village of Berivotra in northwestern Madagascar has been the primary focus of several expeditions, yielding remains of dinosaurian heavyweights and feather-weights, carnivores and vegetarians, as well as various ancient birds, crocodiles, snakes, lizards, and turtles.

An early sequence in the movie *Jurassic Park* depicts a team of paleontologists working feverishly to brush away dirt and uncover the complete, articulated skeleton of a "raptor" dinosaur interred just beneath the surface. Anyone who has spent much time digging up fossils can only laugh at this bit of celluloid science because the reality of excavation differs dramatically. The rock is typically hard, sometimes brutally so, often necessitating an array of tools to remove the overlying sediments, from picks and shovels to jackhammers and rock saws. Then we turn to hammers, sharp awls, and brushes to uncover the fragile bones. Frequent applications of a thin coating of liquid glue to newly exposed fossil surfaces slows the excavation process but strengthens the specimens prior to their removal. The fossils typically leave the site still encased in sediment, because the preparation laboratory is a much safer place to remove the surrounding matrix of rock.

In contrast, site 93-18 is a paleontological marvel. Remarkable, high-quality fossils lie entombed in relatively loose sediments, much of which resembles beach sand. Once the overlying sediment, or OVERBURDEN, is removed, team members can excavate with merely an ice pick and fine-bristled brush. Every few minutes, someone unearths a pristine fossil, exposed for the first time in over 65 million years. In some instances, the sandy coating almost falls away from the fossil, revealing a paper-thin limb bone, a convoluted vertebra, or a shiny, serrated tooth. We still employ the traditional technique of wrapping larger fossils in bandages of plaster and burlap, but the digging is as good as it gets, the closest I've experienced to the Hollywood version of collecting dinosaurs.

In 1995, odd elements of a small carnivore began appearing in the quarry. There was no evidence of a complete skeleton but rather bony parts of multiple individuals scattered over several square meters. Strangest of all was a toothed lower jawbone, the dentary. The few rear teeth preserved in the jaw were narrow, sharp and serrated, suggesting a carnivorous diet. But the frontmost tooth sockets were directed forward instead of upward, demonstrating that this particular meat eater was bucktoothed. During subsequent field seasons, we found the remains of additional individuals that together included about half of the bones of the skeleton. Some bones, like the femur (thigh bone), were represented by numerous examples. A few more bucktoothed jaws even came to light, demonstrating that the frontmost teeth were long and conical, more like spears than knives. Finally, in 2001, I and two other team members published a paper describing this little predator, which would have weighed about as much as an adult German shepherd.

Paleontologists and biologists have the honor of naming any new species that they describe in the scientific literature. They follow a standard established by botanist Carolus Linnaeus in the eighteenth century. Like many humans, every identified biological species—protist, fungus, plant, or animal—is denoted by two names: GENUS and SPECIES. Thus, for example, the first part of the name in both *Homo sapiens* and *Tyrannosaurus rex* is the genus, whereas the latter word refers to the species. Of the two categories, the genus is the more inclusive, frequently containing two or more kinds of species. Each two-part name, however, is unique to one species. Other than avoiding use of an existing name or being so uncouth as to name an organism after oneself, there are few limitations. In practice, the genus-species duo often refers to some unique anatomical feature or behavior, or to the place where the animal lived. *Tyrannosaurus rex*, for example, is the "tyrant lizard king."

We dubbed our new little beast from quarry 93-18 *Masiakasaurus knopfleri*. The genus name combines the Malagasy word for "vicious," *masiaka*, and the frequently used Latin for "lizard," *sauros*. Certainly, *Masiakasaurus* seemed appropriate for a small, apparently agile dinosaur with a presumed penchant for predation. It was in the second part of the name that rock music played a role.

Life in remote field camps tends to be minimalist. We sleep in tents, eat the same food for weeks, and make do without amenities like toilets or running water. So people frequently carry a few comforts from home, including music. The bulk of the *Masiakasaurus* bones were recovered during the 1999 season, when we toted a cassette tape player (remember those?) and a pair of small speakers to the dig site. The playlist included the music of Dire Straits, a favorite of expedition leader David Krause. Coincidentally, we seemed to find bones of the bizarre carnivore only when we listened to the now-defunct British band, led by singer, songwriter, and world-class guitarist Mark Knopfler. One evening in camp, while the crew sat on grass mats consuming the daily fare of rice, beans, and lukewarm beer, team member Cathy Forster suggested that it would be fitting to honor Knopfler's role as musical talisman by naming this dinosaur after him. Although unconventional, it was agreed, and so was borne *Masiakasaurus knopfleri*—the "vicious lizard of Knopfler."

We figured that our colleagues would give us some grief for the name, but none of us anticipated the media onslaught that accompanied publication in the British journal *Nature*. Over a three-day span, I conducted several dozen interviews for newspapers, radio, television, magazines, and Web sites; meanwhile, my coauthors were busy doing the same. The story went global, translated into numerous languages. To our dismay, however, the vast majority of questions had little to do with science and everything to do with our choice of moniker. Indeed, the unlikely marriage of music and monsters proved irresistible. I knew things had spun out of control when phone calls came in from *Rolling Stone* and *Guitarist Magazine*. To make matters worse, the press—in particular, the British tabloids—questioned our motives. Some suggested that we named this buck-toothed predator after Knopfler in reference to his physical appearance or perhaps because he was a bona fide rock dinosaur! Fortunately, Knopfler took the tribute in the spirit intended, stating, "The fact that it's a dinosaur is certainly apt, but I'm happy to report that I'm not in the least bit vicious."

Lest you think scientists are generally a boring lot, Knopfler is by no means the first musician to be so honored. There is a trilobite named *Milesdavis*, a wasp named *Mozartella beethoveni*, and a tiny midge named *Dicrotenipes thanatogratus* (*thanatos* is Greek for "dead," and *gratus* is Latin for "grateful," in honor of the Grateful Dead). Biologists have also honored creative characters both real and fictitious outside the music world; take the snake named *Montypythonoides*, the beetle *Garylarsonus* (after the famous Far Side cartoonist), the fossil turtle *Ninjemys* (after the cartoon *Teenage Mutant Ninja Turtles*), the large crustacean named *Godzillus*, and the menacing mite *Darthvaderum*.

Ever since Linnaeus, biologists have subdivided the natural world hierarchically, recognizing a series of successively more inclusive groups. We encountered the largest of these categories, kingdoms, in the previous chapter. Others, from most inclusive to least, include phylum, class, order, family, genus, and species.[1] Of these groupings, all except one are recognized to be arbitrary units invented by humans to help us sort through life's incredible diversity. For example, whereas most dinosaur genera (plural of genus) contain only one species, a single insect or plant genus may contain dozens or hundreds of species. Such decisions depend largely on the whims and trends of a given discipline.

The single nonarbitrary exception is the species. For most biologists, species are real entities with a unique ecological role, each consisting of individuals that interbreed and produce viable offspring. However, agreement on the reality of this concept in theory has not translated into an agreed-on practical definition. Over the past century or so, an often tedious and frequently trivial debate ensued over what is termed the "species problem." Dozens of species definitions were proposed, generally relating to the bias of the proposing investigator. Thus, the definition of species for a field biologist has differed from that of a systematist (someone who reconstructs historical relationships), which in turn has differed from that of a paleontologist. As a pair of paleontologists once proposed, "a species is a species if a competent specialist says it is."

Assessing breeding potential, let alone the viability, of offspring is highly problematic, for living species as well as for extinct species. And nature's species boundaries often do not match those set by biologists. Lacking solid evidence of breeding potential, biologists generally turn to species-specific features. Such features may be found in soft tissues (e.g., skin, feathers or hair), body size, DNA sequences, chromosome number, vocalizations, or some other behavior, all of which have little or no chance of becoming fossilized. Many species alive today are virtually indistinguishable based on skeletal CHARACTERS alone.

So when it comes to distinguishing species, paleontologists are severely hampered. They cannot observe copulatory successes and failures, and they almost never have access to nonbony tissues, let alone genes. Instead, vertebrate fossil hunters must base their estimates of species diversity on handfuls of fragmentary bones and teeth. In cases where key distinguishing features are not preserved, we're doomed to underestimate the diversity of fossil species. Conversely, every individual organism differs from every other in some subtlety; no two bones are exactly alike. So it is often difficult to sort out which features truly distinguish species. Among vertebrates, bony variations could be the result of growth, sex differences, or other individual differences. Assessing such variation with confidence requires a sufficiently large sample of individuals from a single species, a rarity for dinosaurs.

So how did we determine that *Masiakasaurus* represented a new species? In this case, it was reasonably straightforward to demonstrate uniqueness. Unlike mammals, dinosaurs and most other reptiles[2] tend to be conservative in their dentitions. That is, front teeth tend to resemble those at the rear of the jaw, with most variation occurring in size. Yet the Malagasy predator was unlike anything we had seen before, with specialized, forward-jutting conical front teeth and more standard-issue, serrated, bladelike teeth in back. The little jaws looked so unusual that, at first, we couldn't be certain to which group of animals they belonged. While attending a scientific conference, we showed the first enigmatic jaw specimen to some colleagues, whose identifications ranged from dinosaur to PTEROSAUR (flying reptile) to crocodile. Only later, after further study, did we demonstrate that several specialized features were also present on the lower jaws of *Majungasaurus*, the large carnivorous dinosaur found in the same deposits.

Although you're unlikely to discover one in your backyard, finding a new dinosaur is not as rare as you might imagine. *Masiakasaurus knopfleri* is one of about seven hundred named species of dinosaurs. With all the geographic and temporal gaps remaining in the Mesozoic fossil record, we can be quite certain that thousands of additional dinosaur species existed. A recent study by Steven Wang and Peter Dodson (University of Pennsylvania) suggest that we have only discovered about one-third of the total diversity of dinosaurs. Although a proportion of these unknown forms will never be discovered— either because their remains are not preserved in the fossil record or because they will never be unearthed by paleontologists—new dinosaurs are found and described monthly. In fact, more dinosaurs have been named in the past 25 years than in all previous years combined, largely because more people than ever before are out there searching. Yet identification is only the first step.

In addition to identifying and naming species, equally critical for most studies is an assessment of their evolutionary relationships. Darwin's theory of evolution makes it clear that all life on Earth can be viewed as a vast arborescent tree, with the branches and twigs comprising all species, living and extinct. The branching pattern of this tree, or at least a few twigs of it, is a key factor in most biological investigations, including those relating to fossils. For example, biologists cannot trace the evolutionary progression of a particular feature—say, wings—within a group of organisms such as insects or mammals until they have first reconstructed the approximate pattern of relatedness within the group.

Consequently, SYSTEMATICS, the study of biological ancestry and descent, has always been a major area of interest among evolutionary biologists. Here again, paleontologists must rely on assessments of skeletal characteristics. Instead of searching for features unique to a particular species, however, the focus turns to features shared among two or more species. Investigators apply an artful combination of observation, intuition, and experience to determine which characteristics are most important in delimiting groups of species. A variety of methods have been used to unravel the interrelationships of organisms. Many methods lacked rigor, however, and problems arose when different investigators disagreed about which characters were most important.

Since the 1970s, the rigorous approach most widely used to reconstruct the historical relationships of life-forms has been phylogenetic systematics, or CLADISTICS, originally proposed in 1950 by German entomologist Willi Hennig. Rather than relying on assessments of overall similarity, Hennig argued that determination of relationships should be based solely on shared, specialized characters that could identify natural groups, or CLADES, with a common evolutionary history. Investigators search for the evolutionary sequence of branching that is most strongly supported by the evidence. Such analysis would not be possible without high-speed computers, because alternative branching combinations, referred to as "trees," for a given bunch of species can number in the thousands or even millions. Mathematical programs sort through this maze of alternative trees in search of the simplest one, the one requiring the fewest evolutionary changes or "steps." The assumption here is that the most likely answer is the one that entails the fewest evolutionary changes. A distinct advantage of cladistics is its potential for repeatability and thus testability, allowing investigators to rerun their colleagues' analyses, adding data and making changes as deemed necessary. Cladistics, then, has enabled systematics to become far more rigorous and scientific.

Hypothetical relationships within a group of animals are represented by branching treelike diagrams called CLADOGRAMS, akin to human family genealogical charts (see figure 3.4 later in this chapter and similar figures in subsequent chapters). Cladograms depend on the unidirectional flow of time. Thus, common ancestors of related groups must arise before their descendants, in the same way that parents occur prior to their children. Equally important is the idea that closely related organisms share specialized features—for example, feathers or two-toed feet—that first evolved in a common ancestor. Ultimately, then, the branches of a cladogram cluster together closely related life-forms on the basis of specialized features that indicate shared common ancestry. Cladistics has become an essential tool in the paleontologist's toolbox. Its importance will become more evident as we proceed on our journey.

Before embarking further on the dinosaurian odyssey, however, let's consider what kinds of animals qualify as dinosaurs. How would you know one if you saw one? What were the major groups of dinosaurs? How many kinds of dinosaurs were there? For the purposes of our tale, it is important to be able to answer these questions with some confidence. We will, of course, encounter numerous other ecological players as we unravel these ancient dramas, but it is essential to have a basic understanding of dinosaurs in particular, because they are the focus of our attention.

An important note about terminology. As discussed above, paleontologists are now confident that birds are the direct evolutionary descendants of dinosaurs, and thus dinosaurs themselves in a very real sense. However, so as not to confuse the issue, for the remainder of the book I use the word *dinosaur* to refer only to NONAVIAN DINOSAURS, and then refer to bird members of the dinosaur lineage as necessary.

Dinosaurs occur in rocks of Mesozoic age, between 225 and 65 million years ago. This lengthy duration is divided into three distinct periods: from oldest to youngest, the Triassic, Jurassic, and Cretaceous. During the Mesozoic, dinosaurs underwent a number of phases or "dynasties," each with its own unique repertoire of players. At any given instant, there were hundreds of species of dinosaurs existing on landmasses around the globe. An average dinosaur species, like those of other extinct vertebrates, likely persisted on the order of one million years. So, given the 160-million-year interval of dinosaurian success, you might predict that there were an awful lot of different kinds of dinosaurs. This is exactly what the geologic record reveals.

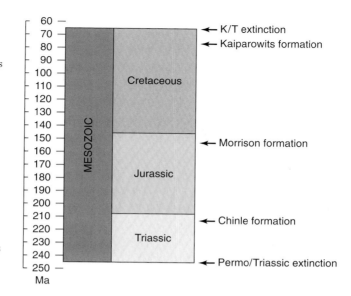

FIGURE 3.1

Mesozoic time scale, show-ing occurrences of two mass extinction events and three geologic formations from the Western Interior of North America. The three formations are featured in chapter 12 (Chinle Forma-tion), chapter 13 (Morrison Formation), and chapter 14 (Kaiparowits Formation); and the extinctions are dis-cussed in chapter 15. The time scale on the left depicts ages in millions of years ago (abbreviated as "Ma").

What unique features unite all dinosaurs and distinguish them from other reptiles, including lizards, turtles, snakes, and crocodiles? You might want to look to their special-ized jaws and teeth, their strange ornamentations, or perhaps their gigantic body sizes. Yet, while many groups of dinosaurs ultimately evolved these features in one form or another, answering this question requires identifying those characteristics common to the earliest dinosaurs and which were then passed along to the bulk of descendants. Strangely enough, it turns out that these defining features reside mostly in the hands and hips, relating to how these animals grasped and walked.

Early dinosaurs and humans have more in common than first meets the eye. One of the hallmarks of the human lineage is our opposable thumb, which enables us to touch the thumb tip to the tips of our fingers. More important, this arrangement allows us to grasp, a critical trait for tool makers and tool users. Although there is no evidence of dinosaurs using tools, they did have an opposable thumb of sorts. This structure differed considerably from that of humans, but it did permit early dinosaurs and many of their descendants to use their hands for grasping, which in turn may have been important for capturing prey or gathering plants, among other behaviors.

How were the hips of dinosaurs different from those of their predecessors? Most rep-tiles have sprawled limbs, with the thigh bone (femur) oriented almost horizontal and the body positioned low to the ground. The HIP SOCKET forms a solid cup that houses the head of the femur. This arrangement resists the sizable horizontal forces required to keep the body off the ground and to propel the animal forward. Think of a lizard in motion; its body and tail swing side to side in an undulating S shape as the animal advances, and the feet are well away from the body's midline.

FIGURE 3.2
Differences in body posture—a sprawled modern lizard (Komodo dragon) contrasted with an upright Cretaceous theropod dinosaur (*Utahraptor*).

In contrast, the first dinosaur possessed a more upright posture, in which the femur was held more vertically. This specialized, erect-limbed conformation, which evolved independently in mammals, elevated the body from the ground. The reorientation also greatly modified the direction of the forces incurred at the hip joint. Rather than generating horizontal forces at the center of the hip socket, an erect posture produces essentially vertical forces directed toward the top of the socket. As a result, the dinosaurian hip socket has a hole in the middle and a thickened, reinforced upper rim. This upright posture placed the left and right feet close together during locomotion, as confirmed by literally thousands of dinosaur trackways showing the limbs held directly beneath the body instead of out to the sides. Although this posture would later prove beneficial for the evolution of gigantic body sizes, the first dinosaurs appear to have been small.

Also unlike most of their reptilian forebears, the earliest dinosaurs were bipedal, walking solely on their hind legs. BIPEDALISM, another hallmark of the human line, carries the great advantage of freeing the hands and forelimbs for tasks other than locomotion. Not surprisingly, just as upright walking has featured prominently in hypotheses of the origins of hominids and human culture, bipedalism has been suggested as a prime mover in the early evolution of dinosaurs. Among carnivorous dinosaurs, for example, the arms and hands became modified for grasping and raking, functions with clear utility for predators. Later, a select group of carnivores modified these forelimbs into wings, permitting one bunch of dinosaurian heretics to take off as birds. Over the course of the Mesozoic, at least four groups of herbivorous dinosaurs independently reverted back to moving on all fours, but the earliest dinosaurs—such as the late Triassic *Eoraptor* from Argentina—appear to have been exclusively small, fleet-footed predators that ran around on their hind legs.

With these features in mind, we can now consider which kinds of animals qualify as dinosaurs and which do not. To begin, let me emphasize a key point. *Being a large, extinct monster is insufficient to confer dinosaur status.* In order to be a card-carrying member of the Dinosauria, an organism must meet several criteria. First (with the exception of birds), the animal must have lived during the Mesozoic. Beasts like the sail-backed carnivore *Dimetrodon*, which roamed during the preceding Paleozoic Era, are frequently included in the pages of dinosaur books, as are mammoths from the succeeding Cenozoic Era. Yet both are more closely related to you and me than to any dinosaur. Second, dinosaurs were exclusively land dwellers. The flying reptiles, or pterosaurs, close evolutionary cousins and contemporaries of dinosaurs, represent an independent evolutionary line. Pterosaurs evolved the capacity for flight separately from birds, adopting an entirely unique solution for becoming airborne (an elongated fourth finger that supported a wing membrane of skin).

The Mesozoic also witnessed the appearance and extinction of various large marine reptiles—PLESIOSAURS, ICHTHYOSAURS, and MOSASAURS—none of which qualify as dinosaurs. The unblinking terrestrial affinity of dinosaurs is remarkably rare among major, long-lived groups of backboned animals (vertebrates). Numerous times in the history of vertebrates, groups that originated as landlubbers evolved descendants that returned to the ocean. The best-known examples occur among mammals—in particular, whales and seals—but the evolution of secondarily aquatic vertebrates has happened numerous times in the last 300 million years. In short, dinosaurs were terrestrial animals with specialized hips and hind legs that lived during the final 160 million years or so of the Mesozoic.

Paleontologists recognize two major dinosaur groups. First are the lizard-hipped forms, or SAURISCHIA, including two major subgroups: the predatory THEROPODS such as *Tyrannosaurus*, and the herbivorous, long-necked sauropodomorphs such as *Apatosaurus* (a.k.a. *Brontosaurus*). Saurischians tend to show the primitive hip condition, with the PUBIS bone directed downward and forward. The second branch, all herbivores as far as we know, is composed of the bird-hipped dinosaurs, or ORNITHISCHIA. In this case, the rodlike pubis bone is rotated rearward to lie alongside another hip bone, the ischium. A similar arrangement of hip bones is found in birds. Yet, just to confuse matters, birds evolved not from the bird-hipped ornithischians but from the lizard-hipped saurischians. One specialized bunch of smaller-bodied theropods, the maniraptors, independently evolved a backward-facing pubis, and it is this group that gave rise to birds. Next I'll introduce the cast of dinosaur characters that will appear in greater detail in subsequent chapters.

The carnivorous theropods are the most famous and, in many ways, the most conservative of dinosaur groups. Whereas at least four groups of herbivorous dinosaurs independently shifted to QUADRUPEDALISM (four-legged locomotion) during the Mesozoic, always in association with large body sizes, the theropods remained bipeds, even truly gigantic

FIGURE 3.3

A selection of large prehistoric creatures frequently confused with dinosaurs. Pterosaurs are Mesozoic flying reptiles closely related to dinosaurs. Dimetrodon is a late Paleozoic member of the synapsids, a group closely related to mammals. Mammoths were late Cenozoic mammals closely allied with living elephants. Mosasaurs, ichthyosaurs, and plesiosaurs were all Mesozoic, seagoing reptiles.

forms like *T. rex*. As a result, the theropod body plan is recognizable by virtually any 5-year-old. With few exceptions, the general structure of theropod skulls remained closely similar to that of many nondinosaurian reptiles such as lizards. In many respects, the skull of *Tyrannosaurus* looks like that of a Komodo dragon on steroids, a statement that does not hold for many plant-eating dinosaurs. Theropods blossomed into a range of sizes, from chicken to tractor-trailer truck. Some specialized members of this clan apparently adopted more omnivorous diets, and a few even became devoted vegetarians. Yet the great majority stuck to eating meat, if their jaws and teeth are any indication. Now, I do not mean to imply that the theropod evolutionary tale lacks intrigue and adventure. Over the course of 160 million years, they spawned a tremendous variety of species, even if one excludes birds. Indeed, nearly half of all named dinosaurs are theropods.

Most theropod dinosaurs were well designed for a predatory lifestyle. Their jaws were lined with knifelike, serrated, and recurved teeth. The lower jaw had a centrally located joint that probably allowed a small degree of bending (presumably to reduce risk of injury during biting). In addition to thin-walled limb bones (a feature taken to the extreme in birds), many bones of the skull and vertebral column (backbone) were hollowed out by air sacs, greatly reducing overall body weight. The fingers and toes were tipped with sharp claws. And the body was held nearly horizontal, balanced over the hip joint, with a long, stiffened tail acting as a dynamic counterbalance to the business end up front.

The oldest-known theropods, *Eoraptor* and *Hererrasaurus*, are also two of the earliest-known dinosaurs. So primitive are their skeletons that some investigators doubt that these two animals are true dinosaurs, placing them instead as very close relatives. Found only in rocks of the Ischigualasto Formation of Argentina, their bony remains erode from rocks of Late Triassic age, about 225 million years old. Many other kinds of vertebrates inhabited these early days of the so-called Age of Dinosaurs, sometimes relegating theropods and their dinosaurian kin to bit roles in Late Triassic ecosystems. By the Early Jurassic, however, about 200 million years ago, dinosaurs had become the dominant large-bodied animals in terrestrial habitats worldwide. The first major success story among theropods was the COELOPHYSOIDS, small to midsized carnivores that originated in the Late Triassic and dispersed around the globe by the Early Jurassic. The small-bodied *Syntarsus* and *Coelophysis* are two well-known examples found, respectively, in Africa and North America.

All other major groups of theropods—CERATOSAURS, ALLOSAURS, COELUROSAURS, and birds—originated during the Jurassic, and many diversified in this period as well. For the first time, some theropods attained giant proportions, exceeding 1,000 kilograms (2,200 pounds). For the remainder of the Mesozoic, most terrestrial ecosystems were dominated by one or more theropods of at least this size. Although the primitive, easily recognizable body type established in *Eoraptor* persisted in later theropods, several interesting variants evolved in each of the major groups. Some ceratosaurs possessed skulls adorned with horns *(Ceratosaurus, Carnotaurus)* or crests *(Dilophosaurus)*. One specialized

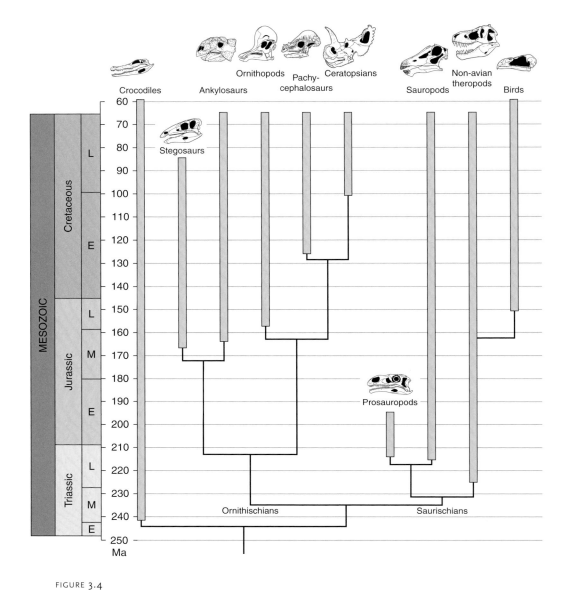

FIGURE 3.4

Evolutionary relationships and temporal occurrences of the major groups of dinosaurs, depicted in a branching diagram (cladogram). The geologic time scale (with numbers depicting millions of years ago, abbreviated as "Ma") is shown on the left. The shaded bars represent time intervals for which the various groups are known. The bird-hipped dinosaurs (ornithischians) are grouped on the left, and lizard-hipped dinosaurs (saurischians) appear on the right. Although not dinosaurs themselves, crocodiles are included to illustrate their relationship (together with birds) as the closest living relatives of dinosaurs.

theropod group, the SPINOSAURS, developed crocodile-like features, including elongate snouts and stout, conical teeth. The strangest forms evolved within a subgroup of the coelurosaurs known as MANIRAPTORA. Maniraptors include sickle-clawed terrors like *Velociraptor* and *Troodon*, informally known as "raptors," as well as OVIRAPTOROSAURS, with strange beaked skulls lacking teeth and often bearing a hollow crest. Also included in the maniraptor gang are THERIZINOSAURS, which appear to have switched from carnivory to herbivory on the way to becoming one of the oddest vertebrate groups ever.

Throughout the Jurassic and Cretaceous, the evolutionary fates of these major theropod clans waxed and waned around the globe as the supercontinent Pangaea fragmented into successively smaller continents. For example, by the close of the Cretaceous, the dominant carnivores on several Southern Hemisphere landmasses were all ceratosaurs; these included larger-bodied terrors such as *Carnotaurus* and *Majungasaurus*, as well as smaller-bodied counterparts, including the aforementioned *Masiakasaurus*. Meanwhile, much of the Northern Hemisphere was home to a diverse bunch of coelurosaurs, including giant TYRANNOSAURS like *Tyrannosaurus* and *Tarbosaurus* together with smaller *Velociraptor* and *Troodon*.

The other major group of lizard-hipped dinosaurs was the SAUROPODOMORPHS—vegetarians with a propensity for gigantism. This extraordinary assemblage of plant eaters appears to have originated and diversified during the Late Triassic. All members have small heads relative to their body sizes, as well as tall tooth crowns, elongate necks, and a pronounced claw on the first toe of the front foot. Paleontologists use the term EVOLUTIONARY RADIATION, or sometimes simply "radiation," to refer to the rapid origin of many species within a single evolving lineage. The first radiation within the sauropod lineage was the PROSAUROPODS, a midsized bunch of long-necked vegetarians that were the dominant large herbivores on land from the latest Triassic through the early part of the Jurassic (a period lasting about 30 million years). Included within the group were the largest herbivores to have evolved on Earth up to that time.

Although some prosauropods achieved giant sizes, most appear to have walked on their hind legs, perpetuating the bipedal legacy of the earliest dinosaurs. However, trackways attributed to prosauropods indicate that at least some forms were capable of moving around on all fours. A recently announced discovery of prosauropod embryos by Robert Reisz of the University of Toronto and others suggests that prosauropods may even have begun life walking on all fours. In all, they were a relatively conservative group, undergoing minimal skeletal modifications once the basic body plan was established. Their small and delicate heads likely included a small BEAK on each jaw. The teeth of prosauropods are relatively broad with coarse serrations, very different from their theropod cousins. Several forms, such as *Plateosaurus*, occur in BONEBEDS containing dozens of individuals, suggesting that at least some prosauropods were gregarious.

The second, more famous evolutionary radiation within this group was the SAUROPODS, sometimes informally referred to as "brontosaurs." This clan of giants includes the largest animals ever to walk the Earth, with the most outrageous examples likely exceeding 100 tons (220,000 pounds) and 33 meters (100 feet) in length. More than a hundred different kinds of sauropods have been recognized, divided into an array of evolutionary groups. Although small compared to their oversized bodies, the skulls of sauropods are wonders to behold. Some—like *Camarasaurus, Brachiosaurus*, and their close kin—have short skulls with blunt snouts and jaws bearing broad, spoon-shaped teeth. These forms also tend to have greatly enlarged bony openings for the nose located up front on the face (though the fleshy nostril was much smaller). In contrast, other sauropod varieties—such as *Diplodocus* and many TITANOSAURS—possessed elongate and slender skulls with narrow, pencil-like teeth limited to the front of the jaws. Many members of the latter groups have smaller, highly specialized bony nose openings, strangely retracted to a position atop the skull.

Sauropods were rare in the Late Triassic and Early Jurassic but exploded into a diverse array of forms by the Middle Jurassic (about 180 million years ago). Popular books often portray the Jurassic as the time of sauropods, including such behemoths as *Supersaurus, Diplodocus*, and *Apatosaurus*. Yet at least one group, the titanosaurs, persisted until the close of the Cretaceous, remaining "large and in charge" in many ecosystems around the globe, particularly in the Southern Hemisphere.

All known sauropods were quadrupedal, carrying their immense bodies on all four legs. Another adaptation linked to gigantic size was columnar limbs, straightened at the elbows, knees, wrists, and ankles. As in modern-day elephants, this arrangement permitted sauropods to support a large portion of their substantial weight directly on the limb bones rather than with muscles. Another example of skeletal change linked to gigantism occurs in the backbone. The neck and back vertebrae of sauropods are some of nature's outstanding engineering marvels, as well as being wondrous works of art. Each vertebra of the neck and back is hollowed out into a masterpiece of thin-walled struts and processes, built for anchoring muscles while minimizing overall mass. The bones were literally carved by air sacs attached to the lungs that invaded and excavated the vertebrae with the help of specialized bone-eating cells. This amazing system, widely employed today among birds, also allowed these Mesozoic vegetarians to evolve long necks through a combination of lengthening and/or increasing the number of neck vertebrae.

By most measures—including body shape, skull shape, teeth, bony ornaments, and locomotion—ornithischians were by far the most diverse of the great dinosaur groups. Included in this vegetarian menagerie were THYREOPHORANS ("shield bearers") or armored dinosaurs, including STEGOSAURS and ANKYLOSAURS; ORNITHOPOD ("bird-foot") dinosaurs, comprising HYPSILOPHODONTS, HETERODONTOSAURS, IGUANODONTS, and HADROSAURS; CERATOPSIANS ("horn-heads") or horned dinosaurs; and PACHYCEPHALOSAURS ("thick-headed lizards") or dome-headed dinosaurs. Pachycephalosaurs moved around solely on their hind legs. Conversely, stegosaurs and ankylosaurs were virtually all quadrupedal, and ceratopsians gave rise to both bipedal and quadrupedal forms. Finally, whereas all smaller ornithopods were bipedal, at least a few of the big hadrosaurs appear to have been locomotory "switch-hitters," moving on all fours at slow speeds and shifting to bipedalism when trotting. Although no ornithischians ever achieved the truly gargantuan proportions of sauropods, some, like the hadrosaurs, came close.

Bony ornaments abound in the bird-hipped dinosaurs. There are the horns and FRILLS of ceratopsians, the crests of hadrosaurs, the domes of pachycephalosaurs, the plates and spikes of stegosaurs, and the full body armor of ankylosaurs. Several of these ornamentation types are extreme; for example, a couple of kinds of long-frilled ceratopsians possessed heads about 3.3 meters (10 feet) long, exceeding skull length in all other terrestrial animals before or since. The possible functions of these bizarre structures are almost as diverse as the ornaments themselves, ranging from defense against predators to control of body temperature to the recognition of and competition for mates. These varied hypotheses will be explored in chapter 10.

All ornithischians bore horny beaks on the upper and lower jaws, presumably for cropping vegetation, and all species possessed an extra bone, the PREDENTARY, up front in the lower jaws. Although bird-hipped dinosaurs lacked the degree of dental variation seen in the herbivorous mammals that came after them, they do include a considerable diversity of tooth types. At one extreme are the stegosaurs, ankylosaurs, pachycephalosaurs, and hypsilophodonts, with relatively simple, leaf-shaped tooth crowns tipped with a variety of bumps and ridges referred to as denticles. These animals evidently carried out minimal food processing in the mouth. At the other extreme are the duck-billed dinosaurs (a.k.a. hadrosaurs) and the larger horned dinosaurs, in which the teeth were arranged into abutting columns so as to form a continuous cutting edge. These so-called DENTAL BATTERIES, composed of hundreds of replacement teeth, evolved independently in horned and duck-billed dinosaurs for efficient processing of large amounts of poor-quality plant matter.

The armored thyreophorans are easily recognized by their prominent body armor, composed of bony growths called OSTEODERMS (literally, "bone skins") that formed within the skin. Primitive forms, such as *Scutellosaurus* and *Scelidosaurus*, possessed parallel rows of osteoderms running down the body. In stegosaurs, this armor is limited mostly to the midline region, taking the form of narrow plates or stout spikes. The North American form *Stegosaurus*, the most famous member of this clan, displayed dramatic pointed plates along its back and lethal-looking spikes at the end of the tail. Yet stegosaurs as a group are known from much of the globe, each species with its own distinct osteodermal patterning. The Chinese form *Huayangosaurus*, for example, has erect plates along the top of the neck that grade into spikes over the back region. Ankylosaurs took an entirely different evolutionary tack. These prehistoric tanks filled in the spaces between rows of osteoderms to form a seemingly impenetrable armored shield over the head and body. One major group, the nodosaurids, is characterized by low, narrow skulls and spikelike bones projecting from the shoulder. The other group,

ankylosaurids, possesses higher, broader skulls, often with pointed osteoderms projecting from the rear corners. Rather than shoulder spikes, ankylosaurids "opted" for bony,
macelike tail clubs.

Among the ornithopods, heterodontosaurs were small (body lengths less than 1 meter,
or 3 feet), fleet-footed bipeds with pronounced lower canine teeth as well as elongate
forelimbs with well-developed grasping hands. The iguanodonts cover a range of body
sizes and tooth types, with the later forms tending toward bigger sizes and jaws with
closely packed columns of teeth. Most famous of these is the European *Iguanodon*, one
of the first discovered dinosaurs. Finally, the duck-billed dinosaurs, or hadrosaurs,
spanned much of the globe and became the dominant large herbivores in most ecosystems where they occurred. This success is often attributed to their highly specialized
jaws and teeth, with a pronounced horny beak up front and those elaborate dental batteries in the rear. Hadrosaurs differentiated into a remarkable array of species, recognized
almost entirely on the basis of unique skull features, including elaborate ducklike bills
and ornate crests. Crested hadrosaurs—bearing elaborate, hollow bony crests that communicate with the nasal cavity—fall within the group known as LAMBEOSAURINES, whereas
noncrested (or solid-crested) forms belong to the HADROSAURINES.

Pachycephalosaurs and ceratopsians have been united by cladistic analyses into a
single group, the MARGINOCEPHALIA. All are known exclusively from Cretaceous rocks
on the northern continents, with most found in deposits of Late Cretaceous age. The
name *marginocephalian* means "margin-headed," in reference to a bony shelf projecting
rearward from the back of the skull. This shelf is modest in pachycephalosaurs, yet
the skull roof above the brain compartment is thickened into a rounded dome, often
rimmed with additional ornamentations resembling outrageous Christmas tree baubles.
Pachycephalosaurs varied greatly in size, from small-bodied *Stegoceras*, at about 1.8 meters
(5 feet) in length, to sizable *Pachycephalosaurus*, about 6 meters (18 feet) long. The
ceratopsians, a diverse and highly successful group, lacked skull domes. Instead, they
extended the bony shelf at the rear of the skull into an elongate crest that was often
ornamented in the largest forms. Smallest and most primitive were the parrot-beaked
PSITTACOSAURS, all placed within a single genus, *Psittacosaurus*. Next in line was a series
of small to medium-sized forms such as the famous *Protoceratops*, with relatively large
heads bearing a pronounced bony crest or "frill." The largest and most specialized
members of the group were the CERATOPSIDS, such as *Triceratops* and *Styracosaurus*, with
giant frills and signature ornamentations over the nose and eyes, as well as along the
frill margins.

So there you have our paleontological playbill. Removed from the context of ecology and
evolution, brief descriptions like these are akin to looking through a family photo album
without knowledge of the stories that linked the people's lives within and between generations. Names and corresponding images, even if accompanied by a family tree, serve

as the poorest of proxies for understanding a given family. All too often, however, discussions of dinosaurs stop at or near this superficial level, satisfied with accounts of who lived when, who was related to whom, and (at least in the popular media) who ate whom. If we are to understand even the rudiments of the dinosaur odyssey, we must probe the depths of their world so as to visualize previously unseen connections, the actual warp and weft of Mesozoic habitats. Nevertheless, armed with a rudimentary understanding of dinosaur diversity, we can take the next step of our journey—an exploration of the physical world of dinosaurs.

Abelisaur theropod dinosaurs from the Late Cretaceous Southern Hemisphere, depicted in association with their landmass of origin. From left to right: *Carnotaurus* (South America), *Rugops* (Africa), *Majungasaurus* (Madagascar), and *Rajasaurus* (Indian subcontinent). The Southern Hemisphere continents are depicted in their approximate Late Cretaceous positions.

4

DRIFTING CONTINENTS AND GLOBE-TROTTING DINOSAURS

DINOSAURS GOT AROUND. Their fossilized remains occur on every continent and from pole to pole, encompassing a tremendous variety of ancient environments: forests, savannahs, deserts, seashores, river floodplains, and mountain valleys, among others. How did these prehistoric landlubbers disperse across oceanic barriers to populate the globe? Did they make marathon swims (particularly challenging for tyrannosaurs, given their tiny arms), or climb aboard vast mats of floating vegetation? No, we now know that earlier dinosaurs spread over great distances simply by walking, whereas many of their descendants hitch-hiked aboard continental rafts. And it turns out that the processes underlying continental movements have influenced life in other profound ways—in particular, through effects on CLIMATE. This chapter explores a pivotal subplot of the dinosaurian saga: the impact of physical processes on life.

During the early portion of the Mesozoic, including the Late Triassic when dinosaurs appeared, all continents were united in a supercontinent called PANGAEA. From the Jurassic onward, the Mesozoic world witnessed the fragmentation of this monster continent into successively smaller landmasses. Beginning in the mid-Jurassic, Pangaea split into northern and southern blocks. A massive tear in Earth's continental crust began in the east and proceeded slowly westward, as if the world were being unzipped. The widening rupture filled steadily with water, giving birth to a new ocean, the TETHYS SEA. The now-distinct yet still-gargantuan supercontinents, known as LAURASIA in the north and GONDWANA in the south, continued to fragment until, by the close of the Mesozoic, Earth

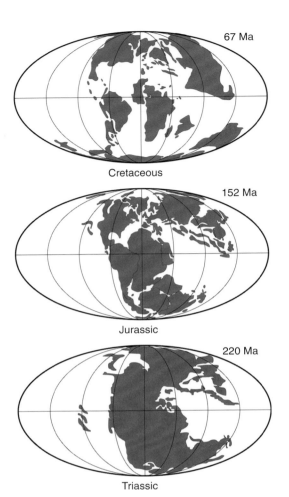

67 Ma

Cretaceous

152 Ma

Jurassic

220 Ma

Triassic

FIGURE 4.1

The sequence of breakup of the supercontinent Pangaea during the Mesozoic,
beginning with Pangaea during the
Triassic (bottom) and concluding with
isolated landmasses in the Cretaceous
(top) that closely resemble the arrangement present today. The abbreviation
"Ma" refers to millions of years ago.

looked very similar to the present-day Earth. Observed in a speeded-up, computer-simulated sequence, the breakup of Pangaea looks like a giant pane of glass shattering, as if smashed by an enormous asteroid. It turns out, however, that the mechanism behind this continental fragmentation was not extraterrestrial but "intraterrestrial."

Prior to the twentieth century, several people, including such luminaries as Francis Bacon and Benjamin Franklin, recognized that the east coast of South America and the west coast of Africa appear complementary, like two connecting pieces of a (very large) puzzle. However, it was an interdisciplinary scientist by the name of Alfred Wegener who, in 1912, first formally proposed the hypothesis of mobile continents, or continental drift, an idea that today we know as plate tectonics. PLATE TECTONICS refers to the concept that Earth's surface consists of large blocks, or "plates," that move across the planet's surface. Wegener's proposed mechanism to account for continental migrations invoked a pair of primary processes. First were tidal effects, essentially the gravitational influence of the moon on Earth. Second was the so-called pole-flight force—a supposed tendency for continents to move away from the poles as a result of centrifugal forces generated by Earth's rotation. Most investigators at the time (and since) regarded these forces to be far too weak to mobilize landmasses. In addition, for most of the twentieth century, Earth's crust was generally believed to be too brittle to allow for such continental fluidity.

However, Wegener's ideas failed initially not only because he lacked a convincing mechanism. In addition, the weight of accumulated evidence was simply inadequate to sway opinions from the traditional view of stable, fixed continents.[1] Along with the puzzle-like fits of certain continental margins, Wegener noted several other patterns that appeared to support his radical idea, including the cross-oceanic occurrences of certain kinds of fossil organisms and nearly identical rock sequences along the east coast of South America and the west coast of Africa. Although these same lines of evidence would later be cited as key data in support of the mobile-continent hypothesis, more evidence of patterns, as well as a more convincing mechanism, was needed to trigger a major paradigm shift within the earth sciences. It wasn't until the late 1960s that technologies advanced to the point that the necessary data could be collected and placed within a theoretical framework.

One key discovery was MIDOCEAN RIDGES and the structure of the rocks that surround them. Like the seam running around a baseball, ridges of basalt traverse Earth's oceans at a depth of about 2,500 meters (8,000 feet). These gargantuan seafloor swellings can be wider than the state of Texas. Were it not for all the water above, midocean ridges would be visible from the moon. The lavas erupting from these ridges form vast expanses of basalts, youngest at the crests and increasing in age with distance from the ridge. This pattern suggests that the seafloor has actually spread outward from the midocean ridges, and that is exactly what has happened. New OCEANIC CRUST forms along the ridges as the molten MANTLE wells up from below. Over time, the older, displaced crust is pushed farther and farther from its starting point, a phenomenon known as SEAFLOOR SPREADING.

An equivalent process occurs when molten hot spots occur below continents. In this case, however, it is the CONTINENTAL CRUST, less dense than oceanic crust, that wells upward and cracks, creating an elongate rift valley. The Great Rift Valley in Africa, famous for its diverse wildlife and fossilized remains of our hominid cousins, is an example of this process in action. In the future, tectonic forces will literally rend the African continent in two, creating a new ocean in the process. A series of such hot spots may have caused the initial breakup of Pangaea.

The recognition of spreading ridges allowed geologists to determine that Earth's crust is divided into approximately a dozen major plates that ride atop a semimolten layer beneath. This 1,000-kilometer- (over 600-mile-) thick outer layer of Earth, known as the LITHOSPHERE, is composed of a mixture of continental and oceanic crust plus a portion of the underlying mantle. Whereas virtually all other natural systems on Earth are driven by solar energy, plate tectonics feeds on energy from within the planet. The heat is generated deep within the core by radioactive decay of three elements—potassium, uranium, and thorium—as well as by residual heat from the planet's formation. The presence of heat at the core sets up a GRADIENT with the cooler surface of the planet, producing heat flow via a process known as CONVECTION.

We are most familiar with convection through boiling water, which creates cells that transport hotter water to the surface and cooler water back down toward the heat source. You might wonder how convection could occur in solid rock. It turns out that the kinds of rocks that occur in the mantle lose their rigidity at very high temperatures and pressures. The heated, supple rock expands, becoming less dense and rising slowly through the mantle until it contacts the lithosphere. Arriving at the crust, the material travels along the crust's surface as it cools, becoming increasingly dense until it finally begins to sink downward toward the core and begins the cycle anew. Think of a planet-sized lava lamp. Just as in the boiling water example, the result of this circulating matter is convection cells that enable heat to dissipate away from an energy source.

We can think of plate tectonics, the movement of crustal blocks at the surface, as merely a side effect of energy dispersal. As the upwelling heat reaches the planet's surface, it drives oceanic crust away from the midocean ridges. When relatively heavy oceanic crust eventually collides with lighter, less dense continental crust, the former is shunted back down into the mantle in a process known as SUBDUCTION. Subduction actually assists in driving plate motion, providing a pulling force that augments the pushing force from the spreading centers. Operating along continental margins, often in association with deep oceanic trenches, this "subduction factory" is an effective recycler, converting raw materials such as oceanic crust and sediments into magma and continental crust. Earthquakes, volcanism, and mountain building are geologic offspring of this process. In all, lithospheric plates engage in three general types of behavior that mark their boundaries: they diverge at midocean ridges, they converge in trenches or subduction zones, and they slide past each other along faults. Thus, for example, in western North America, a variety of phenomena—from Mount St. Helens and other volcanoes to mountain ranges like the Rockies and seismic shakers along the San Andreas Fault—result from the collision of two major plates: the Pacific Plate to the west and the North American Plate to the east.

Moving at a blistering pace of about 4 centimeters (1.6 inches) per year, roughly equivalent to the growth of human fingernails, plate tectonics might seem rather ponderous and ineffectual. But summed over geologic time, this relentless process gobbles up oceanic crust at a gluttonous rate of about 40 kilometers (25 miles) per million years. One complete rock cycle—from extrusion at a midocean ridge to subduction back into Earth's interior and finally back up to a midocean ridge—takes about 100 million years. As a result, oceanic crust is continually renewed and recycled; the oldest-known

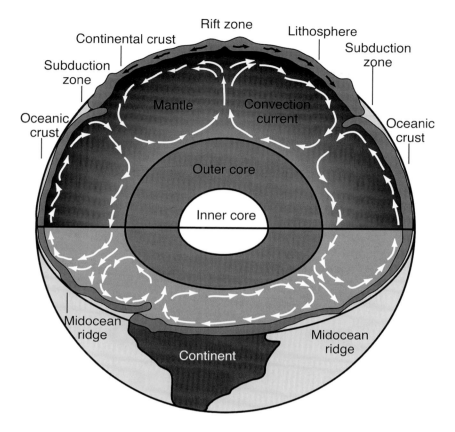

FIGURE 4.2

Simplified diagram showing the inner structure of Earth, as well as the presence of convection currents, or "cells," in the mantle, and the movement of oceanic and continental crustal plates (lithosphere) at the surface. See text for discussion.

crust of this type dates to about 200 million years ago. By contrast, the lighter, more buoyant continental crust is recycled much more slowly and therefore tends to be much older, up to 4 billion years in a few instances. Remarkably, it appears that Pangaea-like supercontinents have formed multiple times in Earth history, apparently cycling on a time scale of about 500 million years. Viewed from a deep time perspective, Earth is not a static body but a dynamic sphere undergoing constant upheaval, both from within and without.

General acceptance of plate tectonics in the 1960s caused a major paradigm shift within the earth sciences, one that fundamentally altered our understanding of dinosaurs and other extinct groups. After all, life could not avoid being caught up in this glacially paced dance of continental cleaving and collision. So it became clear that dinosaurs were a global phenomenon because they originated long before the breakup of Pangaea, allowing ancestral forms to spread over the entire supercontinent. As fragmented continental rafts charted their various courses, the dinosaurian passengers and other fauna and flora on board unwittingly went along for the ride. Once two blocks became separated, the plants and animals on each embarked on separate evolutionary journeys, spawning new forms along the way. Occasionally, two blocks collided or sea levels dropped, establishing land bridges and allowing floral and faunal exchanges between once-separate landmasses. Given this dynamic backdrop, understanding the pattern and timing of continental movements becomes essential to reconstructing the changing world of dinosaurs.

You might predict that the earliest dinosaurs would have been more cosmopolitan—that is, widely distributed—because they would have been able to move throughout much of the supercontinent. In contrast, you might guess that the latest dinosaur faunas would have been more localized or endemic, because many of the landmasses we recognize today had become isolated by the end of the Mesozoic. To a great extent, the record of land fossils supports these predictions, with Triassic and Jurassic forms exhibiting broader geographic distributions (COSMOPOLITANISM) and their Cretaceous descendants restricted to single, smaller landmasses (ENDEMISM). For example, certain varieties of Jurassic sauropods like *Brachiosaurus* (or at least very close relatives) are known from both the Northern and Southern hemispheres, including Africa, Europe, and North America. Conversely, the Late Cretaceous horned dinosaurs, or ceratopsians, are known only from Asia and North America, with the biggest forms (*Triceratops* and its ceratopsid kin) so far restricted to North America.

Some important exceptions, however, contradict this trend toward increasing localization, or endemism, in some instances highlighting persistent connections between continents and in others the reconnection of once-separated landmasses. For example, during the Cretaceous, North America appears to have shared brief land connections with Europe and Asia. Much of the supporting evidence comes not from plate tectonics but from paleontology, including studies of dinosaurs. Dinosaur faunas from the latter part of the Early Cretaceous in North America and Europe share a number of faunal similarities, including titanosaur sauropods, iguanodont ornithopods, and strange ankylosaurs. Then, in the Late Cretaceous, an intermittent corridor connected Asia and North America. A variety of closely related dinosaurs inhabited these two landmasses, including theropods—big-bodied tyrannosaurs and various smaller maniraptors—and plant-munching ornithischians—ceratopsians, pachycephalosaurs, hadrosaurs, and ankylosaurs.

Another unexpected example of Late Cretaceous cosmopolitanism—that is, the occurrence of closely related organisms on different continents—relates to fragmentation of the southern supercontinent Gondwana. In this case, the scientific debate has been contentious, leading to the development of rival hypotheses. Later I discuss this example in some detail because it highlights not only the process of science but also the potential for simultaneous illumination of two scientific fields: paleontology and geology/geophysics. It also happens to be a debate in which I have actively participated.

Eduard Suess, a nineteenth-century geologist and contemporary of Darwin, noted that a certain type of fossil fern, *Glossopteris*, was present in South America, India, and Africa. Suess thought that this pattern pointed to ancient land connections. Based on the fossil evidence, he coined the name "Gondwanaland" for a supposed supercontinent that had at one time included these three landmasses. The name derives from a region called Gondwana in central India, where Suess did some of his key geologic and paleontologic work. With Wegener's ideas still decades away, Suess did not entertain the notion of mobile continents (at least as far as we know) but, rather, thought that oceans had flooded the vast regions separating these landmasses. Despite the misunderstood mechanism, and the fact that Australia, Antarctica, and Madagascar were also part of the southern landmass, the shortened form of the name, Gondwana, has stuck.

According to most plate tectonic models, the major continental blocks we recognize today had separated from one another by the Early Cretaceous (about 120 million years ago). South America and Africa initially broke away as a unit from the rest of Gondwana, which included landmasses that would eventually separate into Antarctica, Australia, Madagascar, and the Indian subcontinent (including present-day India, Pakistan, Bangladesh, Nepal, and Bhutan). The Indian subcontinent and Madagascar later fragmented as a single block that persisted for several million years into the Late Cretaceous. Finally, Antarctica and Australia went their separate ways, and the Indian subcontinent split off from Madagascar to head northward, ultimately ramming headlong into Asia, causing the rise of the highest and most dramatic mountain range on Earth—the Himalayas. This scenario is based largely on geophysical observations yet has remained poorly tested with fossils.

Some new evidence challenges this traditional Gondwanan breakup scenario. Here again, Madagascar plays a starring role. Madagascar's isolation in the Indian Ocean dates to about 88 million years ago. Given such a persistent oceanic gulf, it's not surprising that approximately 80 percent of the current floral and faunal inhabitants of the island are unknown elsewhere. The generation of highly endemic faunas requires two essential ingredients: isolation and time. Madagascar has had plenty of both.

In chapter 1, I recounted the discovery of *Majungasaurus* skull bones on Madagascar. Even before we had finished unearthing the fossils of this ancient predator, we knew that it shared several specialized bony features with a beast named *Carnotaurus*, found far away in Argentina. *Majungasaurus* and *Carnotaurus* belong to a group of theropods known as abelisaurids, which have the same characteristic wrinkled texture on many of the skull bones. Their closest relatives are smaller-bodied, *Velociraptor*-like predators called noasaurids, including Knopfler's vicious lizard, *Masiakasaurus knopfleri*. Noasaurids and abelisaurids belong to a single evolutionary clan that we call abelisauroids or, more

Pan-Gondwana Hypothesis   Africa-First Hypothesis

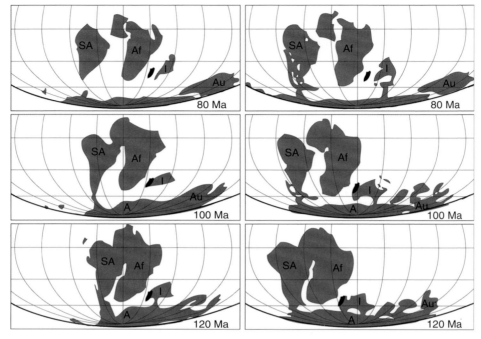

FIGURE 4.3

Diagrams depicting rival hypotheses of the breakup of the southern supercontinent Gondwana during the Cretaceous. Both hypotheses are shown in a sequence of three images, beginning in the Early Cretaceous (120 million years ago; bottom) and concluding in the Late Cretaceous (80 million years ago; top). Abbreviations: A, Antarctica; Af, Africa; Au, Australia; I, Indian subcontinent; Ma, millions of years ago; SA, South America. The island of Madagascar is the solid black region off the southeast coast of Africa. Note hypothesized differences in land bridges in the two scenarios. See text for discussion.

simply, ABELISAURS. To date, abelisaur remains have been found predominantly on Gondwanan landmasses: South America, Africa, Madagascar, and the Indian subcontinent. The only exceptions are a few strays unearthed in southern Europe, which may have shared connections with Gondwana during the Cretaceous. The striking similarity between *Majungasaurus* and *Carnotaurus* suggests that they were the evolutionary equivalent of first or second cousins, with a relatively recent common ancestor, perhaps even in the Late Cretaceous.

The abelisaur evidence—together with parallel patterns of kinship in other groups such as titanosaur sauropods, frogs, crocodiles, and mammals—indicated to our group that current knowledge of Gondwanan fragmentation was, well, fragmentary.

The closest relatives of these extinct species are not found in nearby Africa, as one might predict, but in India and halfway around the world in Argentina. Based on this evidence, our group postulated that many chunks of Gondwana were linked much longer than previously thought. We argued that the most likely corridor for dispersal would have been through Antarctica, which may have retained connections with "Indo-Madagascar" and Argentina long after Africa became isolated from the rest of Gondwana.

At first glance, a migration route through Antarctica might seem far-fetched, conjuring images of dinosaurs trudging like oversized emperor penguins across snow-covered terrain. But Late Cretaceous Earth was a hothouse world. No polar ice caps existed during most of the Mesozoic, and temperatures even at high latitudes rarely dropped below freezing. In fact, the fossil record reveals a broad diversity of Mesozoic plants and animals living in polar regions.

So what does the combined fossil and geologic data tell us about the history of Gondwana? Let's consider the range of scenarios that is consistent with the observed patterns. The presence of abelisaurs and other groups on multiple southern land-masses raises the possibility that their ancestors occurred throughout Gondwana prior to its fragmentation. According to this PAN-GONDWANA HYPOTHESIS, the presence of these animals on southern landmasses involved three events. First, abelisaur theropods (and other groups of organisms) evolved and spread throughout much of Gondwana by the Early Cretaceous, say, 140 million years ago. Second, Gondwana fragmented into smaller landmasses, stranding the life-forms on each. Third, the Mesozoic castaways on the daughter landmasses embarked on independent evolutionary trajectories, ultimately resulting, for example, in divergent groups of abelisaurs in South America, Madagascar, and the Indian subcontinent (and perhaps Africa) at the close of the Mesozoic.

The central problem with this hypothesis is timing. According to traditional reconstructions, most of the key landmasses were separate from one another by about 120 million years ago, fully 50 million years before *Majungasaurus* stalked titanosaurs on the island of Madagascar. This scenario is rendered suspect by the close evolutionary relationship between such forms as *Majungasaurus* and *Carnotaurus*. That is, the evidence suggests that the Malagasy and South American vertebrates present in the latest Cretaceous must have shared a much more recent common ancestor than permitted by generally accepted models of Mesozoic geography. To further complicate matters, we currently have no evidence of abelisaur theropods much older than about 100 million years ago.

Unbeknownst to us, while we were rethinking Gondwanan breakup based on fossil evidence, a group of geophysicists led by William Hay at the University of Colorado independently reached similar conclusions largely from geologic data. Their results indicated that Pangaea had fragmented into three continental blocks by the Early

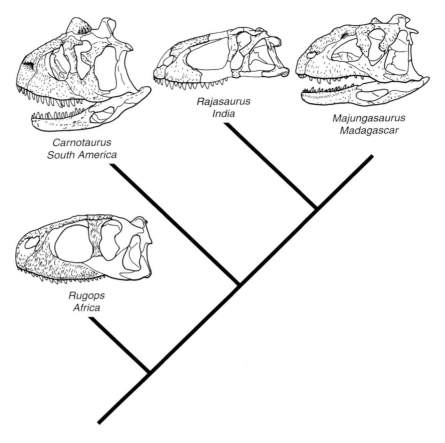

FIGURE 4.4

Branching diagram (cladogram) showing skulls and possible evolutionary relationships of four abelisaur theropods from different Southern Hemisphere landmasses. See text for discussion.

Cretaceous—the first consisted of North America and Eurasia, the second of Africa, and the third comprised South America, Antarctica, Australia, Madagascar, and the Indian subcontinent. They, too, postulated that South America and Antarctica retained terrestrial connections with Madagascar and the Indian subcontinent until sometime in the Late Cretaceous, about 80 million years ago. This lingering connection is thought to have been maintained in part by an isthmus-like land bridge between Indo-Madagascar and Antarctica.

Paleontologist Paul Sereno of the University of Chicago has attempted to salvage the traditional scenario with a modified pan-Gondwana hypothesis. Sereno has led several highly successful expeditions to northern Africa in search of Cretaceous dinosaurs. His discoveries include several abelisaur specimens, including *Rugops primus*, or "first wrinkle-face," which comes from rocks dated to about 95 million years ago. This makes

Rugops the oldest and most primitive known abelisaurid, more distantly related to *Carnotaurus* and *Majungasaurus* than the latter two are to each other. In Sereno's view, brief connections between key southern landmasses were present about 95 million years ago because of three narrow, intermittent land bridges: one between Africa and South America, another between South America and Antarctica, and a third between Antarctica and Indo-Madagascar. Sereno argues that these connections enabled abelisaurs and other Cretaceous beasts to spread through much of Gondwana many millions of years later than previously supposed.

The alternative hypothesis put forward by our group, and supported by the paleogeographic evidence of William Hay and his colleagues, postulates that, following the isolation of Africa, the rest of Gondwana remained connected until the Late Cretaceous, perhaps as recently as 80 million years ago. Connections could have been maintained by a pair of land bridges, one extending from Antarctica to Indo-Madagascar and the other from Antarctica to South America. These lingering links would have created a corridor for plants and animals to move between the eastern and western portions of Gondwana exclusive of Africa. Because this model suggests that Africa was the first landmass to separate from the Gondwanan supercontinent, it has been dubbed the AFRICA-FIRST HYPOTHESIS. Thus, according to us "Africa-firsters," whereas abelisaurs spread over much of Gondwana prior to its fragmentation, specialized members of the group, such as *Carnotaurus* and *Majungasaurus*, evolved after the isolation of Africa, when ancestors of the latter theropods moved between South America and Indo-Madagascar via Antarctica.

The jury is still out on the pan-Gondwana and Africa-first hypotheses, largely because of uneven sampling of fossils in space and time. Africa has yielded plenty of fossils from the middle of the Cretaceous (about 110–90 million years ago), but it has remained virtually mute as to the animals present at the end of the Cretaceous (about 80–65 million years ago). Conversely, India, Madagascar, and South America have generated a bounty of terrestrial fossils from the latest Cretaceous, but, with a few Argentine exceptions, almost nothing is known of their middle Cretaceous faunas. More detailed studies of the evolutionary relationships of key groups may help distinguish between these two hypotheses by showing how the faunas from each of the landmasses are related to one another. In addition, we need more geophysical data to test, for example, whether postulated land bridges actually existed during the periods proposed by each hypothesis. The Gondwana story is a fine illustration of serendipity in science. We never expected to recover fossils in Madagascar that would bring into question accepted notions of supercontinent break-up.

By now, it should be evident that the fragmentation, movement, and collision of Earth's crustal plates had dramatic effects on dinosaurs and other organisms. By continually changing the size, position, and connections of landmasses and oceans, plate tectonic processes provided an ever-shifting stage for life's drama, resulting in profound and irreversible consequences. Yet assessing the size and location of the stage through time is only the beginning. Equally critical in this drama is the climatic backdrop, which, it so happens, is also crucially intertwined with plate tectonics.

Thus far, this chapter has addressed one aspect of Earth's physical processes—the lithosphere or, more generally, the GEOSPHERE. Two other physical elements, or "spheres," deserve attention: air (ATMOSPHERE) and water (HYDROSPHERE). The systems of the geosphere, atmosphere, and hydrosphere are intimately interconnected and interwoven with a fourth partner, the biosphere. Together these interacting systems generate the complex world we see around us, as they did during the Mesozoic. Whereas the activities of the geosphere (and in particular the lithosphere) are driven by energy from deep within Earth, the remaining spheres feed off the sun.

Let's begin with the atmosphere, that astonishingly thin, life-supporting veil of mixed gases that envelops our planet. Compared to the 13,000-kilometer (8,000-mile) diameter of Earth, the atmosphere is positively diaphanous, thinning to nothingness only 100 kilometers (62 miles) above the planet's surface. If Earth were the size of a basketball, the atmosphere would be as thick as an enveloping layer of plastic wrap. Yet this veil of air performs several critical roles, mostly without our notice. The atmosphere circulates ingredients essential for life, such as the oxygen we breathe. Somewhat paradoxically, air is also a critical circulator of water, the universal medium for life on this planet. Through its protective layer of ozone, the atmosphere shields the planet from cosmic radiation. And most important, the atmosphere enabled life to take hold and flourish, occupying virtually every nook and cranny at or near the planet's surface.

Life is sustained by the lowest level of the atmosphere, the troposphere, which extends an average of about 11 kilometers (7 miles) above the surface. The wonderful array of cloud forms attests that the air is continually stirred and churned, in large part by convection. Just as in the lithosphere, atmospheric convection depends on heating from below, which causes air to rise, cool, and mix. However, the driving energy comes not from Earth's interior but from the sun, which generates temperature differences across land, sea, and air. Physical features such as mountain ranges and large bodies of water accentuate these differences. Life also plays a key role. In particular, the metabolic activities of trees release vast amounts of moisture, cooling the surrounding air. The single-greatest temperature difference at Earth's surface, however, is between the equatorial and polar regions, the result of the planet's orientation with regard to the sun.

Faced with such a complex abundance of temperature differences, nature does its level best to reduce them and achieve some sort of steady state or equilibrium. The resulting pattern of flow profoundly affects climate and local weather patterns. It also drives the wind, which generates mixing in both the oceans and the atmosphere. If circulation of the atmosphere resulted entirely from solar heating, energy flow would be relatively simple, involving gigantic convection cells extending from the equator to the poles. Heated air would rise at the equator, flow to the poles, cool and sink, and ultimately flow back along the surface toward the equator. Think of that lava lamp again. Winds in this system would flow only from the equator to the polar regions.

But Earth's rotation complicates atmospheric flow. Spinning like a top, the planet deflects winds and divides air into six latitudinal zones, three per hemisphere. That is, between the equator and poles, the atmosphere flows in *three* separate convection loops or cells: tropical, temperate, and polar. Warm equatorial air, known as trade winds, is separated from the cooler air of temperate regions by faster winds, known as jet streams, generally flowing from west to east. This phenomenon is familiar to anyone who has taken a coast-to-coast flight in North America; an airplane trip from New York to San Francisco lasts more than five hours, whereas tailwinds shorten the return journey eastward by about an hour. Although atmospheric circulation must have varied somewhat during the Mesozoic because of changes in continental positions, we can be confident that the same general patterns and processes applied because Earth has maintained its orientation and direction of spin for over 4 billion years.

In contrast, patterns of water circulation within oceans have been anything but constant. Whereas air moves freely around the atmosphere, landmasses impede the flow of seawater in the hydrosphere; large-scale movements of water, then, depend on the sizes, locations, and interconnections of continents. Nevertheless, many of the same principles and patterns of atmospheric flow also apply to oceans.

Earth is a water planet. Today, about 75 percent of its surface is covered with water, and this figure has been higher many times during the past. So the hydrosphere has a tremendous influence on Earth systems, including those on land. Oceanic circulation is driven by the movement of heat from warmer to cooler areas. The less dense, heated water warmed by the equatorial sun rises to the ocean's surface. It then migrates naturally toward the poles in an attempt to disperse its heat energy. As it progressively nears the poles, the water slowly dumps its heat into the air, warming the surrounding environment. The water, now cold and dense, sinks to a great depth and heads back toward the equator like a liquid conveyer belt completing its circuit.

One such cell incorporates the Gulf Stream, a vast river within the Atlantic Ocean that carries warm water from the Gulf of Mexico northward to Europe, where it has a major moderating effect on coastal climates. The city of London is close to the same latitude as Hudson's Bay in Canada, but the Gulf Stream provides London and the United Kingdom with a much milder climate. Thanks to this warming influence, there are locations in northwestern Scotland where palm trees can grow. If the Gulf Stream conveyer belt were to cease circulating, the north Atlantic would stop receiving an influx of warm southern water, and, within a few years, air temperatures in northern Europe would likely drop by several degrees. (There is some evidence that global warming has already caused a slowing of the Gulf Stream, something that has Europeans understandably concerned.)

Because continental positions and connections differed during the Mesozoic, we can be confident that the oceanic circulation cells during the Age of Dinosaurs were significantly different from those active today. To give the most obvious example, the pole-to-equator circulation of ocean currents would have been very different during the time of Pangaea, when there was one giant "superocean" referred to as Panthalassa. More recently, Antarctica was attached to South America until about 50 million years ago, long after the major extinction of the dinosaurs. This configuration of continents impeded circulation of circumpolar currents, undoubtedly with major effects on global climates. The initial breakup of Pangaea resulted in formation of the Tethys Sea, but well-developed circulation cells for ocean currents in the Atlantic likely were not possible until the Late Cretaceous. So it appears that oceanic circulation shifted throughout the Mesozoic in response to changing continental configurations. These fundamental shifts in ocean flow would have had profound effects on climate and therefore on the terrestrial world of dinosaurs.

Just two parts hydrogen and one part oxygen, water has remarkable heat-dispersing properties. It requires ten times as much heat to raise the temperature of water 1°C as it does to do the same for iron, so water is an excellent heat sink. Because fluids are mobile, heat spreads out and is stored more easily than in solids. Nature tends to distribute heat energy as evenly as possible. When heated, the surface temperature of solids tends to increase quickly, often becoming hot to the touch; yet just below the surface the substance remains cool because solids are typically poor conductors. In contrast, the tendency for fluids to disperse heat and reduce temperature differences means that heat distribution is more uniform in water. In general, an entire body of water must be heated for it to feel warm. These basic properties of heat flow explain why the beach is often hot to walk on during the summer, whereas the ocean only a few steps away feels cool or even frigid. The key point is that continents and oceans have radically different responses to heat, with dramatic consequences for weather and climate. Because the expansive oceans are much slower to heat up and to cool down than landmasses, their temperature-stabilizing effects tend to mitigate the rapid thermal shifts characteristic of continents. This phenomenon applies particularly to coastal regions. Landlocked Wichita, Kansas, near the center of North America, experiences extreme seasonal swings in temperature, with the average daily temperatures in summer and winter varying by about 20°C (50°F). Conversely, coastal San Diego, California, enjoys minor fluctuations of about 7°C (15°F) in average temperatures.

Like everything else in the world of dinosaurs, we can't observe Mesozoic climates directly. So we employ a variety of techniques to assess climates indirectly. Fortunately, climates modify sediments, leaving a variety of clues for paleosleuths. For example, we're confident that Canada and the northern United States experienced an ice age in the not-too-distant past because we can see where glaciers scarred mountains, where vast lakes once filled continental basins, and where ancient soils preserve chemical signs of water saturation in areas that are dry today. The fossils tell the same story; remains of Pleistocene plants and animals (e.g., woolly mammoths) indicate adaptations to cold weather. Similar types of clues can be found for the Mesozoic, revealing how climates back then differed from those of today.

Let's now consider the breakup of Pangaea from a hydrosphere perspective. Rather than focusing on landmasses, contemplate for a moment the changing sizes and positions of oceans. We'll start in the Late Triassic, about 225 million years ago, near the time of the first dinosaurs. With all landmasses united in Pangaea, temperatures over land fluctuated between much wider temperature extremes than today. Lacking the moderating effects of oceans between continental chunks, Pangaea underwent rapid and severe temperature swings. In coastal regions, this pronounced seasonality included hot and dry periods followed by extremely wet, often monsoonal conditions. Temperatures may have oscillated more than at any other time in the last 500 million years. During the Northern Hemisphere summer, the northern part of Pangaea was sweltering hot, while the southern portion stayed cool. Moisture from the oceans, particularly the growing Tethys Sea, was pulled into gigantic low-pressure cells, producing extensive monsoon rains in the coastal regions. This pattern reversed during the Southern Hemisphere summer.

Jumping forward in time to the Late Jurassic, about 150 million years ago, we find that Pangaea has fragmented into smaller landmasses surrounded by oceans. The close proximity of large bodies of water moderated the vast land temperature swings of earlier times, an effect amplified by rising global sea levels. Relatively warm conditions initiated in the Triassic continued, so there was no sign of polar ice or glaciers. Global sea levels were much elevated relative to the present day, in part because there was no water tied up in massive ice sheets. The greatly increased volumes of seawater led to flooding of several continents, including North America, Europe, and North Africa. Shallow seas submerged low portions of these landmasses. High sea levels and inland waterways had a strong stabilizing influence on temperatures and decreased seasonality, so the Late Jurassic saw warm, equable climates.

Moving on to the Late Cretaceous, about 75 million years ago, we find strong evidence of exceptionally warm and mild climates, even at high latitudes. Equatorial temperatures were likely close to modern values at the equator, whereas temperatures at the poles were much balmier than those of today. Polar temperatures during the Cretaceous have been estimated at 0°–15°C (32°–59°F), with a temperature difference between the equator and the poles of about 17°–26°C (63°–79°F), as compared with 41°C (105°F) today. The major moderating effect during this interval was exceptionally high sea levels; just as in the preceding Jurassic, many continents were flooded. Europe became an island archipelago. In North America, the Western Interior Seaway flooded the central plains for about 30 million years, separating the continent into eastern and western landmasses. Had humans been present, Late Cretaceous real estate agents would have had a grand time selling spectacular, subtropical beachfront homes in Utah and Arizona (though they might have been hesitant to disclose the need for electrified fences to keep out the tyrannosaurs). One investigator aptly described the Earth of this time as "wall-to-wall Jamaica," in stark contrast to today's Midwest weather of smothering summers and snow-draped winters.

Clearly, then, the varying sizes of continents and oceans through time have had a tremendous effect on global climates. Yet climate at any given moment is a complex phenomenon, involving interactions among Earth (geosphere), air (atmosphere), water (hydrosphere), and life (biosphere). In particular, tectonic activity influences Earth systems in ways that go far beyond effects on landmasses. To give one pertinent example, global sea levels are closely tied to plate tectonic activity. During intervals of active tectonism—that is, periods of rapid seafloor spreading—the spreading centers at midocean ridges become elevated, pushed upward by the pressure of molten rock beneath. This elevation of the seafloor, in turn, shrinks the absolute sizes of ocean basins, causing sea levels to rise. In extreme instances, rising sea levels exceed the limits of marine basins, causing great volumes of water to spill out onto the continents as inland seas.

Periods of active seafloor spreading such as the Late Cretaceous are also associated with have an increased rate of subduction of oceanic crust along plate margins. More subduction results in additional volcanism, as oceanic crust is thrust down into the mantle for recycling. Volcanic eruptions dump vast quantities of carbon dioxide into the air, raising atmospheric levels of this greenhouse gas and reducing the amount of solar radiation that can bounce back into space from Earth's surface. This captured radiation results in global warming, a phenomenon we are battling today for an entirely different reason (the release of greenhouse gases from human activities such as the burning of fossil fuels). During the Mesozoic, global warming melted the polar ice caps entirely, causing dramatic sea-level rises.[2] In short, more tectonism means higher rates of seafloor spreading, which in turn translates into raised spreading ridges, smaller oceanic basins, higher sea levels, higher subduction rates, increased volcanic pumping of carbon dioxide into the atmosphere, and global warming. Talk about interconnections!

Under the influence of seafloor spreading, landmasses have fragmented, dispersed, and amalgamated through deep time, with profound effects on global climate change and the diversity of life. The dinosaurian odyssey closely overlapped the breakup of the supercontinent Pangaea, a fact reflected in the global distributions and evolutionary history of the group. We have seen that the remains of dinosaurs and other extinct organisms can even be used as independent tests of hypotheses of continental breakup and collision. We have also begun to explore Earth's four major "spheres." For eons, circulation within Earth, the oceans, and the atmosphere has created complex systems driven by the movement of heat. Because the primary sources of energy—radioactive decay in the center of Earth and the sun—have never been exhausted, the gradients have persisted, allowing the dynamic systems of the lithosphere, atmosphere, hydrosphere, and biosphere to persist as well. Importantly, these systems do not work in isolation but in unison. Here, then, are the beginnings of an integrated view of life with consequences for how we regard not only the world of dinosaurs but our own place in nature as well.

While one *Tyrannosaurus* claims a *Triceratops* victim, others chase the remaining members of the horned dinosaur group across a clearing within a forest dominated by flowering plants.

5

SOLAR EATING

SUNBEAMS BREAK THROUGH THE FOREST CANOPY, revealing a group of *Triceratops* nearing a cascading stream. Swollen by seasonal rains flowing from mountain canyons to the west, the stream heads eastward toward a distant inland sea. Wispy fog dances through thick groves of deciduous and evergreen trees. The cool, long shadows of dawn give little hint of the blistering heat just hours away. It's springtime in Late Cretaceous Montana, about 67 million years ago. The *Triceratops* grouping consists of seven small juveniles, two mid-sized subadults, and three enormous adults. Their relative ages are evident not only in differences in body sizes but in the variable lengths of the facial horns and massive bony crest. The old male's skull exceeds 3 meters (10 feet) in length, among the largest of any land-living creature before or since. The animals seem intent on consuming the abundant fresh greenery, aided by a turtle-like beak and a dense battery of slicing teeth. The big male frequently scans the surroundings in search of predators.

Eventually the chunky herbivores amble to the stream bank and take deep gulps of cold water. The juveniles huddle together, flanked by bulky adults. A dark shape passes unnoticed on one side of the group. Suddenly the specter, a stalking *Tyrannosaurus*, explodes and closes in on one of the subadults with menacing strides. Before the other horned herbivores can react, the tyrannosaur gapes widely—exposing broad arcs of thick, serrated teeth—and plunges its jaws into the victim's chest and throat. The predator's massive head easily tears into the flesh, delivering a fatal wound. As the three adult *Triceratops* turn to face their attacker, two other tyrannosaurs emerge from the forest. One raises its head and lets out a deep-throated roar. Intimidated, the herbivores turn and flee, snorting loudly. The tyrannosaurs settle restlessly around the downed and dying *Triceratops* as it takes desperate gulps of air. Within minutes, the carcass is opened wide, and the tyrannosaurs are immersed in a bloody feeding frenzy.

Sound familiar? It should. Such scenes of prehistoric horror have been reenacted in movies, television documentaries, and books, fueled by audiences' twin appetites for dinosaurs and violence. One could be forgiven for concluding that dinosaurs did little else than kill or attempt to avoid being killed. Although life-and-death dramas were regular occurrences on Mesozoic landscapes, dinosaurs, of course, engaged in many more common and less exciting behaviors such as sleeping, defecating, and searching for food. The predator-prey scenario provides a myopic (albeit dynamic) view of dinosaurs. So let's step back and consider the same scenario from the broader perspective of ecology.

Ecosystems are complex entities in which matter and energy are in constant motion. A key maxim of ecology is *energy flows and matter cycles*. Rather than breaking down living systems into ever-smaller parts, ecologists attempt to understand these systems in their entirety, including interactions of organisms with one another and with the environment. Whereas an anatomist might make observations of an eagle's wing to assess its flight capabilities, and a systematist might focus on minute details of wing anatomy to unravel the eagle's evolutionary relationships, an ecologist would be more interested in the eagle as top predator in its ecosystem. The same spectrum of interests applies to those of us who study ancient life. Whereas some paleontologists use anatomy and computer simulations to test whether or not *T. rex* could sprint, and others investigate the evolutionary relationships of tyrannosaurs and birds, paleo-detectives with an ecological bent examine the role that *Tyrannosaurus* and other dinosaurs played in their respective ecosystems. Let's have a look, then, at who else lived alongside *T. rex*.

Tyrannosaurus is best known from the Late Cretaceous Hell Creek Formation, broadly exposed in Montana and the Dakotas. The plants of the Hell Creek ecosystem, like most plants before or since, provided the major energy source for their ecosystems. Thanks in large part to research by paleobotanist Kirk Johnson of the Denver Museum of Nature and Science, we know that the Hell Creek flora included more than 300 different kinds of plants. Angiosperms, or flowering plants, are by far the dominant and the most diverse group, comprising about 90 percent of the total. But a range of nonangiosperm trees were present as well, including evergreen conifers, ginkgos, and cycads. This ancient forest superficially resembled some living mixed deciduous and evergreen broad-leaved forests, although closer examination would show it to be distinct from any modern community. For example, lobe-leaved angiosperms occur in greater abundance here than in any living forest. And in stark contrast to modern-day Montana, the presence of palm trees indicates a warmer climate.

Through photosynthesis, these plants transformed sunlight into their "bodies," including trunks, stems, roots, branches, leaves, seeds, and flowers. In doing so, they converted solar energy into something edible by plant-eating consumers like *Triceratops*. Many horned dinosaurs, in turn, became fodder for carnivorous consumers (often called "secondary consumers") like *Tyrannosaurus*. Here, then, is a simple prehistoric food chain, with energy passing from plant to herbivore to carnivore.

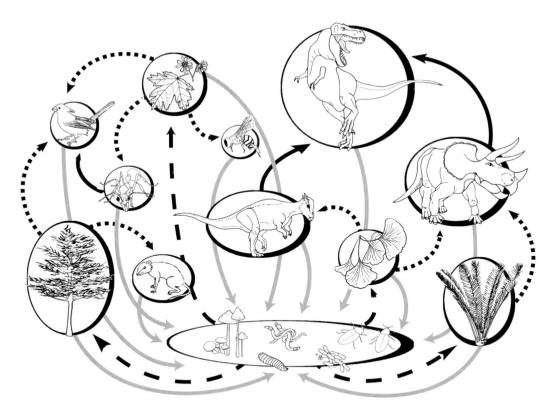

FIGURE 5.1

A partial Late Cretaceous food web, showing postulated ecological relationships in the Hell Creek ecosystem of western North America (about 67 million years ago). Major steps in the transfer of energy are as follows: (1) from plant producers to herbivores (dotted arrows); (2) from herbivores to carnivores (solid black arrows); (3) from plants, herbivores, and carnivores to decomposers (gray arrows); and (4) from decomposers to plants (dashed arrows).

But things were not so simple. While energy can be traced through such food chains, there are many more ecological players—additional chain links—to consider.

Though we lack direct fossil evidence, studies of modern ecosystems give us confidence that Late Cretaceous plants participated in cooperative interactions with many other species. Some of these life-forms were critical for plant survival. For example, fungus and bacteria helped draw nutrients from the soil. Other organisms were necessary for plant reproduction. Fruit- and seed-eating mammals and birds probably dispersed seeds that passed through their guts. Meanwhile, insects (and perhaps birds and other animals) did their part through pollination of flowering plants. So the energy flow captured by the plant producers depended on a range of life-forms.

Moving up the food chain to the next rung or TROPHIC LEVEL, the host of Hell Creek herbivores included *Triceratops* and at least one duck-billed hadrosaur *(Edmontosaurus)*. Among other devoted vegetarians were bipedal pachycephalosaurs—including little *Stegoceros* and

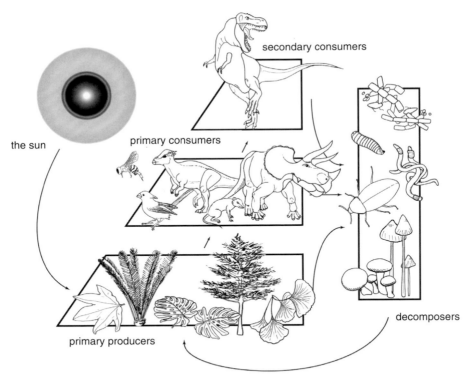

FIGURE 5.2

A "trophic pyramid" for the Hell Creek ecosystem, showing the flow of solar energy from plants (primary producers, of which there are many) to herbivores (primary consumers, of which there are fewer) and then to carnivores (secondary consumers, still fewer). Decomposers dispose of the remains of producers and consumers and ultimately participate in feeding valuable nutrients back to plants.

big-bodied *Pachycephalosaurus*—plus several small ornithopods from a group known as thescelosaurs. Quadrupedal herbivores included a couple of heavily armored anky-losaurs, one of which was the group's tanklike namesake, *Ankylosaurus*. There may have been one or two other horn-headed behemoths besides *Triceratops*. Finally, the menagerie of plant munchers was likely rounded out by a few theropods, including an ostrich-like ORNITHOMIMOSAUR and a maniraptor cousin of *Velociraptor* called an oviraptorosaur. Many paleontologists think that ornithomimosaurs and oviraptorosaurs made the dietary shift from meat to plants (a topic we will return to in chapter 8); alternatively, it's feasible that they were omnivores, eating plants and meat. Finally, there was a long roster of nondinosaurian herbivores, including many species of turtles, mammals, and particularly insects.

Despite their diminutive sizes, plant-eating insects, by force of sheer numbers, may have rivaled or exceeded herbivorous dinosaurs in ecological impact. Although insect body fossils are exceedingly rare in the Hell Creek Formation, abundant fossilized leaves preserve traces of chewing by insects. Conrad Labandeira of the Smithsonian Institution and his col-leagues closely studied these traces, which occur on the leaves of most major plant groups.

The researchers documented dozens of distinct feeding behaviors, placed into such descriptive categories as "leaf mining," "margin feeding," "hole feeding," and "piercing and sucking." Many of the trace types are associated with a single plant species, suggesting that insect varieties were specialized to feed on particular plants. Remarkably, some of these traces closely mimic modern examples, allowing the researchers to infer the identities of many insect trace makers. Overall, this ecosystem was home to a great diversity of plant-devouring insects, rivaling modern temperate ecosystems. In addition to keeping the plants in check, herbivorous insects provided an abundant food resource for many other species, likely including birds, mammals, pterosaurs, other insects, and perhaps even a few dinosaurs.

So a diverse range of species, from the gargantuan to the microscopic, participated in the flow of energy from plants to herbivores. Complicating matters further, whereas some of these prehistoric vegetarians were undoubtedly specialized feeders, many likely derived their share of the energy flow from several different plant species. Our *Triceratops* victim may well have been made of solar energy that had passed through multiple kinds of plants. You can begin to see that the ecosystem is complex, with more links in the food chain.

Next are the carnivores. While certainly the king of the beasts in its ecosystem, *T. rex* was by no means the sole meat eater. Many other dinosaurs sought the same high-quality fleshy resources. Most conspicuous were the "raptor"-like forms such as *Troodon*, *Ricardoestesia*, and *Dromaeosaurus* (or a near relative). Only a fraction the size of *T. rex*, these predators likely pursued smaller prey, from baby dinosaurs to birds, mammals, fishes, and lizards. They may have scavenged the kills of tyrannosaurs as well. Finally, there were a number of other nonbird and nondinosaur carnivores, both predators and scavengers, which might have partaken of a dead *Triceratops*, with crocodiles topping the list. Still more chain links.

The *Tyrannosaurus*-eats-*Triceratops*-eats-plants scenario also ignores an entire ecological category: decomposers. Predators never consume all energy available in their prey, frequently leaving behind plenty of meat and other tissues. Leftovers rarely last long, however, because a host of decomposing life-forms waits in the wings, including bacteria, fungi, and many insects (detailed in chapter 9). Tyrannosaur leftovers were fodder for flies, ants, beetles, and termites, among others. Over weeks and months, our hypothetical horned dinosaur carcass will be consumed by a multitude of microbes and insects. Decay and decomposition will dispatch teeth, bones, sinew, fat, blood vessels, nerves, and organs. A substantial portion of the remains will ultimately be converted into soil nutrients that, in turn, will sustain future generations of plants and start the cycle anew. The same fate awaits all players in this drama.

Upon closer examination, then, our simple food chain closes to form a loop: producers to consumers to decomposers, back to producers, and so on. Moreover, the chain is merely a few links in an intricate, solar-powered web that connects all organisms within an ecosystem. Plants and other photosynthesizers form the web's productive core. All others must consume the energy provided by producers—either directly, by eating producers, or indirectly, by feeding on other consumers. *Triceratops* and its herbivorous cohort are primary consumers, whereas *Tyrannosaurus* and other meat-eaters are secondary consumers.

All dinosaurs occupied one or the other (and, likely in a few instances, both) of these ecological roles. As a unique category of consumers, decomposers specialize in breaking down organic matter and recycling its nutrients. Together, this web and its components are sustained by two constant, integrated processes: flowing energy and cycling matter.

This is a good point to reflect on the ecological role of *Tyrannosaurus rex*. Thanks to an abundance of fossils (nearly forty articulated skeletons) and a barrage of recent studies employing an array of techniques, *T. rex* offers the best window into the ecology and evolution of giant predatory dinosaurs. Lately, there has been much discussion of the dietary habits of *Tyrannosaurus*. Indeed, this animal's status as "Tyrant King" has even been challenged. The controversy boils down to a single question: was *T. rex* a voracious predator, as long assumed, or did this famous carnivore make its living by scavenging the dead? The question, which has been around for about a century, was raised most recently by Jack Horner, a paleontologist at the Museum of the Rockies in Bozeman, Montana. Several other paleontologists countered with evidence supporting the predator hypothesis. The media, ever in search of new, easily digestible angles on science, quickly anthropomorphized the debate into "proud hunter" versus "lowly scavenger," with the implication being that demonstration of the latter would effectively relegate *Tyrannosaurus* to the status of prehistory's biggest maggot. News of the debate spread widely within the general public, as evidenced by the fact that the predator-scavenger question has become one of the most frequently asked of dinosaur paleontologists.

Let's take a brief look at the evidence. Horner has argued that multiple bony, telltale clues are inconsistent with a predatory lifestyle for *Tyrannosaurus*, suggesting instead a scavenging niche. These clues include (1) a relatively small eye that would have prevented long-distance spotting of potential meals; (2) an enhanced sense of smell, useful for sniffing out distant rotting carcasses; (3) limb proportions suggestive of a slow, lumbering gait, inconsistent with chasing down prey; (4) tiny forelimbs apparently useless for holding, let alone dispatching, prey; and (5) broad teeth unlike the narrow, bladelike teeth of most theropods.

To my mind, each of these points has been convincingly refuted by other paleontologists, including Jim Farlow (Indiana University–Purdue University) and Thomas Holtz (University of Maryland). These authors note that a predator does not need a large eye in order to see well. Indeed, although small for the gargantuan size of the animal, the eye of *T. rex* was large in absolute terms, with increased light-gathering capacity and perhaps a high level of acuity. The bony evidence is consistent with acute smelling ability, yet it seems reasonable that this sense could have served equally well to detect the living as well as the dead. With regard to foot speed, a number of studies support Horner's contention that *Tyrannosaurus* was closer to the lumbering end of the spectrum than the sprinting end. However, other studies demonstrate that the likely prey animals, including hadrosaurs and ceratopsids, were even slower (see chapters 7 and 8). So the giant predator appears to have been fully capable of catching its probable prey. It's hard to argue against the fact that *T. rex* had ridiculously small arms for its size; at 6 feet,

2 inches (188 centimeters) tall, my arms are roughly the same length as those of the 12-meter- (40-foot-) long dinosaur. Yet many carnivores—think of wolves, orcas, and secretary birds—are highly effective predators without using their arms as weapons. Similarly, crocodiles and toothed whales (together with a host of extinct, presumably predatory animals) have managed to remain at the top of their respective food chains for millions of years despite having broad teeth.

It's also important to consider the ecology of the predator-scavenger question and search for clues among the living. Today, carnivore species regularly compete for kills, often with the predator having its meal usurped by larger or more numerous challengers. *Tyrannosaurus*, many times larger than other predators in its ecosystem, would have been the undisputed bully in conflicts over carcasses. Yet there remains the question of quantity. Was there sufficient carrion available to sustain populations of 5,000-kilogram (11,000-pound) carnivores? It's difficult to answer this question in the face of important unknown variables such as the metabolic rate of *T. rex* (which influenced daily caloric needs) and the sizes of herbivore populations. Nevertheless, a study by Graeme Ruxton and David Houston at the University of Glasgow attempted to estimate all the relevant parameters, modeling the Late Cretaceous ecosystem closely after the modern Serengeti. The authors' tentative conclusion, based on a number of untestable assumptions, is that there may have been sufficient carrion available to support populations of gargantuan carnivores like *Tyrannosaurus*.

Yet very few living carnivores are exclusive scavengers. Several species long regarded as scavengers, such as hyenas, obtain much of their diet from predation. The vast majority of carnivores turn out to be opportunistic, killing when necessary yet content to feed on a fresh carcass killed by some other animal. Among vertebrates, the only animals that subsist almost solely on carrion are twenty-three species of vultures, which are able to soar from carcass to carcass, expending a minimum of energy in the search for food. In contrast, *T. rex* had to carry its substantial bulk overland, a nontrivial task even on level, open terrain. Interestingly, in a follow-up paper, Ruxton and Houston concluded that obligate scavengers must be large, soaring fliers like vultures. Clearly, this strategy was not available to *T. rex*.

In my view, nothing about the skeleton of *Tyrannosaurus* precludes a predatory lifestyle. Because devoted scavengers are extremely rare in ecosystems today, it seems conservative to conclude that the biggest land-dwelling carnivore of all time mimicked the majority of modern meat eaters in being both a vicious predator and an opportunistic scavenger. Horner's scavenging hypothesis was important in that it forced us to reconsider some basic assumptions. Yet I think we can be quite certain that *T. rex* sat alone atop the food web in latest Cretaceous North America.

Nevertheless, life at the top was no picnic for the Tyrant King. Ecosystems, like all natural systems, run on energy. In the case of Earth systems, this energy comes almost exclusively from sunlight, wind, or moving water.[1] Although most life-forms are entirely dependent on the sun for their energy supply, only a few select groups of organisms—plants, certain bacteria and algae—can convert sunlight into usable energy. These organisms, the only truly productive life-forms, act as reservoirs of carbon and generators of oxygen. Through the metabolic process known as photosynthesis, green plants and some bacteria capture energy from sunlight and transform it into sugar—the universal food of life.

The cells of photosynthetic organisms contain chlorophyll, a green pigment that absorbs packets of sunlight (photons) and uses them to fragment water into hydrogen and oxygen. Whereas much of the oxygen escapes into the air, the hydrogen is bound up with available carbon to generate sugars, the building blocks of carbohydrates, fats, proteins, genes, and other biomaterials. Plants also use sugars to make cellulose, which stiffens plant cells and allows them to grow skyward. Thus, producers generate food in the form of their tissues, as well as oxygen as a metabolic by-product, sustaining all other kinds of life. Nonphotosynthetic organisms reverse this process, "breathing" oxygen via respiration, burning carbohydrates, and releasing carbon dioxide. With the exception of life's origin almost 4 billion years ago, arguably the single-most important event in the history of life took place about 2 billion years ago with the evolution of photosynthesis (see chapter 2), because photosynthetic organisms provide the basis for the balance of biological diversity.

From the shapes of individual leaves and trees to the architecture of an entire forest, plants seem to be optimally "designed" to "grab" as much of the sun's energy flow as possible. What do plants do with it? Remarkably, only about 1 percent of the captured energy goes into making more plant matter. A considerably larger portion of the incoming radiation is simply reflected back into the atmosphere, which heats up the plant's exterior. About two-thirds of the absorbed energy moves water through the plant to the leaves and then out to the air, where it transforms from liquid to gas. The interface between plant and atmosphere occurs through microscopic pores called stomata, on the undersides of leaves. This physiological process, with the lengthy moniker of "evapotranspiration," dissipates a lot of heat. Collectively, forests cool themselves by dumping the heat of water vapor into the atmosphere.

Forests are planetary air conditioners. Like invisible fountains, they spray prodigious volumes of vapor into the air. The rising water vapor forms rain clouds, which in turn reflect sunlight, cool the surrounding region, and return moisture to the forest. Thanks to the near-ubiquity of plants on land and the incredible rates of energy flow, evapotranspiration is a key factor controlling local, regional, and global climates.

Plants do not gorge themselves on available energy. They show great sensitivity to environmental conditions and react accordingly, controlling this flow with remarkable exactitude. If soil moisture drops below critical levels, plants avoid drying out by throttling back on rates of water uptake, photosynthesis, and evapotranspiration. In other words, plants regulate the flow of energy according to changing external conditions. They hunker down in the bad times and thrive in the good times, making maximal use of available energy.

This trend toward energy dispersal continues further along the food chain. Herbivores are able to convert and use only about 10 percent of the plant energy they consume. Remember, this is 10 percent of the 1 percent used by plants to make their bodies, equivalent to a meager 0.1 percent of the original pool of energy from photosynthesis. The loss of energy to heat continues at the next trophic level, with carnivores eating herbivores, and again at the following level, with carnivores consuming other carnivores. The energy is concentrated as it moves along the food chain, so meat is a higher-quality source of energy (with more usable energy per unit weight) than plants. Yet because these energy transfers up the food chain tend to occur at less than 15 percent efficiency (with most being lost to heat), consumers at successively higher trophic levels must live off an ever-shrinking pool of energy.

This dramatic reduction in available energy as we climb the TROPHIC PYRAMID explains why ecosystems contain many more herbivores than carnivores, and also why the total BIOMASS of plants greatly exceeds that of plant eaters (see figure 5.2). Rampant heat loss constrains most of an ecosystem's available energy to the lower levels. Imagine 1 square mile (2.6 square kilometers) of Serengeti grassland. That chunk of savannah might sustain one hundred wildebeest, such that each animal lives off about 1 percent of the grass that is consumed. Yet, a standing crop of one hundred wildebeest might be necessary to feed a single lion, so the lion gets only about 1 percent of the herbivores. Consequently, in order to sustain itself, a pride of lions requires a territory on the order of 100 square miles (260 square kilometers).

The perpetual drop in useable energy limits trophic pyramids to no more than five or six levels. Assuming that a given kind of plant diverted only 10 percent of the energy it captured toward making new tissue, and that each level of consumer could use only 10 percent of its energy intake, the carnivores at the fifth tier of a trophic pyramid would be subsisting on 0.00001 percent (one hundred thousandth of 1 percent) of the energy originally diverted by the plants. Given such dramatically diminishing returns, there is simply insufficient energy for additional levels of consumers.

We're all familiar with at least a few of nature's cycles—for example, sunrise and sunset, the lunar month, the passing of seasons, and the twice-daily ebb and flow of ocean tides. These earthly events are driven by extraterrestrial processes. However, Earth has its own cycles, many of which form closed loops.

It has been said that life loves loops, because most biological processes feed back on themselves. A one-way process lacks control and can run away on itself until resources are exhausted. Hardly a recipe for survival. But turning the process into a FEEDBACK LOOP lets the system monitor and regulate its own behavior. Think of a home thermostat or the way your body adjusts its temperature (shivering when cold, sweating when hot). The regulation of flow does not require any intelligence or purpose. Rather, feedback loops employ circular control. The system monitors and takes in information about its performance, and this input is then used to self-correct as necessary.

The most important Earth-bound cycles are founded on the continual movement of a handful of elements through air, earth, water, and life. Indeed, the biosphere and atmosphere are regulated by the flow of six elements: carbon, hydrogen, nitrogen, oxygen, phosphorus, and sulfur. This elemental "cocktail," easily remembered by the acronym CHNOPS (pronounce the C as an S and think about the German liqueur), moves through every organism and ecosystem, underscoring the close ecological and evolutionary bonds shared by all life-forms. For example, DNA molecules, which house the biosphere's genetic library, are composed of five of the six elements, lacking only sulfur. The panoply of proteins that exist on Earth today are formed by chains of amino acids, of which there are only twenty different varieties; all of these chains contain carbon, oxygen, nitrogen, and hydrogen, and two of them include sulfur. Over the course of Earth history, the living world has become deeply embedded in these cycles so as to become integral to each. In fact, it's not exaggeration to state that the biosphere has persisted for billions of years on this planet *because* it has inserted itself into these loops, making use of the continually recycled elements.

Let's look at carbon. Life is sometimes referred to as "carbon based." About one-fifth of the body mass of any backboned animal is made of carbon, and this element is critical for building cells. As we've seen, photosynthesis involves plants removing carbon from the air in order to construct cells. Plants burn a small portion of this carbon, but much of it is passed through the food web to make the bodies of other organisms and fuel their activities. Plant and animal consumers then release or exhale carbon, which combines with oxygen to form carbon dioxide. Finally, decomposing organisms recycle leftover carbon through the ecosystem. As noted earlier, carbon dioxide is a greenhouse gas important in regulating global climates. When organisms, from dinosaurs to diatoms, are fossilized, much of the carbon stored in their bodies becomes trapped for millions of years within the planet's surface rocks. A portion of this carbon is later released into the air as carbon dioxide through plate tectonics, volcanism, and erosion. In this way, life and nonlife work in unison to form a loop or cycle.[2]

Nitrogen is another key ingredient in CHNOPS, comprising 80 percent of the air we breathe, and about 1 percent of the cells in all life-forms. That one percent turns out to be critical, because nitrogen is present in all complex protein molecules, including DNA. The nitrogen cycle, too, is intimately tied to earth, air, water, and life, and includes some unusual twists and turns. Plants acquire the nitrogen they need to grow and thrive from the soil, which is porous to the air. But plants can't use nitrogen in its gaseous form, so many millions of years ago they formed a partnership with fungi that we refer to as MYCORRHIZAE ("fat roots"). The fungal side of this union, which occurs as swellings on plant roots, improves the plant's ability to absorb water and minerals from the soil. The fungus in turn receives carbohydrates from the plant. In the Mesozoic, as today, this SYMBIOSIS was critical in sustaining terrestrial ecosystems. Similar kinds of loops cycle the remaining CHNOPS elements—hydrogen, oxygen, phosphorus, and sulfur.

So an ecosystem, whether modern or Mesozoic, comprises a characteristic community of species and its surrounding environment, all organized by the flow of energy and the cycling of nutrients. Each species has its own NICHE or role, including strategies for obtaining food and water, reproducing, and avoiding predation. Plant-eating ceratopsians and meat-eating theropods are examples of coexisting species with defined roles. As previously noted, many species pairs compete, with each side continually attempting to gain the upper hand. This leads to an ongoing jostling of niches, with advances on one side countered by gains on the other. Despite this ongoing flux, the typical result is a standoff, a fluctuating balance of power among the ecosystem players. In this way, the species of a given community divide up the portion of the solar energy captured by plant and bacterial producers.

Why are ecosystems stable? What prevents them from disintegrating? It turns out that part of the answer involves strength in numbers. That is, a high diversity of species is one of Mother Nature's chief survival strategies. More players results in more paths for energy flow, building a kind of redundancy into the system. So if one or a few species disappear locally, the ecosystem can adjust. In addition, greater diversity makes it less likely that an invasive species can take hold and wreak havoc on an ecosystem, because many predators will be present. Over the long term, the processes of ecology and evolution interact, allowing ecosystems to adapt, persist, and occasionally undergo major reorganizations.

Throughout nature we find multilevel structures of systems nested within systems—for example, cells within organs within organisms within populations. Each level forms an integrated whole while at the same time being part of a larger whole. Each level also has its own peculiar, or "emergent," properties, such that every whole far exceeds the sum of its parts. For example, consciousness is an emergent property at the level of the organism. Notwithstanding the efforts of modern geneticists, such emergent phenomena will never be understood simply by conducting ever-more detailed studies of genes and their interactions. The different levels within a hierarchy interact with one another via feedback loops that regulate their function.

Ecosystems are no different, although their boundaries can be difficult to define. Forests are often regarded as ecosystems, but within them are miniecosystems—creeks, ponds, meadows, canopies—that also tend to be inhabited by unique subsets of species. Moving up the scale, local ecosystems are united into regional ecosystems or bioregions—say, the Western Interior of North America—and these in turn can be linked until one finally ends up with the planetary biosphere. Viewed as a multilevel hierarchy, the biosphere at any given moment in time is composed of nested sets of ecosystems. There is always some transfer of energy and exchange of nutrients across ecosystem boundaries, but typically at a much slower rate than occurs within a given system. Ecologists regularly observe changes in species diversity at the local level; a bird species, for example, may disappear for a while, only to return at some later time. Regional ecosystems, in contrast, tend to be much more stable over long time spans, sometimes persisting with only minor changes for thousands or even millions of years.

Going back to our Hell Creek example from the Late Cretaceous, we see that the ecological perspective enables us to regard the environment as far more than a stage for titanic struggles between predatory tyrannosaurs and their horned dinosaur prey. Rather, such confrontations are properly relegated to side plots in an epic saga of flowing energy, cycling matter, and shifting environments.

As the primary source of energy, plants are the core of any terrestrial ecosystem. So our quest to understand the odyssey of dinosaurs must include an exploration of Mesozoic plants. Paleontologists have learned more than you might suspect about plants from the Age of Dinosaurs. This information comes primarily from ancient plant remains—including fossilized tree trunks, branches, leaves, and flowers, as well as microscopic pollen grains—augmented by observations of living plants. Although most of the species were different, many of the major plant groups in the Mesozoic have persisted to the present day, giving us insights into the biology of their ancient ancestors. Other Mesozoic plants, however, appear to have no living descendants, so we must make inferences based on more distant relatives among modern floras. Further information is gleaned from the types of rocks that entomb plants, giving us insights into environments where the plants lived and died.

We tend to think of the Mesozoic as very similar to the present day, except with dinosaurs roaming around. This is a mistake. Mesozoic Earth was a hothouse world marked by high sea levels and a lack of polar ice caps. The problem is that, as children of the Pleistocene, we tend to view the past through "icehouse eyes," to borrow a phrase from Kirk Johnson. Most of the major terrestrial biomes (large-scale communities of plants and animals) we are familiar with today—rain forests, tundra, boreal forests, and temperate and tropical grasslands—were entirely absent during the Mesozoic. Instead of lush vegetation covering much of the tropics, equatorial plants were patchily distributed and adapted to high seasonality. Whereas flowering plants dominate modern global floras, they arrived late into the world of dinosaurs, playing a significant role only from the Late Cretaceous onward. Grasses, one of the most successful groups of flowering plants, are the primary fuel source for many terrestrial ecosystems today. Yet grass did not evolve until the very end of the Late Cretaceous. So what did Mesozoic biomes look like? At this point, the best we can offer are vague guesses.

Dinosaurs originated in a Triassic world dominated by two major groups of land plants: the spore-bearing pteridophytes and seed-bearing gymnosperms. The angiosperms, or flowering plants, were a specialized group of seed plants that first appeared in the Early Cretaceous and became widespread by the end of that period. Both pteridophytes and gymnosperms are vascular plants, transporting water and food products through internal conducting tissues. In contrast, other kinds of plants such as various mosses and seaweeds are limited in size because they lack vascular tissues.

The spore-bearing PTERIDOPHYTES occur in three main types—FERNS, HORSETAILS, and LYCOPODS—all of which lived among dinosaurs. Most of us are familiar with ferns and horsetails, but few know about lycopods. Today lycopods are best known as club mosses—small creeping plants that often grow on other plants. During the latter part of the Paleozoic Era, some lycopods grew to be large trees more than 30 meters (100 feet) high. In the succeeding Mesozoic Era, these plants were limited to much more modest sizes, and indeed the majority of spore bearers that lived with dinosaurs were small, low-growing, herbaceous plants lacking well-developed stems.

Because their reproduction is based on spores, which require at least seasonally available water, pteridophytes tend to be rare in dry environments. They propagate by the windblown spreading of spores, a portion of which land in moist settings and are able to germinate. Sperm are then produced, which swim in search of eggs formed in nearby germinated plants. Most pteridophytes are able to circumvent this laborious reproductive cycle by sprouting new branches from a single underground stem; in this way, one reproductive event can generate an entire field of plants.

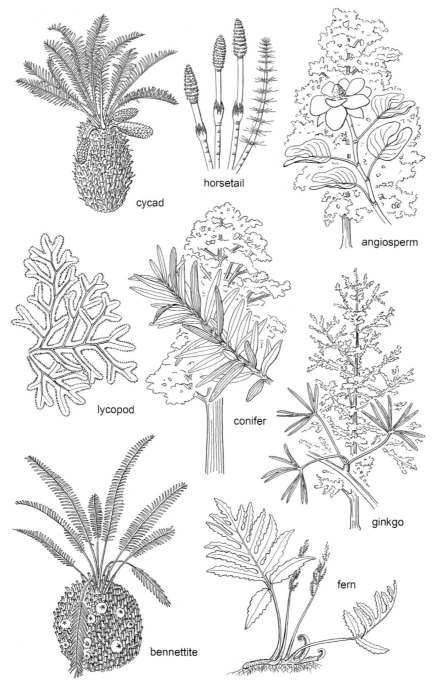

FIGURE 5.3

Representative varieties of the major Mesozoic plant groups, including three spore-bearing pteridophytes (ferns, horsetails, and lycopods) and five seed-bearing gymnosperms (cycads, ginkgos, conifers, angiosperms, and bennettites). The plant genera represented are as follows: fern, *Onoclea*; horsetail, *Equisetum*; lycopods, *Selaginella*; cycad, *Encephalarctos*; ginkgo, *Ginkgoites*; conifer, *Agathis*; angiosperm, *Archaeanthus*; bennettite, *Cycadeoidea*. All of the depicted plants are based on Mesozoic forms, although close relatives of most (all except the bennettite) have persisted to the present today. A number of smaller plant groups—for example, the seed-plants known as gnetales—are not depicted here.

Spores are a successful strategy, but seeds were a pivotal vegetative innovation because they freed plants from standing bodies of water. Seed plants, the GYMNOSPERMS, were thus able to radiate away from water sources and colonize drier terrain. The four major groups of seed plants alive today are cycads, ginkgos, conifers, and angiosperms (the flowering plants), all of which were present for at least part of the Mesozoic. CYCADS are evergreen plants with short, stout trunks and long, pinnate leaves that bear a general resemblance to palm trees. GINKGOS are mostly trees with characteristic, fan-shaped leaves; although represented today only by a single species, ginkgos were quite diverse during the Mesozoic. CONIFERS are a much more familiar group of woody plants, with seeds borne in cones and leaves that usually take the form of needles. Modern conifers include cedars, cypresses, firs, junipers, pines, and redwoods. Mesozoic conifers were dominated by two other groups, podocarps and araucarians (monkey puzzle trees), both restricted today to small pockets on Southern Hemisphere landmasses. These three groups of seed plants first evolved in the Paleozoic and were present throughout the Mesozoic.

The fourth group, angiosperms, are the most recent addition to the seed plant family. Flowering plants first appeared in the Early Cretaceous, perhaps starting out as small, herbaceous forms, but soon diversifying to fill a vast range of ecological niches. Today, angiosperms are by far the most dominant plant group, with about 275,000 recognized species, compared to about 1,000 conifer species; 200 species of cycads and ginkgos; and 11,000 species of spore-bearing pteridophytes. Modern angiosperms encompass everything from oaks and magnolias to grasses and herbaceous shrubs. A major factor in angiosperm success was flowers, a reproductive strategy involving partnerships with other species. Mesozoic flowers were likely pollinated most frequently by insects, though various birds and perhaps other kinds of organisms may have served this role as well. Angiosperms are also notorious for their rapid growth. Combined with high rates of reproduction and tolerance to disturbance, weedy growth enabled angiosperms to colonize areas quickly and apparently more efficiently than other seed plants.

A fifth group of seed plants, the BENNETTITES, are also worthy of mention. The now-extinct bennettites came in a variety of sizes, but many had large, palmlike leaves that resemble those of some flowering plants. Like most members of the seed plant family, they also first joined terrestrial ecosystems in the Paleozoic and were widespread throughout most of the Mesozoic. However, unlike the other gymnosperm groups, bennettites have not persisted to the present day, going extinct in the middle of the Late Cretaceous. Perhaps the flowers were simply too much for them.

Plant communities, like dinosaur communities, changed throughout the Mesozoic. During most of the latter Paleozoic, spore-bearing plants were widespread and dominant. But the formation of Pangaea during the Permian Period changed all that, resulting in dry, highly seasonal conditions that were inhospitable for water-loving ferns, horsetails, and club mosses. However, one group's tragedy was another's success story. With their greater reproductive freedom, seed plants survived the "xeric bottleneck" triggered by the formation of Pangaea, diversifying into a multitude of forms well adapted to more arid conditions.

Triassic floras were ruled by such seed plants as cycads, conifers, and bennettites, with gink-gos also holding their own. The fortunes of these four groups waxed and waned throughout the Jurassic. With the Early Cretaceous appearance of flowering plants, the diversity of other seed plants began to decline. By the close of the Cretaceous, angiosperms were the most common plants in many regions worldwide, retaining this dominance to the present day.

Depictions of Mesozoic Earth vary mightily. Some show densely vegetated environ-ments, much as we see today. Many others look more like dirt parking lots, with a few monkey puzzle trees tossed in the background. The true story was almost certainly more variable—and more interesting. Some investigators have argued that vegetation, partic-ularly in equatorial regions, was concentrated near waterways such as rivers, lakes, and seashores, with much more patchy distributions inland and in the uplands. At higher latitudes, vegetation may have been distributed more evenly because of higher moisture levels. There is some evidence that dinosaurs mimicked this pattern, with the greatest abundance and diversity of species at midlatitudes, where plant life was the most abun-dant. Such conclusions are suspect, however, and may be reflective of biases in sampling more than any Mesozoic reality. Although much remains dimly understood, we can be quite certain that dinosaurs and plants coevolved along with insects and other life-forms throughout the Mesozoic, with each influencing the evolutionary fates of the others.

In general terms, the workings of Mesozoic ecosystems would have been very similar to their modern counterparts. During the Age of Dinosaurs, plants were the major produc-ers on land, the only life-forms capable of tapping into that boundless energy source—the sun. Herbivorous dinosaurs were part of a cadre of primary consumers, eating a variety of plants. Carnivorous dinosaurs were among the secondary consumers, feeding on the her-bivores and on each other. Finally, the feedback loop was closed by decomposers such as insects and bacteria, which fed on both plants and dinosaurs. All the major players of today were present in this ancient drama—bacteria, protists, fungi, plants, and animals. Given that these loops were energized entirely by the sun, all dinosaurs were "solar eaters," con-suming sunlight either directly as plants or indirectly, through the bodies of animals.

As we continue on the dinosaur odyssey, I invite you to look past the blood-and-guts imagery of *Tyrannosaurus* battling *Triceratops*, focusing instead on how scenes like this played out as minor acts in a much larger drama of flowing energy and cycling nutrients. Step back and shift your gaze away from a few trees (or dinosaurs) to focus instead on the entire forest, on the ebb and flow of whole ecosystems. Intricate webs of relationships provide the foundation for ecosystems to withstand disturbances, making minor adjust-ments as needed—bending but not breaking. Periodically, however, ecosystems are faced with large-scale events that require a more drastic response. Sometimes the aging actors in the ecological play must be replaced by younger upstarts, who then establish new roles and relationships. Like a long-running Broadway show, the cast changes reg-ularly (at least in a deep time sense). This occasional overhaul involves an entirely different set of processes best encapsulated in a single word: *evolution*.

A scene from Early Cretaceous Asia: a feathered maniraptor theropod, *Microraptor*, landing in a forest conifer, *Brachyphyllum*, is distracted by a passing pair of *Confuciusornis* birds.

6

THE RIVER OF LIFE

AS THE FIRST ORANGE-YELLOW HUES OF DAYLIGHT illuminated the California coast, a great horned owl flapped hurriedly past the window, presumably headed to daytime sanctuary. Stars were still plentiful. Off in the east, Venus sat perched like a beacon just above the horizon, looking more like an oncoming 747 than a planetary member of our solar system. A short time later, the sun climbed above the Marin headlands, revealing the Pacific surf pounding the rocky coastline. Above this grand juxtaposition of land and sea, six turkey vultures soared in broad, sweeping arcs, seemingly in silent celebration of the new day. Two coal-black ravens squawked at each other from a nearby Monterey pine. From the corner of my eye, I detected movement and shifted focus to just beyond the window. An Anna's hummingbird, replete with emerald green back and scarlet head and throat, had arrived to sip from the feeder. After consuming its fill, at least for the moment, this 3-inch ball of feathers and energy lifted off and hovered, its tiny wings beating something like 50 times per second. Suddenly, the iridescent apparition was gone. Were I to venture down to the nearby estuary on this day, I would see many other birds, potentially dozens of species, with evocative names like black-necked stilt, Caspian tern, belted kingfisher, marbled godwit, and long-billed dowitcher.

The remarkable truth of the matter is that all of these creatures are flying dinosaurs. To the best of our knowledge, every bird species past and present can be traced to a single common ancestor that evolved perhaps 160 million years ago from dinosaurian stock among small, bipedal carnivores akin to *Velociraptor*. Contrary to popular belief, then, dinosaurs didn't go extinct 65.5 million years ago. Represented by about ten thousand living species, avian dinosaurs still rank among the most successful groups of vertebrates. Despite millions of years of separation, today we are directly connected to the world of dinosaurs through these wondrous volant creatures. I still find this idea stirring, not just intellectually but on a deeper level of awareness. And it makes me chuckle to contemplate all of those binocular-toting members of the Audubon Society as dinosaur lovers.

But how do we know? How can we paleontologists be so sure that birds are linked through direct evolutionary ties to a particular group of animals that lived millions of years ago?

Archaeopteryx, the oldest and most primitive bird, was discovered in 1860, just one year after the publication of Darwin's famous treatise on evolution, *On the Origin of Species*. *Archaeopteryx* is presently known from ten specimens, including several nearly complete skeletons, all recovered from the Solnhofen limestones in Germany, which date to the Late Jurassic (about 155 million year ago). These same deposits have revealed a diverse variety of animals, among them a small theropod dinosaur named *Compsognathus*. *Archaeopteryx* and *Compsognathus* are so similar that specimens of each have been confused for the other multiple times over the years. Particularly given the timing of its discovery, *Archaeopteryx* made an ideal transitional fossil, providing strong support for Darwin's revolutionary theory that all life is related. On the one hand, this Jurassic bird has several primitive, reptilian features, including teeth, three separate clawed fingers, and a long bony tail. On the other, it possesses specialized bird features such as feathers, fused clavicles, and an elongate forelimb modified into a wing. Various nineteenth-century scientists picked up on the transitional nature of this ancient avian and argued that birds are closely related to dinosaurs, perhaps even their direct descendants. Chief among these people was the famed biologist T. H. Huxley.

Yet for most of the twentieth century, an alternative view of bird origins prevailed. In 1926, Danish artist and paleontologist Gerhard Heilmann published a landmark work titled "The Origin of Birds." Heilmann agreed that *Archaeopteryx* and other birds were clearly similar to theropod dinosaurs, but he dismissed the notion of bird ancestry from dinosaurs because dinosaurs apparently lacked clavicles (the collarbones of humans). In birds, the clavicles fuse into a single element, the FURCULA or wishbone. In keeping with the prevalent idea that evolution could not retrace its steps, Heilmann argued that, once lost, these elements could not have reevolved. He concluded that bird origins were to be found among more primitive members of Archosauria—the larger, more encompassing group that includes dinosaurs, crocodiles, and birds. It was presumed that fossil representatives of the specific group that gave rise to birds had simply not yet been discovered.

Enter John Ostrom, whose work in the 1960s and 1970s provided the catalyst for the dinosaur renaissance described in chapter 1. Following his description of the "raptor" dinosaur *Deinonychus*, Ostrom reignited the bird origins debate through several papers that detailed specialized anatomical features shared between birds and certain raptorlike carnivorous dinosaurs. Subsequent workers have added many more characteristics to this list, such that the total number today exceeds one hundred. Indeed, we now know that many features traditionally associated with birds evolved in the prebird theropod ancestor. For example, like birds, carnivorous dinosaurs have thin-walled limb bones and air-filled bones within the vertebral column. Additional evidence supporting the dinosaur-bird connection includes growth patterns, nesting behavior, and even sleeping behavior.

But what of Heilmann's objection regarding the absence of clavicles in dinosaurs? Over the past few decades, clavicles have been found in a broad range of theropod specimens, including those thought to be closest to the origin of birds. So this objection has been removed.

By far the most convincing evidence supporting the notion that birds are directly descended from dinosaurs are the numerous transitional fossil forms, the so-called dino-birds, discovered over the past 15 years. Most spectacular of the recent discoveries are numerous forms unearthed from Cretaceous-aged rocks in Liaoning Province in northeastern China. Like the specimens found in the Solnhofen quarries of Germany, the Chinese fossils show exquisite preservation, not only hard tissues like bones and teeth but also soft tissues such as skin impressions and feathers.

Biology has few absolutes, but even kids know that fishes have gills, mammals have fur, and birds have feathers—at least so the story goes. In a world where all species are created independently of one another, this might be true. But in a Darwinian world, we expect transitional forms that blur the lines. Many of the Liaoning fossilized feathers occur on exquisitely preserved birds such as *Confuciusornis*. Yet many others are associated with the skeletons of nonbird dinosaurs. To date, more than a dozen different kinds of feathered nonavian dinosaurs have been found in China, and that number increases with every passing year. One of the most exciting examples is *Microraptor gui*, a diminutive dromaeosaur theropod bearing feathers not only on its forelimbs, but on its hindlimbs as well. In the initial scientific paper describing *Microraptor*, Xu Xing, a paleontologist at the Institute of Vertebrate Paleontology and Paleoanthropology in Beijing, and his colleagues resurrected an old idea, hypothesizing that the origin of flight in birds included a four-winged gliding stage before evolution honed the system to a pair of wings up front. So even feathers, previously the quintessential avian feature, now cloud the boundaries between two major groups of organisms.[1]

Evolution can't make something from nothing. Instead, it modifies existing structures. For example, both mammal hair and bird feathers are highly modified scales. The Liaoning dino-bird discoveries supply important insights into the evolution of feathers, from very simple, hairlike structures to complex flight feathers. Moreover, the discovery of feathered, ground-dwelling dinosaurs indicates that feathers originally evolved to serve some function other than flight. The most likely alternatives are control of body temperature and courtship/display. Together with the evolution of feathers, the origins of flight in birds entailed wholesale transformation of the forelimb into an elongate wing capable of flapping flight. Related changes included loss of the bony tail, reorganization of the hindlimb muscles, and modification of the foot for perching instead of running.

Archaeopteryx and the many recent dino-bird discoveries are often referred to as "missing links." This unfortunate term is a double misnomer. Most trivially, because they are now in our possession, these finds are clearly no longer missing. More fundamentally, few if any of these fossils qualify as links in the sense of being part of the ancestor-descendant lineage leading from dinosaurs to modern birds. In the previous chapter, we found that the term *food chain* was a poor descriptor of the flow of energy through an ecosystem; rather, the degree of life's interconnectedness means that the web metaphor is much more useful for understanding ecological dynamics. The term *missing link* similarly invokes a linear chain—in this case, a chain of relationships, with primitive ancestral forms evolving through a ladderlike progression into more advanced descendants. Yet reality is once again much more messy and interesting. The history of virtually any major group, and for life as a whole, is better conceived as a densely growing bush than a series of step-by-step, linear progressions. The new dino-bird discoveries are filling in some of the branches near the base of the avian family tree, showing that this particular part of the foliage was bushy indeed.

The dinosaur-to-bird story is one of thousands of narratives relating to the history of life on Earth. All life on this planet belongs to a single extended family, descended from a common ancestor that lived between 3.5 billion and 4 billion years ago. Darwin's great contribution was to firmly establish these evolutionary ties in a reasoned, well-supported theory, resulting in a fundamental shift in our perception of the world and the human place within it. The underlying tenet of the theory of evolution is that species are not fixed or immutable. All species have ancestors and relatives, and many pass on descendant forms. For example, domestic cats and panthers come from a common stock of feline mammals, and among their close relatives are extinct forms like saber-toothed cats. I examine this idea in some detail later because knowledge of a few evolutionary basics is prerequisite to understanding the dinosaurian odyssey.

Darwinian evolution forms the conceptual bedrock of the life sciences and is resoundingly accepted by the scientific community. However, outside the realm of science, considerable doubt and misunderstanding persist about this revolutionary idea. A number of recent polls indicate that only about a third of the U.S. population regards the theory of evolution to be well supported by the evidence. These same polls show that about two-thirds of respondents favor teaching alternatives to evolution—for example, intelligent design—along with evolution in science classrooms. Yet no scientific alternatives account for the diversity of life through time. Evolution is often maligned as "just a theory," as if to say that it is no more than our current best guess. In contrast to its vernacular usage, however, the word *theory* as applied in science does not imply mere speculation; it refers instead to an interconnected set of hypotheses that have withstood numerous attempts at falsification. Descent with modification—the notion that all living and extinct species share a common ancestry—is one such theory, accepted with the same degree of confidence as the theories of gravity and relativity.

Why are scientists so confident about the evolutionary ties that link all life on Earth? Because over the past one and a half centuries, biologists have amassed a veritable mountain of evidence in support of Darwin's "dangerous idea." In brief, this evidence includes the following:

Biological structures, from genes to gross morphology, support the same major evolutionary groupings. With few exceptions, organisms thought to be close relatives on the basis of gross anatomy are also closely similar in their genes (DNA sequences)—for example, horses and zebras, or lions and tigers. Sometimes these genetic similarities are even closer than first anticipated; for example, chimpanzees and humans share about 95 percent of their genes, which means that we are more closely related to each other than either of us is to gorillas! The reverse pattern is typically true, too; the more dissimilar the body types and presumed relationships, the more dissimilar their DNA (genomes). Occasional exceptions to this genetics-anatomy linkage do not disprove evolution; on the contrary, they point out interesting and unexpected relationships that further our understanding.

Organisms within a given group have been modified from a single ancestral form. All bodily structures represent the culmination of deep time interactions between organisms and environment. Evolutionary theory predicts that related organisms will share features derived from common ancestors. Such shared characteristics passed on from an ancestor to multiple descendants are known as HOMOLOGIES. For example, frogs, birds, rabbits, and lizards all have different forelimbs, each reflective of a unique lifestyle. But those different forelimbs all share the same set of elements—a single upper arm bone (the humerus) and a pair of lower arm bones (radius and ulna)—inherited from a common ancestor.[2] This same trio of arm bones is present in the vast majority of all living and extinct land-living vertebrates, providing further evidence of common ancestry. The few exceptions are groups like snakes that secondarily lost their limbs.

Organisms have numerous features that make no sense in terms of independent creation or functional design. Because evolution proceeds by modifying preexisting forms, adult organisms possess traits that reflect their evolutionary history. These include vestigial structures, or evolutionary "leftovers," such as hip bones in whales, nonfunctioning eyes in cave-dwelling creatures, and the appendix in humans. The history of life is replete with examples of evolutionary jury-rigging, in which evolution made the best of a bad situation by transforming raw materials present in ancestral organisms. A classic example is the panda's thumb. These Asian bears lack a true thumb, which is found only in a few primates. However, in response to a need for stripping shoots from bamboo stalks, evolution modified one of the panda's wrist bones into a thumb substitute.

Geographic distributions of organisms are best explained through a combination of evolution and physical events, such as continental movements. Perhaps the best known example of this pattern is the distribution of marsupials, with the bulk of modern forms restricted to Australia. The island of Madagascar, featured in chapter 1, is another example, with a largely unique flora and fauna resulting from its ancient isolation in the southern

Indian Ocean. In groups of creatures with distributions that encompass two or more continents, such as the primate and deer families, the isolated subgroupings possess many unique, shared features, indicating that ancestral forms migrated long ago and the descendants then evolved along independent paths. On the flip side, comparable ecosystems in different parts of the world often include ecologically similar species that appear physically alike, suggesting that evolution has occurred in parallel within entirely distinct evolutionary lineages. A classic fossil example of such CONVERGENT EVOLUTION relates to a pair of distantly related saber-toothed carnivores among mammals—a placental saber-toothed cat from North America and a saber-toothed marsupial from Australia.

Key processes of evolution have been extensively documented by experiment and observation. These processes include mutation, natural selection, and even the origin of new species. Substantial evolutionary changes have now been documented in a variety of organisms. One of the best studies of natural selection in action relates to Darwin's finches on the Galápagos Islands. Within a given finch species, there is considerable variation in beak sizes. When a drought alters the availability of certain kinds of seeds, individuals with beaks best able to feed on the remaining seeds are able to outcompete and outreproduce their peers in a single generation. In times of plenty, the fortunes of beak types can be reversed. In both instances, the result is a significant shift in average beak size within the population. Equivalent kinds of changes have been observed in a variety of groups, including bacteria, moths, fruit flies, and fishes.

Transitional forms abound in the fossil record. Antievolutionists often decry the lack of transitional fossils. Yet the truth of the matter is that many sequences of transitional, or intermediate, fossils are known, among them the dino-birds described earlier. As noted, although very few of these fossils contributed to the direct line of descent leading to modern forms (remember, the history of life is a bush, not a ladder), they do record key stages in the evolution of their particular groups. Finely preserved examples of such transitional creatures have multiplied rapidly in the past two decades and are now scattered throughout the range of life's diversity, from plankton and clams to horses and horned dinosaurs. Key exemplars among vertebrates include intermediate forms linking fishes to amphibians, land-dwelling ungulates to whales, and apelike ancestors to upright human bipeds. Not surprisingly, it is this last transition that most troubles antievolutionists, despite the fact that we now have on the order of two dozen species of bipedal hominids that include a wonderful array of forms intermediate between chimpanzee-type apes and modern humans. Today our planet is home to only one species of upright hominid, *Homo sapiens*. Over most of the past 5 million years, however, at least two, and sometimes four or five, bipedal, humanlike forms existed at any one time. Like the dino-bird example, the human family tree is surprisingly bushy.

The order of appearance of fossils in the geologic record is comprehensible only from an evolutionary perspective. The fossil record reveals a progression from simple to complex—from single-celled bacteria all the way to whales, redwoods, and humans. This pattern cannot be explained by a single creation event or catastrophe, such as a flood. Instead, the appearance and disappearance of species in the fossil record supports the idea of ongoing origins and extinctions. That is, only the simplest organisms are found in the oldest rocks and the most complex life-forms occur only in much younger rocks. Yet simpler forms have not simply disappeared. Instead, representatives of all of life's major groupings (bacteria, protists, fungi, plants, and animals) have persisted to contribute to the present-day biosphere.

In short, overwhelming evidence confirms that all organisms on Earth—starfish, dinosaurs, mushrooms, petunias, and humans alike—trace their origins to a humble single-celled bacterium perhaps one ten-millionth of a meter in diameter. This fact is astounding and difficult to grasp. Recall, however, that you and I began our lives as a single cell. If nine months is sufficient to transform one cell into a highly complex 7-pound baby, surely it's conceivable that life's diversity could arise from a single cell in several billion years!

Charles Darwin was by no means the first evolutionist. Others before Darwin, including his grandfather Erasmus, argued for the changeability of species. Most famous of these early evolutionists was Jean-Baptiste de Lamarck who, in the early 1800s, proposed that species could take on new forms in response to their needs. The textbook example of this hypothesis relates to the giraffe neck. This is a book about dinosaurs, however, so let's use a more appropriate example—sauropods. As the story goes, primitive, short-necked sauropods were unable to reach succulent leaves in tall trees. Driven by the need for food, sauropods stretched upward to higher and higher branches, producing slightly longer necks during their lifetimes. Those elongate necks would have been passed on to off-spring, which continued the stretching trend in an effort to reach ever higher into the canopy. Over many generations, sauropod necks would become progressively longer until they reached the exaggerated form we are familiar with in animals like *Brachiosaurus*.

This hypothesis, known as "the inheritance of acquired characteristics," might seem at first to be an attractive solution. Yet it has a fatal flaw. To use a human example, imagine a man and a woman working out regularly at a gym, lifting weights for years on end. Both decide to become professional bodybuilders, ultimately bulking out to massive proportions. Now suppose that the muscle-bound couple marry and have children. Do you think that their offspring would emerge from the womb looking like mini–Arnold Schwarzeneggers? Would their kids even inherit a greater affinity toward "pumping up" later in life? No, no such simple correlation exists between one's actions in life and the inherited characteristics of one's children. The chief lesson here is that evolution operates only on heritable features—in modern parlance, this means features coded in DNA.

Charles Darwin provided the first convincing mechanism for evolution, a process he termed NATURAL SELECTION. Darwin compared selection in nature with artificial selection conducted by breeders of pigeons, plants, and livestock. Breeders preferentially combine varieties that have characteristics of particular interest—for example, larger size in pigs, more elaborate plumage in pigeons, or bigger seeds in cereals. Over time, the breed

Long-necked
descendant

Original
short-necked
ancestor

FIGURE 6.1

The evolution of long necks in sauropod dinosaurs. Lamarck hypothesized that evolution of this sort accrued through the inheritance of features acquired in the lifetimes of individuals. In contrast, Darwin recognized that populations rather than individuals evolve. In this example, sauropod dinosaurs with slightly longer necks proved to be more successful than individuals with shorter necks, tending to live longer and pass on more offspring. Over time, this process of change in populations resulted in increasing neck length for species within this lineage.

changes in direct response to this selective force. Darwin argued that nature does the same thing, demonstrating an unconscious bias toward animals best able to compete for limited resources. Offspring vary in characteristics that relate to survival and reproduction, and those variations that are most successful are the ones most likely to be passed on to offspring. Over many generations, numerous features of that species will change in response to this unrelenting external pressure. So, in contrast to the Lamarckian view, common ancestry plays a pivotal role in evolution.

Darwin's theory involves two major themes, and it's important to distinguish between them. The first is a pattern, descent with modification, or the common ancestry of all life on Earth. The second major theme, natural selection, is a process—an explanation for how evolutionary change occurs over time. The great bulk of the scientific community embraced descent with modification within a few decades after the publication of Darwin's *Origin*, and today virtually no professional biologists question the veracity of evolution as a guiding principle of their field. (Ironically, descent with modification is the most controversial part of the theory of evolution among the general public, largely because of its implications for the origin of humans.) What scientists continue to examine is Darwin's second theme, natural selection. Although there is strong agreement that natural selection is a key factor directing evolutionary change, lively debate persists about its efficacy and the possible roles of other factors.

Darwin's second theme, natural selection, is founded on three fundamental premises. First, individuals within a population vary in their expressions of numerous heritable traits. Variation within species has long been known to breeders of plants and animals who establish new varieties—of cats, pigeons, wheat, cotton, corn, and so on—based on breeding particular variants. You can easily observe this pattern yourself. Go to a park and check out the dogs, or turn your attention to their human companions. Or look closely at a particular species of plant in the forest or in your own garden. If you observe a bunch of individuals closely, you will find plenty of variability within species. Paleontologists see this kind of variation as well. Every dinosaur skeleton is different in some way from every other member of its own species. Variation, it turns out, is the sine qua non of evolution, the fodder that natural selection feeds on.

Second, organisms produce more offspring than the environment is able to support. One oak tree drops thousands of acorns each year. A female salmon produces about 30,000 eggs during each spawning. One oyster can generate 114,000,000 eggs in a single reproductive event. Even among elephants, one of the slowest-breeding animals known, if all the young of a single female survived and reproduced at the maximal rate, after 750 years the descendants of this single mother would number 19,000,000. Clearly, only a tiny fraction of all these individuals grows to adulthood. And this is a good thing, because otherwise the world would soon be overrun by oak trees, salmon, oysters, and elephants!

Third, organisms must compete to survive and reproduce, and individuals possessing variations that best fit them to their current environments are the ones most likely to survive, reproduce, and pass along those desirable variations to the next generation. This process is called ADAPTATION. Almost all biological activities can come under the scrutiny of natural selection and adaptation, from catching and evading predators to acquiring mates, avoiding disease, and forming symbiotic alliances with other species. Adaptation tends to reward successful traits and eliminate less successful variants. For example, if a desert plant inherited some feature that favored the retention of water, such a trait would likely be passed on to future generations, and individuals lacking this feature might not be able to compete. Over time, such favorable traits accumulate. If the lineage spawns new species, these successful traits are likely to be passed on. This three-step process is evolution by natural selection.

Returning to our sauropod example, let's suppose that the ancestral forms had short necks. Although all had short necks, neck length would have varied within the population, and those with slightly longer necks might have been able to browse on higher foliage. As a result, these fortunate animals would be more likely to live longer lives and produce more offspring. Succeeding generations would then inherit slightly longer necks, and so on, until, after many thousands of years, sauropod necks became very long. In truth, the actual course of events was much more complex and interesting. Neck elongation likely occurred independently in several groups of sauropod dinosaurs, with change occurring in fits and starts over millions of years across numerous species. In some forms, neck lengthening involved adding more neck (cervical) vertebrae, whereas other forms paralleled the giraffe solution, increasing the lengths of existing vertebrae. Nevertheless, the Darwinian scenario based on natural selection gives a relatively accurate account of neck elongation within this group.

As envisioned by Darwin, natural selection modifies the traits of plants and animals by making use of heritable variation that arises naturally within all populations. These variations are neither inherently good nor bad. They are simply there, and the rigors of life ensure that only the fitter combinations of variations are inherited by future generations. An inevitable consequence is a change in the genetic information stored within the population.

You can think of natural selection as a sieve. In each generation, all members of a given population are tossed into the sieve. Those that are less fit tend to fall through the holes and make no contribution to the next generation. Those that are most fit for that environment remain on the mesh—that is, they survive and pass on offspring. The offspring then grow up to comprise the next generation, which receives the same harsh sieve treatment, resulting in more evolutionary winners and losers. Over time, the genetic makeup of the population shifts, rewarding the features that are most successful. Because environments change periodically, a set of features that confers success, or fitness, at one ecological moment may send an organism tumbling through the sieve at some later time.[3] In short, individuals do not adapt or evolve—only populations do. Let me restate this fundamental and frequently misunderstood point. *Populations, not individuals, evolve.* Evolution occurs because certain kinds of individuals have greater success than others in survival and reproduction, resulting in shifts in the heritable variation of the population as a whole.

In a nutshell, natural selection is driven by two kinds of events. First, nature randomly alters the content of genetic information through such processes as mutation and recombination, resulting in variations at the level of organisms. Second, organisms interact with their environment, and those best able to survive and reproduce tend to pass on more offspring to the next generation. Over time, the information content of the entire population shifts. Evolution happens.

While Darwin was contemplating the workings of natural selection, he soon recognized a problem. He noted that many characteristics of animals did not seem to confer any advantage in the fight for survival. The famous evolutionist soon realized, however, that these same features increased an organism's odds of winning the reproductive lottery. In particular, males with elaborate signaling structures had a greater success rate when it came to competing for mates. Darwin's answer was to label a new selective process: SEXUAL SELECTION.

Like natural selection, sexual selection weeds out less successful traits in favor of those that are more successful. In this case, however, success is measured on the basis of reproductive success rather than survival. Sexual selection often favors the evolution of elaborate weapons and mating signals—horns, crests, colors, calls, and the like—that increase the chances of acquiring mates and passing on genes to the next generation. Sometimes, the bizarre features that prove most successful in reproduction actually decrease the odds of survival, working contrary to natural selection. Peacock feathers and deer antlers, for instance, are costly to build and maintain. Other features, like the calls of many frogs and insects, make these animals more vulnerable to predators.

Darwin reasoned that if such elaborate signaling structures resulted in improved mating success, they would be passed on to subsequent generations, even if these same features were detrimental to survival. In short, when it comes to evolution, sex trumps death, at least to a point. If female preference results in the evolution of more elaborate horns or more intricate calls, sexual selection might then lead to a runaway effect, causing succeeding generations to become increasingly extreme and bizarre. Numerous examples of bizarre structures among dinosaurs may have evolved at least in part under the pressure of sexual selection. Examples include ceratopsian horns and frills, hadrosaur crests, pachycephalosaur domes, stegosaur plates and spikes, and theropod crests and horns. The origin and functions(s) of these strange bony appendages have spurred a long and ongoing debate, but that's a topic for a later chapter.

If energy is the currency of ecology, information is the currency of evolution. Life can be likened to a vast river of information flowing through time. Information arises, flows from generation to generation within the bodies of organisms, and branches into myriad tributaries. Over time, as environments shift—for example, becoming hot or cold, dry or wet—some of these waterways dry up (extinction), while others become raging torrents that top their banks and spawn new sets of tributaries (speciation).

Darwin's view of evolution was founded on unending gradual change. According to this perspective known as GRADUALISM, organisms are forever under the close and merciless scrutiny of natural selection. Consequently, variable species are continually shifting from one form to another as they become ever-more adapted to their (frequently unstable) ecological surroundings. Over geologic time, these progressive, generation-by-generation heritable changes accumulate until they result in major evolutionary transformations. Following this view, species boundaries must be arbitrarily defined, because no discernible breaks occur in the ongoing flow of change. In keeping with the Darwinian perspective, if we could trace the evolutionary lineage of, say, bald eagles thorough an unbroken chain of fossil ancestors back to some small Jurassic theropod dinosaur, the boundaries between species would be arbitrary, a matter of personal preference because one form would grade seamlessly into another.

Placing gradualism into the context of our river metaphor, the information content of the flow is in constant flux; a sample of water taken at one point in the river would differ from other samples taken farther downstream, with the degree of difference proportional to the distance between sampling sites and the rate of flow. If you were to take a series of samples between these two well-separated sites, there would be no spot where you could

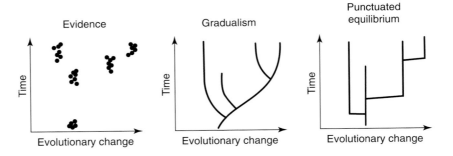

FIGURE 6.2

Two modes of evolution compared: *gradualism*, or evolutionary change within an unbranching lineage; and *punctuated equilibrium*, or long periods of minimal change within species followed by brief bursts of evolutionary change associated with the origin of new species. The graph on the left depicts a series of data points that represent fossil samples. The remaining two graphs show how these data might be interpreted to represent either gradual evolution (continuous change) or punctuated equilibrium (long periods of stability punctuated by short bursts of change). In all three graphs, evolutionary change is shown on the horizontal axis and time is shown on the vertical axis. See text for discussion.

pinpoint a major transition. The river would fork occasionally, producing new tributaries, but transitions in the composition of the flow would remain gradual.

Over the past few decades, the gradualist view has come under attack. Today, many, perhaps the majority of, evolutionary biologists consider species to have individual histories, or "lifetimes," bounded by discrete births (SPECIATION) and deaths (EXTINCTION). As early as the 1930s, the population geneticist Sewall Wright presented a view of the natural world that should have made all biologists pause. He described species in terms of semi-isolated populations whose members periodically interbreed and thereby exchange genetic material. If these populations are spread over a large area, every population would tend to undergo minor adaptive changes in response to local conditions. Given this structure, it's difficult to envision how significant, directional change could accumulate across an entire species, as Darwin thought, because any evolutionary modification in one population would tend to be swamped by periodic interbreeding with individuals from different populations. One could imagine an occasional mutation—leading to better eyesight, improved digestion, or faster running ability—that might spread widely because it proved beneficial to members of all of the populations. However, in general, beneficial changes would likely improve fitness only locally, and distinct populations might even evolve in opposite directions.

The gradualist view of Darwin has also been challenged by direct observation of patterns in the fossil record. When a paleontologist walks up a hillside littered with fossils, she is moving through time. Because strata are deposited in layer-cake fashion, fossils eroding out at the bottom of the hill tend to be older than those at the top. In some cases, millions of years separate the deaths of organisms preserved in adjacent layers. Paleontologists have long observed that fossils of a given species tend to persist with minimal change through long time intervals, sometimes millions of years, only to be replaced abruptly by some distinct but related form. This pattern, which clearly conflicts with the Darwinian notion of constant gradual change, was traditionally attributed to gaps in the fossil record rather than to evidence of evolutionary process. It was thought that a more complete record would surely reveal continuous changes consistent with gradualism.

Then, in 1972, Niles Eldredge of the American Museum of Natural History and Stephen J. Gould of Harvard University turned the paleontological world on its head. Suppose, they said, that the fossil record is more complete than we thought, that observations of long-term stasis and relatively brief bursts of change reflect the real, dominant pattern of evolution rather than deficiencies of the fossil record. Suppose that evolutionary change is generally concentrated around the origin of new species and that species then persist largely unchanged until their eventual extinction. Eldredge and Gould cited abundant evidence in support of their idea. They gave the hypothesis the name PUNCTUATED EQUILIBRIUM, because it entailed punctuations of change layered on a background of equilibrium, or stasis. Punctuated equilibrium allows for small-scale evolutionary change within species and within populations, but it predicts that directional change in the sense of evolving entirely new forms requires evolutionary branching events—that is, the origin of new species.

Over the past several decades, this hypothesis has been the subject of intense testing and debate. Paleontologists have scoured old data sets and generated new ones in a concerted effort to reveal patterns and check them against contrasting sets of predictions. Today punctuated equilibrium is supported in one form or another by most paleobiologists. However, its profound ramifications have yet to be appreciated by biologists generally, who still cling to a gradualist view of evolution. Several exceptions to punctuated equilibrium have been documented, yet species stasis (i.e., minimal change within species over deep time) has emerged as an overwhelming pattern in the fossil record. This finding marks a dramatic departure from Darwinian gradualism. In general, directional change does not accumulate continually within local populations. Local modifications to the information flow occur, but these are generally swamped by interbreeding, which results in the exchange of information between population tributaries.

Most of the time, then, it seems that species don't do much of anything, at least from an evolutionary standpoint. Contrary to the gradualism perspective, they appear to persist relatively unchanged for most of their histories. But if so, how do evolutionary changes build up over time to lead to new forms? The answer is to be found in the origin of new species, or speciation. Speciation is pivotal because it sets in place a reproductive barrier that prevents interbreeding and genetic mixing. For most sexually reproducing organisms, once that reproductive barrier is in place between two populations, the flow of information thereafter remains isolated, whether the populations-turned-species persist for one year or millions of years.[4]

Returning to the river metaphor, populations within a species act like criss-crossing tributaries. The information flow is separated into distinct channels some of the time but is reunited at other times by the act of interbreeding. Within regional ecosystems, this pattern of branching and reconnecting enables a limited amount of information change (evolution) to occur, but little in the way of directed, transformational change that affects the entire species. A portion of this new information cascades through the entire river (via matings between members of different populations), but many other bits are swamped by exchange between and among the various tributaries. However, once two tributaries (populations) become isolated from each other (i.e., the members of each no longer interbreed), the resulting (reproductive) barrier generally ensures that the flows travel in separate channels, and the result is at least one new species. In the process, new information—the evolutionary changes accumulated in the local population—is injected into the long-term flow of evolutionary lineage. And the longer those streams are isolated, the less likely it is that they will commingle in the future, because evolutionary changes in information content accrued along the way increase the likelihood that the two streams will be unable to intermix. Speciation, then, is a fork in the river, a permanent branching in the flow of information. It promotes the origin and capture of new information, leading to new forms, which in turn can produce dramatic changes to the overall information flow of an ecosystem. Given such a revised perspective, speciation takes on heightened importance. In particular, any factors that influence the pace of species births and deaths are likely to have a major effect on patterns of diversity.

How do ecology and evolution interact? As we've seen, ecological processes deal with the day-to-day running of ecosystems—capturing and distributing solar energy on the one hand, and cycling key nutrients (think CHNOPS) on the other. Conversely, through natural selection and a host of other processes, evolution ensures that life is able to adjust to the inevitable changes that occur over deep time. Whereas ecology can be thought of as nature's short-term memory, maintaining the flow of energy to handle the immediate tasks at hand, evolution, tapping into vast stores of genetic information, provides the long-term memory, making midstream course corrections as necessary.

Life's episodic pattern of lengthy stasis and brief bursts of change demonstrates that there is much more to evolution than the generation-by-generation battle for genetic dominance. Indeed, evolution in the sense of generating new species is perhaps better regarded as a "last resort" than an ongoing survival strategy. When faced with changing environmental circumstances—for example, the onset of an ice age or hothouse conditions—species and their populations do not merely stay put, grimly determined to change their form to meet the new conditions. Instead, where possible, organisms track their preferred habitats. Often it is only when habitat tracking is no longer an option that we see much evidence of extinction and speciation, the removal and addition of ecosystem players.

Several recent paleontological studies, from both the marine and terrestrial realms, suggest that the origins and extinctions of species are not arrayed haphazardly throughout Earth history; rather, these events tend to be clumped and cross-genealogical, impacting many distantly related groups of organisms over short durations. In one of the best-documented examples, Carlton Brett (University of Cincinnati) and Gordon Baird (State University of New York at Brockport) documented a sequence of eight assemblages of invertebrates (e.g., trilobites, clams, and corals) from the Middle Devonian (about 380 million years ago) of North America. Brett and Baird found that each assemblage formed a community that could be characterized by its own unique composition of organisms. And while there was an occasional isolated appearance or disappearance of a species, the great bulk of species changes were concentrated in brief turnover events that completely restructured the ecosystem.

The bursts of species turnovers documented in this and other studies have been linked to major environmental shifts. The apparent synchronicity between evolutionary and environmental change (although questioned by some investigators) suggests that evolution is frequently triggered by external factors that affect entire ecosystems. So not only do single species remain in stasis over geologic time; entire groups of species within regional ecosystems may do so as well. This is not to say that species originate and go extinct only in large-scale turnover events; a regular background of such events continues during even the most stable intervals. At present, debate continues as to how such ecological widespread stability might be maintained over deep time. Nevertheless, the dominant pattern appears to be periods of long-term stability dominated by the forces of ecology, paired with brief episodes or pulses of upheaval, dominated by the forces of evolution. Entire ecosystems become established over (geologically) short durations, and their constituent species then persist relatively unchanged for extended periods (up to millions of years). Ultimately, some environmental change forces species turnover and a major reorganization, resulting in the formation of a new ecosystem.

This is an exciting time in the history of evolutionary biology. Whereas Darwin's first theme—descent with modification—has been confirmed beyond all reasonable doubt, his proposed mechanism for change—natural selection resulting in continual gradual change—no longer explains all of the patterns that biologists (and particularly paleontologists) observe in nature. The next decade or so is likely to yield important new insights regarding the integration of ecology and evolution. I'm quite certain that Darwin would have appreciated the discussion.

Whether evolution occurs gradually or in fits and starts, the central message of Darwin's legacy is that change is ubiquitous. Observed on the scale of a human lifetime, the world may sometimes appear permanent and unwavering (although less so in an age of rapid global warming). And, as we've seen, species can persist relatively unchanged for hundreds of thousands, even millions, of years. Yet viewed from the perspective of deep time, change is a constant, and nothing escapes its touch. Armed with the dual perspective of ecology and evolution, let's return to the Mesozoic and unravel some of the threads that linked dinosaurs to one another and to their shifting environments.

Contrasting feeding strategies in a pair of Cretaceous herbivores. The crested, bipedal duck-billed dinosaur *Lambeosaurus* feeds on overhead browse, while the four-footed horned dinosaur *Chasmosaurus* concentrates on cropping low plants. Mixed-species associations like these may have occurred as a means of avoiding predators. At present, however both the behavioral differences and the mixing of species are based on speculation.

7

THE GREEN GRADIENT

WILDEBEEST CROWDED THE RIVER'S EDGE, snorting loudly and jostling nervously for position. The tension in the air was palpable. It was clear that, at any moment, the herd would leap into the murky water and swim to the opposite shore. It was equally evident from their agitated behavior that no one wanted to be first.

In 2004, following a successful field season hunting dinosaurs (or at least their bones) in the remote Turkana region of northwest Kenya, a few of the crew members decided to travel to the Masai Mara National Reserve to experience the famed Mara ecosystem firsthand. Situated in the southwest corner of the country, Masai Mara comprises the northernmost portion of a vast, contiguous protected area, the Serengeti-Mara Ecosystem, the bulk of which is located in Tanzania. This is a land of shimmering grasses and breathtaking skies interrupted only by an occasional rocky escarpment or verdant riverine forest. Encompassing 25,000 square kilometers (9,700 square miles), this ecosystem is also one of the last great refuges on Earth for free-roaming large animals, predator and prey alike.

That makes Masai Mara a fine place to think about dinosaurs.

For most of the year, a vast assemblage of ungulates—dominated by wildebeest, zebra, and gazelle—inhabit the southern portion of this reserve. They live off the abundant short-grass plains, feeding, rutting, giving birth, and earnestly avoiding large predators. During June or July, with the onset of the dry season, the animals begin a long, northward trek of about 800 kilometers (500 miles). Many wind up in Kenya's Mara region, which remains wetter at this time and thus offers more food. This magnificent event, referred to simply as "the migration," culminates with the arrival of the wet season several months later, which triggers a return en masse southward.

We had arrived in August, near the migration's peak. At one point, our guide, a Kenyan man named Ken Kamau, drove the Toyota Land Cruiser slowly over the crest of a hill into the midst of a vast mixed throng of zebras, wildebeest, and gazelles. Ken turned off the engine, and we found ourselves immersed in a sea of hoofed animals stretching to the horizon in all directions. I squinted, and the entire landscape seemed to move. We marveled at the cacophony of hundreds of jaws consuming solar energy masquerading as grass. I felt like a voyeur in another world and couldn't help but imagine similar scenes played out with dinosaurs many millions of years ago.

In order to reach the lush promised land of tall waving grass in the north, the wildebeest must first ford the Mara River. Months later, they will do so again on the return trip to Tanzania. These brief water crossings, each requiring less than a minute for most animals, nevertheless comprise some of the most frightening moments in a wildebeest year. In terms of distance—perhaps 50 meters (160 feet)—the swim seems trivial, though you need to imagine dog-paddling in a strong current while being pushed and bumped by dozens of panicked companions.

Why the mayhem, fear, and tension? In a word . . . *crocodiles*. Mara crocodiles gorge themselves annually during the wildebeest migration, consuming enough meat to satisfy their cold-blooded metabolic needs for months to come. In addition to those taken by predators, some of the hoofed herbivores drown in the frenzied crossing.

As we stood in the open-air truck on the far side of the river, the herd of several hundred wildebeest ambled slowly toward us, munching on grass along the way. We asked Ken whether they might be preparing for a river crossing. He expressed doubt, stating that he had conducted game drives in Masai Mara for several years and had yet to witness this famous event. However, he happily agreed to stop and wait, just in case. Over the next 15 minutes the herd moved ever closer to the short, steep slope that descended into the muddy waters of the Mara. Suddenly, one adult wildebeest stepped slowly, hesitatingly, off the grassy plain and headed down the slope toward the water. Before reaching the river, however, it turned and ran back up to rejoin the herd. Then another came down the hill, followed by another. Each time the nervous animals ventured closer to the water, and each time they returned to the growing mass of herbivores above. This anxious sequence continued a while, until crowding from the rear of the herd literally forced the animals up front to descend the slope. Now there was no longer any option to return to the safety of the plains above; a wildebeest wall blocked the way. The herd's front guard continued to advance until several adults teetered at the muddy water's edge.

Meanwhile, in the river, several massive hippos lounged in the shallows, doing what hippos do during the day—pretty much nothing. Nearby, a large crocodile perhaps four meters (12 feet) long sat motionless, partially submerged, while the herbivores gathered. After a few minutes, the croc casually roused itself and, with long powerful strokes of its armored tail, set off slowly downstream. Perhaps, we speculated, the oversized reptile sensed what was about to happen and was preparing for an underwater attack.

Finally, suddenly, the first wildebeest leapt high into the air, splashed down midstream in the river, and made for the opposite shore. Other animals quickly followed until there was a frenzy of ungulates leaping and swimming for their lives. The resulting commotion was staggering. Soon a long line several wildebeest wide stretched from one side of the river to the other. The reaction of the hippos lounging nearby seemed to be a combination of boredom and disgust that their quiet day had been so rudely interrupted. In less than ten minutes, the entire wildebeest herd had negotiated the crossing. One frightened youngster made the trip almost three times, paddling most of the way across, heading back in confusion, and then performing the swim again. In the end, all made it safely to the south side of the river. Perhaps the crocodiles had fed recently enough so as not to be tempted by this particular glut of meat fording their realm.

Wildebeest are one of many varieties of plant-eating ungulates that inhabit the Serengeti's grasslands and woodlands. Smaller-bodied examples include impala, as well as Thompson's and Grant's gazelles. Also present are several similarly sized herbivores—zebra, topi, and hartebeest—and a handful of larger forms—buffalo, eland, giraffe, and elephant. Rhinos venture into the open savannah on occasion, though they prefer the cover of bushland, also home to diminutive antelope like dikdik and bushbuck. Although one occasionally glimpses the long, curving horns of a waterbuck poking above the grass, these majestic ungulates favor the lush riverine forest that snakes through this landscape. Given such a bounty of prey species, it's not surprising that the Serengeti is also predator central, with lion, leopard, cheetah, hyena, wild dogs, jackals, and foxes, among others. Of course, all of these glamorous members of the fauna are accompanied by a spectrum of smaller animals such as rodents, turtles, fishes, birds, lizards, and insects. Finally, as in any ecosystem, most of the biodiversity is tied up in microbes that remain hidden from view.

People often think of the Age of Dinosaurs as a strange and unique period in Earth history when giant animals were common worldwide. Yet for the vast majority of time since the dinosaurs first diversified in the Early Jurassic—that is, for the better part of 200 million years—terrestrial ecosystems over most of the globe have typically included numerous big-bodied animals. Although post-Mesozoic faunas may have lacked animals approaching body sizes of the largest dinosaurs, various mammals such as elephants and huge rhino relatives regularly filled the role of land-dwelling giants. The few exceptions to the persistent presence of bigness have occurred in the wake of mass extinctions, including the one that wiped out the dinosaurs. Today, due primarily to human-induced extinctions, we are in the midst of another of these exceptional intervals, one in which giant animals are increasingly restricted to small, protected reserves. With the exception of Africa, the cradle of our bipedal ancestors, the arrival of humans on virtually every other landmass has precipitated the rapid disappearance of megafaunal species. Consequently, today you must travel to places such as Kenya to glimpse habitats full of glorious big beasts.

The highly diverse Serengeti ecosystem is fueled largely by grass. The dominant varieties here are red oat grass, thatch grass, and sweet pitted grass. With so many herbivore species feeding on this seemingly homogenous plant community, you might wonder how the grasslands manage to survive at all. The answer involves a mixture of specialized anatomy, partitioned resources, and a division of labor. Zebra feed on the coarse, high-fiber, and generally unpalatable portions of the grass, mostly seed heads and stems. They are the first wave of herbivores, exposing the greener, high-protein grass leaves that are the preferred fodder of the second wave—wildebeest and several other large ungulates. The feeding activities of wildebeest in turn expose fresh shoots and herbs favored by a third wave of smaller ungulates such as Thompson's gazelles. Other players in this sequence include topi and hartebeest, with long pointed faces and narrow mouths ideally suited for selectively picking out the most nutritious grasses. African buffalo, with their broad mouths and massive molars, adopt the opposite strategy, munching unselectively through grasslands like a battalion of horned John Deere mowers. Finally, a few animals such as elephant, eland, and Grant's gazelle are even more generalized in their feeding habits, not only grazing on grass but also browsing on a variety of broad-leaved plants.

Surprisingly, it appears that the grasses thrive in response to this multipronged, coordinated assault. Studies have shown that the succession of grazers—first zebra, then wildebeest, then smaller ungulates like Thompson's gazelle—passing through a grassland actually stimulates growth of grass species and produces a greater diversity of low, herbaceous plants palatable to a range of herbivores. This is not to say that the system is impervious to overgrazing. But, in general, the spectrum of carnivores helps to keep the total number of plant consumers in check. Similar types of relationships interlinking plants, herbivores, and carnivores almost certainly existed in dinosaur-rich Mesozoic ecosystems.

The Mara woodlands, as well as the grasslands, take a thrashing from the activities of herbivores. While driving around, we noticed that the larger acacia trees were bare throughout most of their heights yet were topped by neat, broad, bell-shaped canopies of vegetation. One could almost imagine large teams of Serengeti gardeners trimming the vast woodlands, a vision not far from the truth. Ken explained that the acacias are regularly pruned by giraffes, which use their long tongues and prehensile upper lips to great effect, stripping leaves to a height of about 6 meters (18 feet). This activity produces a "browse line," above which the tree is able to spread beyond the reach of even these long-necked herbivores.

Much more devastating are the elephants. These behemoths frequently topple mature trees and indiscriminately trample thickets of bush. It's amazing to watch elephants tear off large tree limbs with their muscular trunks or literally snap tree trunks merely by leaning on them. In a short time, a few rambunctious proboscidians can decimate large expanses of woodlands. Sauropods and other giant dinosaurian herbivores may well have generated giraffe-like browse lines. And, given that many dinosaurs such as hadrosaurs and ceratopsians approached elephantine proportions, and that the sauropods exceeded this limit by many times, it's difficult to conceive that dinosaurs were not ancient precursors to elephants in their treatment of woodlands. It was likely trivial for sauropods to knock down small to medium-sized trees.

The end result of these varied herbivore-plant interactions in the Serengeti is a continuous, abundant flow of energy and cycling of nutrients, with each species having a unique role that perpetuates the ecosystem as a whole. The energy gradient generated by the sun is transformed by photosynthetic plants into a green gradient, which herbivorous animals attempt to dismantle. Transformed into the flesh of herbivores, the green gradient then feeds a variety of carnivores and, ultimately, decomposers.

Animal diversity, then, is inextricably tied to plant diversity. The green gradient defines the capacity of an ecosystem to support consumers of varying sizes and trophic levels. Of the millions of species currently living on Earth, it's estimated that about 50 percent, and perhaps as much as 90 percent, of that diversity occurs in the exceptionally productive tropical rain forests. As one moves away from the equator north or south, there is a progressive drop in biodiversity, with the frigid polar regions supporting the fewest number of species.

In the Mesozoic, things were very different. Remember, this was a hothouse world lacking polar ice caps, with elevated sea levels and greatly reduced temperature differences between high and low latitudes. The poles supported a range of plants, dinosaurs, and other animals; however, just as today, dramatic seasonal variations in solar energy would have limited the potential for life at the highest latitudes. In contrast to the present day, the aridity associated with Mesozoic global warming appears to have severely restricted plant productivity in the equatorial regions as well. The world of the dinosaurs lacked tropical rain forests and all of their associated biomass, with peak levels of biodiversity perhaps occurring instead at the wetter midlatitudes. Thus far, the fossil record bears out this prediction, with diversity and abundance very high at the midlatitudes (e.g., North America and Asia) and much more limited in the polar and equatorial regions (Alaska and North Africa, respectively). However, we still have relatively few fossil localities from outside temperate latitudes with which to test this idea.

Herbivores come in all shapes and sizes, from gargantuan elephants and sauropods to tiny beetles and caterpillars. Despite their diminutive nature, by virtue of sheer numbers, insects usually have a much greater overall impact on plant communities than do vertebrates. Even in savannah grasslands like the Serengeti—replete with vast herds of wildebeest, zebra, and impala—vertebrates are responsible for less than one-third of the plant matter consumed, whereas plant-eating insects carry out closer to two-thirds of the herbivory. Like mammals and dinosaurs, insects evolved a variety of strategies to help them process plants, including a wondrous array of jaw structures and vast numbers of symbiotic gut microbes. Just as in us, those gut microbes participate in digestion by breaking down food and allowing nutrients to be absorbed into the body.

Among vertebrates, food quality and herbivore body size are inextricably interwoven. Any kid knows that big animals eat more food than small animals. No surprise there.

However, plant foods are extremely variable in quality, even within the same plant. And plants, like animals, vary greatly in their occurrence within ecosystems. Small animals, with their lesser dietary requirements, can afford to search out high-quality, easily digestible items, such as fruits, seeds, and fresh shoots, which typically have patchy distributions. Consequently, small herbivores tend to have more specialized diets, consuming not the entire plant but the most nutritious parts. Think about squirrels storing nuts for the winter. Conversely, because highly nutritious dietary items such as seeds and fruits tend to occur in smaller quantities with more limited distributions, they simply cannot support populations of bigger animals. Instead, large herbivores must consume not only greater quantities of food but also food of poorer quality, high in fiber and low in fat and protein. Serengeti wildebeest must migrate hundreds of miles annually because they can't make a living off stored nuts or other highly nutritious plant parts.

This intimate relationship between body size and food quality results in highly divergent strategies for large and small herbivores. Unable to find sufficient amounts of easily digestible fruits and seeds, larger herbivores tend to be "whole-plant predators," consuming substantial portions of plants rather than select parts. Yet by themselves, herbivores generally lack the enzymes to break down the cell wall compounds of plants, such as cellulose and lignin. So they unknowingly enlist the help of microorganisms, which reside in fermentation vats within the hind gut. With big appetites, big populations, and unselective diets, these animals tend to have a major impact on their native habitats. Here, then, is a pattern with clear implications for a Mesozoic world of giants.

Yet, past or present, plants do not take abuse from herbivores lightly. Despite their relative immobility and seemingly passive approach to life, plants have evolved an array of effective defenses to thwart the best efforts of herbivores. Most obvious is a range of tough, nonnutritious outer coverings such as bark and spines, though leaves, too, can be highly fibrous or covered in a waxy, unpalatable coating. Faced with such tough, low-nutrition (not to mention bad-tasting) fodder, digestibility is typically a major obstacle for herbivores. In response, plant consumers tend to adopt one of two strategies. The first is mechanical breakdown, using such "front-end" tools as jaws, teeth, and stomach stones. The second strategy involves back-end modifications, such as large, elongate guts and prolonged passage times of food; these features in turn enable bacteria living within the gut to ferment the plants and thereby gain access to nutrients. Over time, plants and herbivores engage in evolutionary "arms races," each attempting to achieve ecological one-upmanship over the other.

To extend the military metaphor, some of these arms races escalate into chemical "warfare," with plants producing biotoxins, even cancer-causing agents, as a means of combating herbivores. Much of this biochemical production is aimed at herbivorous insects, which often have a greater overall impact on plant communities than vertebrates, despite dramatic size differences. Some of these chemicals are merely distasteful, whereas others can be downright deadly, resulting in damage to the reproductive cycle or even the demise of the consumer.[1] Remarkably, plant-generated insecticides may be the most potent force limiting the numbers of herbivorous insects.[2]

Some plants, rather than resorting to intimidating armor or poisonous cocktails, adopt an alternative defense strategy, fending off voracious herbivores through symbioses with other species. For example, the bull's-horn acacia, which inhabits the lowlands of Mexico and Central America, teams up with ants. Like most acacia, the bull's-horn is thorny, but the thorns are swollen and hollow, providing a home for the ants; in addition to shelter, the plant provides sugars, fats, and proteins to its ant partners. In return, the ants swarm the surface of the acacia, biting, stinging, and in general deterring animals of all sizes that come into contact with the plant.[3]

The last and perhaps most "ingenious" strategy that plants have evolved for dealing with herbivores is to welcome them to the dinner table, even going so far as to put out a sign advertising a free meal and then put the unwitting animals to work. Flowering plants in particular take advantage of consumers' need to feed, producing a variety of irresistible delicacies. Less than two centuries ago, it was thought that plants made beautiful flowers because the Creator had put them there to please humans. Later it became clear that flowers serve an important ecological function that has nothing to do with bipedal primates (no surprise in retrospect, given that the first flowering plants preceded humans by over 125 million years!). Flowers turn out to be the siren song of the plant world, the means of signaling pollinators. Drawn like flies to dung, these pollinators—from butterflies and bees to bats and birds—spread pollen from flower to flower and, in doing so, facilitate the proliferation of angiosperms. Today, the great majority of angiosperms (up to 98 percent in the highly diverse lowland tropical rain forests) are pollinated by animals. Flowering plants literally depend on these coevolved symbioses in order to disperse and reproduce.

Pollination partnerships are the result of millions of years of coevolution, and many are finely tuned. For example, interspecies relationships are often highly specific, with a single animal pollinator linked to a single plant. Moreover, both plant and pollinator typically have specialized structures that facilitate their relationship. Thus, bees possess color vision, sensitivity to certain odors, and specialized pollen- and nectar-carrying structures, all linked to the flowers they pollinate. Insect-pollinated flowers tend toward hues of blue and yellow, because most insects cannot see red. Red flowers, in contrast, are largely the domain of hummingbirds. Paleontologists often assume that many herbivorous dinosaurs during the Cretaceous fed on angiosperms. It's important to remember, however, that these giant herbivores depended on insects to aid in the reproduction of their food supply, just as we humans do today.

Mesozoic plants and herbivores coevolved for almost 200 million years, with dinosaurs making up the dominant large-bodied plant eaters for most of that era. As with modern examples, Mesozoic plants and animals evolved a range of shifting strategies for dealing with each other. Here I briefly describe innovations on both sides of this coevolutionary dance, and then I address some best guesses as to how this dance progressed through time. This endeavor requires that we try to envision plants from the point of view of herbivorous dinosaurs and contemplate herbivorous dinosaurs from a plant perspective. As we proceed, I invite you to keep the Serengeti in mind. Although we can currently reconstruct Mesozoic ecosystems with only the broadest of brushes, like the African example, they also were undoubtedly complex and tightly interwoven.

Chapter 5 introduced eight major groups of plants that dominated the world of dinosaurs. First were three clans of spore-bearing pteridophytes—ferns, horsetails, and lycopods—most of which were low-growing, water-dependent herbaceous plants. Next came four groups of seed-bearing gymnosperms that were widespread from the Triassic through the Early Cretaceous: cycads, ginkgos, conifers, and bennettites. Finally, angiosperms, or flowering plants, were the latest-appearing group of seed plants, originating in the Early Cretaceous and diversifying globally by the end of that period. These Mesozoic plants faced the same kinds of challenges as their modern counterparts. And they, too, evolved a variety of mechanical and chemical defenses to ward off herbivores, from insects to dinosaurs. Indeed, paleobotanists think that many of the defenses we see in living plants arose during the Mesozoic, when both insects and dinosaurs were proliferating.

Plants respond to herbivores in a variety of ways, with the preferred "strategy" dependent in large part on the evolutionary raw materials inherited from ancestral forms. The foliage of spore-bearing pteridophytes tends to be succulent and lacking in any major mechanical defenses such as spines or tough bark. However, living examples like ferns and horsetails frequently produce poisonous chemicals. It's feasible, and perhaps likely, that Mesozoic spore bearers also possessed chemical defenses with similar effects on dinosaurs. From the perspective of a dinosaur herbivore, pteridophytes may have offered a potentially fast-growing, renewable resource. Yet, this resource would have been restricted in abundance by the need for aquatic reproduction, and chemical defenses may have generated considerable obstacles for digestibility.

Among nonangiosperm seed plants, we find a good news/bad news situation for the dinosaurs. Cycads, gingkos, conifers, and bennettites were able to spread widely and somewhat independently of bodies of water, creating diverse habitats and providing an abundant food source. Yet many living representatives of these groups possess tough outer coverings and/or resistant foliage, sometimes with spines. In addition, the cells often have thick walls, making them difficult to break down and digest. Finally, a large proportion of seed plants are rich in indigestible chemicals and resins. Despite these many obstacles, members of all five of these gymnosperm families may well have been consumed by dinosaurs during most of their tenure.

In contrast to the spore bearers and other seed bearers, angiosperms may have offered a welcome alternative for plant-eating dinosaurs. If modern flowering plants are any indication, Mesozoic examples had succulent foliage and fewer chemical defenses. Like other seed plants, angiosperms grow in a wide range of habitats largely independent of standing water. Unlike most seed plants, a damaged angiosperm can often recover quickly by regrowing from underground buds. Thus, they had the potential to spread widely, exploiting (and cocreating) a diverse range of habitats. In short, flowering plants may have been a dinosaur herbivore's dream: abundant, succulent, and fast growing, with fewer indigestible chemicals, tolerance to intensive herbivory, and the ability to recover quickly following cropping. Keep in mind, however, the fact that angiosperms were Mesozoic latecomers, first appearing in the Early Cretaceous. Jurassic herbivores such as *Brachiosaurus* and *Stegosaurus* never saw a flower. Some paleobotanists have argued that the terrestrial world became much more vegetated after the appearance of flowering plants, allowing animals to disperse much more widely as well.

The most important thing to remember about plant-eating dinosaurs is that most were really big. The largest sauropods were on the order of 70 tons (150,000 pounds), greater than ten times the weight of the biggest elephants. Even the smallest known dinosaur herbivores were still pretty large; *Lesothosaurus*, a primitive form from South Africa, was

FIGURE 7.1

Size variation in dinosaurian herbivores. Top row from left to right: *Lesothosaurus* (an Early Jurassic primitive ornithischian), *Psittacosaurus* (an Early Cretaceous ceratopsian), and *Argentinosaurus* (a Late Cretaceous sauropod). Second row from left to right: *Protoceratops* (a Late Cretaceous ceratopsian), *Euoplocephalus* (a Late Cretaceous ankylosaur), *Pachycephalosaurus* (a Late Cretaceous pachycephalosaur), and *Triceratops* (a Late Cretaceous ceratopsian). Bottom row from left to right: *Edmontosaurus* (a Late Cretaceous ornithopod) and *Stegosaurus* (a Late Jurassic stegosaur). Human in top row added for scale.

about 1 meter long and would have weighed somewhere in the range of 4–7 kilograms (9–15 pounds). This is still many times larger than the smallest examples among birds and mammals.

If you're interested in dinosaur feeding behavior, teeth are the best place to start searching for clues. These resistant bits of enamel and dentine yield valuable indicators as to their owner's diet. When it comes to teeth, dinosaurs were much like sharks. Sharks replace their teeth continuously and are constantly generating new teeth. Employing a marvelous, conveyer belt–like system, sharks eject old teeth and rotate fresh replacements into position. Break a few teeth attacking a sea lion? No worries— there's more on the way. Although the jaw mechanisms of sharks and dinosaurs are radically different in many respects, both groups generate(d) new teeth and replace(d) old teeth continuously.

In contrast, humans and most other mammals have just two sets of teeth, commonly referred to as "baby" and "adult." When the adult teeth wear out, humans (at least those who can afford them) turn to dentures. Old individuals of other mammal species, like elephants and horses, have been known to grind their teeth smooth; unable to feed, these animals quickly succumb to starvation. Why would evolution curse mammals with a mere two sets of teeth, whereas the remainder of the vertebrate world happily pumps out replacements indefinitely? The answer is specialization. Because their teeth do not continuously fall out, mammals are able to have precise and varied kinds of contacts between upper and lower dentitions. Look in the mouth of your average mammal (or simply look in the mirror and open your own mouth), and you'll find a variety of tooth types, each crafted over millions of years to serve a particular function. Depending on the mammal in question, there may be pointed teeth for stabbing, bladelike teeth for slicing, and broad molar teeth for grinding and pulping.

Now check out the mouth of a dinosaur (or, say, a shark, lizard, or frog), and what you'll see is sameness. The teeth at the front and back of the jaws tend simply to be larger or smaller variations on a single theme. Importantly, however, the specific characteristics of each tooth type vary considerably, depending on the group in question and on the animal's diet. In other words, though dinosaur teeth tend to be homogenous from back to front within a single species, evolution crafted a diverse set of solutions to the challenge of feeding. In addition, much of the plant processing undertaken by dinosaurs occurred farther along the gastrointestinal tract.

The closely related prosauropods and sauropods—together known as the sauropodomorphs—represent one the most successful "experiments" in the history of land animals. They also gave rise to the earliest radiation of dinosaur herbivores. First appearing in the late Triassic, these dinosaurs smashed all size records set by previous large-bodied land animals, and few animals have come close since the Mesozoic. They were the dominant terrestrial herbivores globally during the Jurassic, and they persisted in this role over much of the planet—particularly in the Southern Hemisphere—during the Cretaceous as well. People have long marveled at the incredible sizes of sauropods

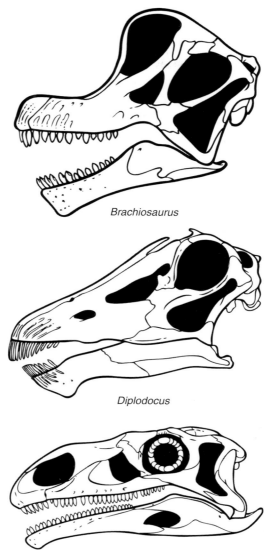

Brachiosaurus

Diplodocus

Plateosaurus

FIGURE 7.2
Variation in the skulls of
sauropodomorph dinosaurs:
Brachiosaurus (a sauropod), *Diplodocus*
(a sauropod), and *Plateosaurus*
(a prosauropod). Not to scale.

and wondered how such animals could have swallowed sufficient food to keep their internal furnaces stoked. Regardless of metabolic rate, animals of such great mass could not have been overly selective in their dietary preferences.

Sauropod skulls come in two general types, with plenty of odd variations. The first skull type, epitomized by *Camarasaurus* and *Brachiosaurus*, is short front to back, but tall, with thick spoon-shaped teeth and the bony external nostril positioned up front. The second type, characteristic of *Diplodocus* and a number of titanosaurs, is longer and lower with narrow, pencil-like teeth restricted to the front of the jaws. It is generally thought that sauropods fed by raking their teeth across a branch to remove the foliage. Many forms had tooth-to-tooth occlusion, with upper and lower tooth crowns meeting in a relatively precise fashion when the jaws were closed. Yet sauropods had little if any ability to chew as mammals do. Some researchers speculate that, like giraffes, they relied on powerful tongues to manipulate food items for slicing. Much has been written about how such giant animals were able to pass enough food down their gullets to sustain their enormous bulks, and we will return to this fascinating problem in chapter 13.

Sauropods are typically regarded as high browsers with graceful, upright, swanlike necks. With the neck held nearly vertical, the largest of these animals would have been able to peer, periscope-like, into a fourth-story window. More extreme reconstructions depict these animals rearing up, using a tripodal stance composed of tail and hind legs in order to fight, mate, or feed. In recent years, various arguments have been presented against these behaviors. Some say that the articulation of bones within sauropod necks prohibited upright, giraffe-like neck positions. Others point to a lack of muscular leverage to lift the head and neck. Still others argue that these animals would have been unable to generate sufficient blood pressure to pump blood from the chest to the head. At present, there is little consensus on this problem. Nevertheless, the evidence at hand indicates that most sauropods tended to hold their necks in a more horizontal posture and that different species and groups varied in their ability to raise the neck to an erect position. Some, like *Brachiosaurus*, may have habitually positioned the neck as giraffes do; others, like *Dicraeosaurus*, likely lacked the ability to raise the head much at all; and still others, like *Diplodocus*, probably possessed some abilities for horizontal and vertical movements. As for rearing up onto the hind legs, this behavior seems intuitively unlikely for such giants. Yet the act of copulation demanded that at least some males engaged in this behavior on occasion. Like elephants living in the wild (as opposed to the circus variety), I imagine that sauropod rearing, though possible, was not a common sight on Mesozoic landscapes.

Other than sauropodomorphs, all major groups of dinosaur herbivores fall within the bird-hipped clan, Ornithischia. Members of this group possessed toothless beaks presumably covered in life with a horny sheath. Feeding strategies, however, apparently varied greatly among ornithischians. The simplest teeth—mostly spoon- or leaf-shaped structures topped with coarse ridges—are found in animals such as ankylosaurs, stegosaurs, and pachycephalosaurs. Whereas the former two groups were low-browsing quadrupeds, the dome-headed pachycephalosaurs were bipeds, capable of feeding somewhat higher in the canopy. The largest examples within these herbivorous clans were on the order of 10 meters (30 feet) long and weighed as much as 6,000 kilograms (13,000 pounds). Thus, here also we must wonder how the animals managed to ingest sufficient foliage to maintain such massive bodies, particularly given their wimpy teeth.

The most specialized teeth and jaws among ornithischians are found in the duck-billed dinosaurs (hadrosaurs) and horned dinosaurs (ceratopsids). These two groups

Stegosaurus

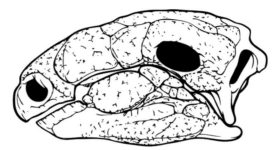

Panoplosaurus

FIGURE 7.3
Variation in the skulls of
thyreophoran (armored)
dinosaurs: *Stegosaurus*
(a stegosaur), *Panoplosaurus*
(a nodosaurid ankylosaur),
and *Euoplocephalus* (an
ankylosaurid ankylosaur).
Not to scale.

Euoplocephalus

independently evolved powerful dental batteries, each composed of closely packed columns of teeth that together formed a continuous cutting edge. With up to forty rows of teeth on each side of both upper and lower dentitions, and multiple replacement teeth in each column, the jaws of a single adult *Triceratops* or *Edmontosaurus* contained literally hundreds of teeth. In carnivores, the jaw joint of the lower jaw is generally located at the same level as the tooth row, causing the jaws to close in a simple, scissor-like fashion. However, the scissor design is less effective for processing plants. In ceratopsids and hadrosaurs, the jaw joint is located below the level of the tooth row, a pattern seen in other herbivores before and since. The offset arrangement allows the entire upper and lower

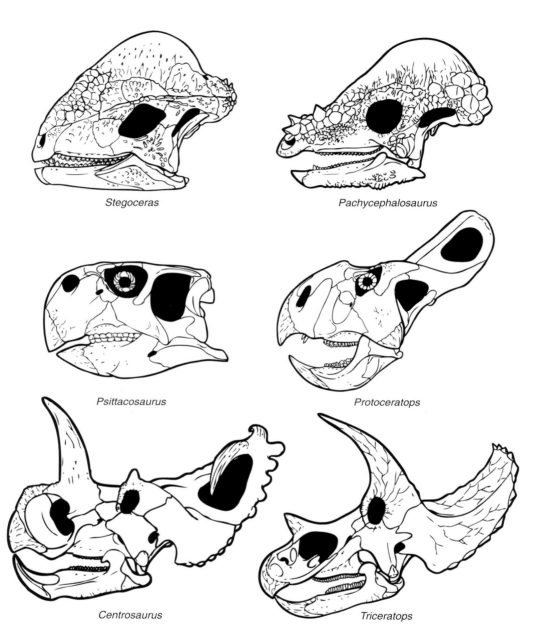

Stegoceras

Pachycephalosaurus

Psittacosaurus

Protoceratops

Centrosaurus

Triceratops

FIGURE 7.4

Variation in the skulls of margin-headed (marginocephalian) dinosaurs: top row, the pachycephalosaurs *Stegoceras* and *Pachycephalosaurus*; middle row, the ceratopsians *Psittacosaurus* and *Protoceratops*; bottom row, the ceratopsid ceratopsians *Centrosaurus* and *Triceratops*. Not to scale.

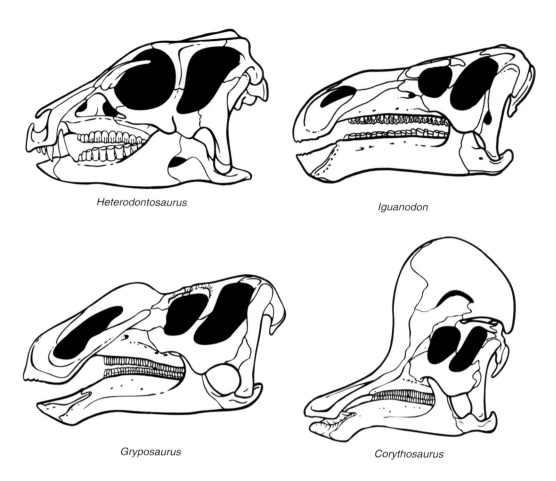

FIGURE 7.5
Variation in the skulls of ornithopod dinosaurs: *Heterodontosaurus* (a heterodontosaur); *Iguanodon* (an iguanodont); *Gryposaurus* (a noncrested hadrosaur); *Corythosaurus* (a crested hadrosaur).

dentitions to contact each other almost simultaneously, a great advantage when dealing with a mouth full of plants. The lower teeth passed inside the uppers, allowing the upper and lower tooth rows to come into contact and slide past each other like two elongate blades.

These elaborate food-processing systems, which evolved independently in hadrosaurs and ceratopsids, were not well suited for chewing, as in mammals. Ceratopsids jaws simply sliced food items up into smaller chunks, whereas an oblique contact between upper and lower dentitions in hadrosaurs suggests some ability to crush and grind as well as slice. In both groups, the tooth-ridden jaws were augmented up front by an extremely thick, reinforced beak, narrow and parrot-like in ceratopsids, broader and more ungulate-like in hadrosaurs. Presumably the beak was used to break off plant matter, which was then sliced by the teeth prior to swallowing. With a keratin-covered beak, a formidable set of dental batteries, a lowered jaw joint, and plenty of muscle power, these advanced food-processing systems were the Cuisinarts of the Cretaceous, equipping horned and duck-billed dinosaurs with finely tuned feeding apparatuses as intricate as any that large herbivorous mammals would evolve more than 30 million years later. Based on their capacious gut regions, we can be quite sure that all of these giant herbivores also possessed elongate intestines full of fermenting microbes.

When it comes to pairing up particular kinds of herbivorous dinosaurs with their preferred plant diets, we remain entrenched in the dark. This unfortunate situation represents a major gap in our knowledge that future research may be able to fill. Studies of microscopic wear patterns on teeth could help, because different plants can leave varying traces of wear. Tooth enamel can also preserve isotopes, allowing researchers to investigate at least large-scale questions regarding diet—for example, meat versus plants, and open versus closed canopy forest settings. At the back end of the gastrointestinal (GI) tract, exceptionally rare instances of preservation sometimes include gut contents with processed plant materials. Similarly, there are a growing number of fossilized dung specimens (COPROLITES) sufficient in size to implicate dinosaurs as the poop makers. Examples of the latter two lines of evidence—gut contents and dung—suggest that hadrosaurs consumed conifers at least on occasion. Conversely, a few other coprolite specimens attributed to dinosaurs appear to include partially digested bits of angiosperms. Mostly, however, we are currently limited to sweeping generalizations and bald speculations.

Yet, given that plants were the dominant source of energy on land, and that dinosaurs were the dominant large-bodied herbivores in terrestrial ecosystems, we can be quite confident that coevolution between these groups occurred. Plants must have evolved various techniques for curtailing dinosaur herbivory, and plant-eating dinosaurs undoubtedly countered with a number of strategies to thwart the evolutionary innovations of plants. We must also keep in mind that there were many other kinds of smaller herbivores in these ecosystems, including mammals, birds, and insects. As noted earlier, insects in particular are known to have major effects on plant communities.

In lieu of direct evidence linking specific kinds of dinosaurs and plants, we must turn to indirect sources of information. One of the strategies most commonly employed by paleontologists is to search for patterns of evolutionary change in two coexisting groups, and then see whether these patterns coincide in meaningful ways. So, let's take a look at the entire Mesozoic tenure of dinosaurs and plants and consider some possible examples of coevolution. Although the details of these plant-herbivore interactions remain obscured by millions of years of deep time, we can begin to construct testable hypotheses. And the known evolutionary patterns on both sides of the plant-dinosaur equation lead to some provocative scenarios. Many of the hypotheses summarized here come from the work of two paleobotanists, Scott Wing of the Smithsonian Institution and Bruce Tiffney of the University of California, Santa Barbara.

When dinosaurs first appeared, during the Late Triassic, the plant world was dominated by a range of seed plants. Remember that terrestrial ecosystems had undergone a dramatic transformation after the formation of Pangaea, with the water-loving spore bearers displaced in large part by arid-adapted seed bearers. Some of these, particularly among the conifers, reached heights far beyond those of any previous land plants, concentrating much of the foliage high above the ground. In contrast to their predecessors, Late Triassic plants also tended to have tough or thorny outer coverings, which meant that big herbivores now had to get by on relatively low-quality diets.

As previously noted, the Late Triassic rise of conifer trees, in both an evolutionary and an absolute height sense, may have been linked to coevolution with herbivorous dinosaurs. Conifers reached new heights at approximately the same time as the prosauropods. Long-necked prosauropods such as *Plateosaurus* in Europe and *Riojasaurus* in South America emerged as major players in Late Triassic ecosystems, greatly exceeding the sizes of all previous land-dwelling herbivores. They were accompanied by primitive sauropods and, whereas the prosauropods went extinct in the Early Jurassic, the sauropods remained the dominant large herbivores globally through the Jurassic. So it's possible that the Late Triassic shift in defensive tactics seen in plant communities may have occurred partially as a response to dinosaurian consumers. Sauropodomorphs, with their newly evolved giant sizes and elongate necks, would have been able to browse at greater heights. With high and low browsers present, Triassic ecosystems appear to have been the first to include partitioning of plant resources by herbivores on the basis of height. This multitiered approach to herbivory is a pattern that evolved again and again, becoming a standard feature of terrestrial ecosystems during and after the Mesozoic.

The sauropod-conifer connection is thought to have persisted through the Jurassic and, at least in some areas, into the Cretaceous, with both groups diversifying into a broad range of forms around the globe. The largest Jurassic sauropods were capable of reaching more than 10 meters (30 feet) into the canopy. If certain kinds of sauropods were able to rear up on their hind legs, these animals would have been able to access even greater heights. Although some sauropods were likely unable to lift their heads high above their bodies, others probably used their large bodies and elongate necks to browse the crowns of tall trees. So long necks and tall trees could have evolved in tandem. Yet, if we are to be scientific in this investigation, we must consider all other possible reasons why trees became tall. For example, there may have been increased competition among the plants themselves for access to sunlight, with the advantage going to species that achieved greater heights.

Herbivore height, if it was a factor at all, was only part of the equation. Another important evolutionary factor was body mass—sheer bulk. That is, herbivores with greater body masses may have been more successful because Late Triassic floras were tougher and thus offered a lower-quality diet. In a world dominated by increasingly tough, fibrous vegetation, bigger animals with more elongate guts may have had a distinct advantage because they would have been better able to digest these problematic plants. Some have even argued that the feedback loop exemplified by this coevolutionary race provides the solution to an enduring mystery. Why were dinosaurs so big? Perhaps because herbivorous dinosaurs evolved increasingly large sizes in order to feed on taller, poorer-quality plants. Carnivorous dinosaurs, in order to be effective predators in a world of giants, simply followed suit, forced to mimic this prevalent pattern in their herbivorous prey. Although reality was undoubtedly more complex and less reminiscent of a "Just-So" story, this simplistic scenario may encapsulate an important kernel of truth, revealing a set of linked events that forever changed the face of terrestrial ecosystems.

Let's now turn to the Cretaceous, during which the biggest event in the floral realm by far was the origin and spread of angiosperms. Flowering plants first appear in the fossil record in the Early Cretaceous, about 125 million years ago, and became the dominant land plants by the close of the Late Cretaceous, 65.5 million years ago. The Cretaceous also witnessed the diversification of many groups of dinosaur herbivores. This list includes duck-billed hadrosaurs (a subgroup of ornithopods) and the horned ceratopsids, with their highly specialized, food-processing jaws. Recognition of these patterns led to the hypothesis, first suggested in the mid-1980s by Robert Bakker and independently by Scott Wing and Bruce Tiffney, that herbivorous dinosaurs triggered the evolution of flowering plants. Bakker, in characteristically provocative style, went so far as to claim that "dinosaurs invented flowers." According to this idea, during the Late Jurassic and Early Cretaceous the pervasive feeding activities of plant-eating dinosaurs placed great pressure on communities of seed bearers like cycads, gingkos, and conifers. This profound disturbance, they argued, resulted in greater evolutionary success (thanks to natural selection and adaptation) for plants with certain characteristics: rapid growth, tolerance to disturbance, and the ability to regrow from underground buds. With spore bearers and most seed bearers unable to respond to this pressure, the stage was set for the spread of a new group of seed plants that possessed all of these features—angiosperms.

This hypothesis has received a pretty thorough thrashing from other paleobiologists, who note several contradictory points. First, there is minimal evidence of angiosperm-dinosaur interactions in the Early Cretaceous (and in the Late Cretaceous, for that matter). Second, while flowering plants became widespread by the Early Cretaceous, the fossil record suggests that they were minor ecosystem players, and that conifers and other nonangiosperm seed plants remained the dominant plants in terms of actual biomass. Third, many of the complex jaw mechanisms and other major feeding innovations in herbivorous dinosaurs appeared well before or long after the origin of angiosperms, suggesting that these features had little impact on the origin of flowering plants. In other words, the key evolutionary patterns in angiosperms and dinosaurs were not coincident, suggesting that other factors must have been involved in the "invention" of flowers.

Nevertheless, it is certainly conceivable that angiosperms effectively "underwrote" the Late Cretaceous radiations of herbivores such as hadrosaurs and ceratopsids. About half of all dinosaurs are known from the last 20 million years of the Mesozoic—thus, on the order of 50 percent of all dinosaur species are restricted to the final 12 percent of their 160-million-year Mesozoic tenure. Although this observation may result, at least in part, from problems of preservation and sampling, it has prompted some to argue that latest Cretaceous ecosystems were truly distinct from anything that had come before. Flowering plants provided an abundant, accessible energy source for plant-eating dinosaurs, and the capacity of these plants to exploit a wide range of environments meant that the dinosaurs could follow along into these novel settings. Angiosperms lacked many of the deterrents—particularly nonnutritious tissues and poisons—that made pteridophytes and earlier seed plants unappetizing by comparison.

This is not to say that flowers "invented" Cretaceous dinosaurs, any more than the opposite. Rather, plants and dinosaurs, together with the many other life-forms, coevolved—they cocreated each other; that is, the diverse parts of the ecosystem responded to one another via evolution. The danger with this kind of an explanation is that one is simply left with "everything caused everything"—not an enviable position for a scientist because, at least stated in this broadest sense, it precludes testing. Nevertheless, this kind of dependent co-arising of forms is likely a much more accurate description of the events and, indeed, of the way evolution works in general.

Although evolution's complexity can be daunting, the situation paleontologists face today is exciting rather than bleak. Now that we are better able to identify patterns and their timing, we can begin to refine our ideas. And while we are still a long way from understanding Mesozoic ecosystems like we do the Serengeti, it's important to remember that paleontologists have only begun to seek answers at the higher, more integrative level of ecosystems. Once this new, more integrative, coevolutionary perspective really takes hold, numerous unexpected discoveries are likely to follow. Of course, the dinosaur odyssey included carnivores as well as herbivores, and it is to the meat eaters that we now direct our attention.

Representative theropod dinosaurs, including probable omnivores and herbivores, as well as carnivores. From left to right: the ornithomimosaur *Ornithomimus* (omnivore?); the tyrannosaur *Tyrannosaurus* (carnivore); the abelisaur *Carnotaurus* (carnivore); the therizinosaur *Therizinosaurus* (herbivore); the spinosaur *Suchomimus* (carnivore); the oviraptorosaur *Khaan* (omnivore?).

8

PANOPLY OF PREDATORS

WITH FEW EXCEPTIONS, wherever you find plant-eating animals, you'll also find meat eaters. In fact, almost all ecosystems have at least two levels of consumers—that is, consumers that eat other consumers—with feeding, or trophic, pyramids sometimes topped by such renowned beasts as lions or orcas. This ecological structure enables the flow of solar energy to pass from plants to herbivores to carnivores, with each succeeding link in the consumer chain limited to a drastically reduced pool of available energy.

The world of dinosaurs was no exception to this rule. During the latter two-thirds of the Mesozoic, theropod dinosaurs were the predominant large-bodied carnivores on land. After originating on the supercontinent Pangaea during the Triassic, this family of predators blossomed into a remarkable global diversity of forms. Theropod remains have been recovered from all major continents, and extend from Antarctica to the extreme arctic. Of the seven hundred or so distinct dinosaur species recognized today, about 40 percent are theropods. Although this estimate is likely inflated by several biases, it's safe to say that theropods were a diverse bunch. When contemplating ecosystems, however, it's important to distinguish between the total number of individual animals and the total number of species present. So, even though the total number (and thus biomass) of individual carnivorous animals is typically much less than that of herbivores, the number of carnivore species may approach that of the plant eaters.

How do we know that most theropods were flesh eaters? Direct evidence of a carnivorous diet comes from multiple lines of evidence. Any 5-year-old will tell you the most obvious indicator—pointed teeth. Although tooth shapes show substantial variation across theropod species, the great majority are bladelike, recurved, and serrated. When viewed under a microscope, each serration tends to resemble a miniature tooth with a blunt tip. Like a steak knife, these tiny structures helped separate muscle fibers in the meat of their prey. Serrated teeth are so useful as meat-processing tools that they have evolved a number of times in carnivorous vertebrates, including sharks, Komodo dragons, and saber-toothed cats. It's been suggested that the tooth serrations of at least some theropods, including *T. rex*, took on a different role, trapping fibers of flesh that putrefied on the teeth and enabling these predators to deliver lethal, bacteria-laden bites like those of living Komodo dragons.

As in other dinosaurs, theropod teeth were disposable, erupting in waves that replaced older teeth with fresh recruits. Over the course of a long life (say, about 30 years of age), a single *Tyrannosaurus* likely shed hundreds of teeth. Some of these were dislodged prematurely (e.g., following a violent impact with bone), but most were simply shed on a regular cycle. Isolated theropod teeth, including absolutely pristine examples, are common finds in Mesozoic fossil deposits.

Tooth-marked dinosaur bones are also pretty common discoveries, providing additional evidence of theropod diets. These bite traces range from simple, conical holes or deep furrows to parallel lines left by serrations as the tooth dragged across the bone. Some sauropod bones have been found in Colorado with healed bite marks, suggesting that these animals survived attacks from Late Jurassic theropods. Similarly, punctured and healed tailbones in a Late Cretaceous hadrosaur suggest to some investigators that the herbivore managed to survive an attack from *Tyrannosaurus*, the only giant predator present in that particular ecosystem. Returning to Madagascar, the Late Cretaceous abelisaur *Majungasaurus* appears to have been the only medium- to large-bodied theropod present on the island at this time. Numerous *Majungasaurus* bones have been recovered with theropod tooth marks that match those of *Majungasaurus*. No, this doesn't mean they were biting themselves. Nor could the tooth marks have resulted from brief donnybrooks between competing adults, because many of the bites occur on limb bones that would have been inaccessible in a living animal. Instead, these specimens are the best evidence to date of cannibalism within theropods.

Even more definitive dietary evidence comes from extremely rare preservation of the remains of theropod meals, as either fossilized gut contents or dung. Little coelurosaurs have been found with the bones of lizards or small mammals in their guts. A finely preserved specimen of the spinosaur *Baryonyx* included fish remains in its gut region, fueling speculation that its long, crocodile-like snout was specially adapted for catching fish. (Often ignored is the fact that this same specimen includes remains of a partially digested juvenile ornithopod dinosaur.) And a remarkable, 44-centimeter (17-inch-) long fossilized coprolite (turd) attributed to *Tyrannosaurus rex* included numerous pulverized chunks of bone from one or more juvenile bird-hipped dinosaurs.

Of course, theropods shared their world with other carnivores. In the Late Triassic, dinosaurs frequently fell prey to a variety of distantly related archosaurs (the group of reptiles that gave rise to dinosaurs), a topic we shall return to in a subsequent chapter. During the

Jurassic and Cretaceous, carnivorous crocodiles aplenty lived alongside theropods. Some Cretaceous examples grew to "supercroc" sizes, easily big enough to prey on dinosaurs. Paleontologist Paul Sereno estimates that the African monster crocodile *Sarcosuchus* reached up to 12 meters (36 feet) in length and a whopping 8,000 kilograms (17,640 pounds), more massive than any predatory dinosaur. In North America, the similarly sized *Deinosuchus* competed with tyrannosaurs for meat. Evidence of this ecological relationship includes bones of both duck-billed dinosaurs and tyrannosaurs with bite marks attributed to *Deinosuchus*. Although we cannot say for certain whether tooth marks are the result of predation or scavenging, it's likely that, even in a world of dinosaurian predators, such crocodilian nightmares reigned supreme around their watery realms.

Yet, whereas freshwater crocodiles in the Jurassic and Cretaceous were restricted to wet environments, theropods occupied a spectrum of Mesozoic terrestrial ecosystems— from sand-dominated deserts to lush subtropical forests to frigid polar regions (not nearly as cold as the present, but chilly enough to be snow covered some of the year). As far as I know, every reasonably sampled Mesozoic ecosystem that has yielded plant-eating dinosaurs also includes one or more varieties of theropods. Many of these ancient hunters were multiton giants, but theropods varied greatly in size. Gigantism evolved multiple times within the group, as did the opposite trend, miniaturization. At the giant end of the spectrum is none other than the 15-meter (45-foot) *Tyrannosaurus*, which likely tipped the scales at 5,000–6,000 kilograms (11,000–13,200 pounds).[1] At the dwarf end is raven-sized *Microraptor*, about 40 centimeters (16 inches) long as an adult. With most of that length devoted to tail, this little animal weighed less than 1 kilogram (2 pounds) and could have sat comfortably in your hand. If we include living birds in the mix (because they, too, are

FIGURE 8.1
Variation in body size within theropod dinosaurs: (from top to bottom) the tyrannosaur *Tyrannosaurus*; the hummingbird *Stellula*; the dromaeosaur *Microraptor*; the dromaeosaur *Velociraptor*; the ceratosaur *Ceratosaurus*; the spinosaur *Spinosaurus*. Human added for scale.

theropods), hummingbirds qualify as the smallest members of the group. Given that a hummingbird may weigh about 0.001 kilogram, the Tyrant King was about 5 million times heavier!

To refer to the biggest theropods as giants doesn't begin to convey the radical departure that these predators took from other land-dwelling meat eaters before and since. The biggest terrestrial predators today are polar and Kodiak bears weighing about 800 kilograms (1,760 pounds). If we look to the fossil record beyond dinosaurs, most land-dwelling carnivore heavyweights are mammals that lived long after the Cretaceous mass extinction. These include the 880-kilogram (1,940-pound) hyena-like creodont *Megistotherium osteothlastes* from the Miocene of Africa, the 750-kilogram (1,650-pound) bear *Agriotherium africanum* from the Pliocene of Africa, and the 600- to 900-kilogram (1,320- to 1,980-pound) mesonychid *Andrewsarchus mongoliensis* (a relative of living whales) from the Oligocene of Asia. A remarkable Pleistocene predatory giant was *Megalania priscus*, the Australian terror lizard. Although estimates vary, it appears that the average weight of this goanna on steroids was 320 kilograms (700 pounds), with rare individuals perhaps approaching a colossal 2,000 kilograms (4,400 pounds). Notice that outside the dinosaur realm, no fully terrestrial carnivores had adult body masses averaging greater than 1 metric ton (1,000 kilograms, or 2,200 pounds). In contrast, over the course of the Mesozoic, at least six theropod lineages spawned numerous species with average body masses of 1,000 kilograms or greater, with several exceeding 3,000 kilograms (6,600 pounds). And *T. rex* was six times larger than any nondinosaurian land predator!

With few competitors and plenty of herbivore meat lumbering around the landscape, you might guess that *Tyrannosaurus* and its bloodthirsty brethren had it easy. Not so. Big carnivores face formidable physiological and ecological obstacles. Extremely large animals—whether carnivore or herbivore, warm- or cold-blooded—must consume prodigious amounts of food. By necessity, then, giants tend to occupy huge home ranges in order to obtain sufficient calories. Such problems are especially acute if those giants happen to be meat eaters. Because the great bulk of available energy at each successive trophic level, from plants to herbivores to various carnivores, is lost to heat (see chapter 5), top meat eaters must subsist on a tiny slice of the ecosystem's energetic pie. Increased home range size also means that a given area can support fewer individuals of giants than non-giants, which translates into lower population densities (number of animals per habitat area). Low population densities, in turn, present profound challenges to the longevity of a species; if the density of animals becomes too low, the species will be excessively prone to extinction. A single environmental event—for example, drought—or a rapid series of such events might be sufficient to wipe out the remaining animals. In other words, when it comes to species longevity, there is strength in numbers, at least up to a point. Maximum body size in giant carnivores, then, reflects an evolutionary balance between maintaining population densities low enough to avoid overexploitation of prey yet large enough to reduce the probability of extinction. Based solely on these two barriers—minimal available energy and low population densities—you can begin to see why big, fierce animals tend to be rare in the history of life.

Taking a closer look at the theropod family tree, you may remember that theropods and sauropods comprise the lizard-hipped group known as Saurischia. Historically, theropods were divided into two major groups. Large-bodied forms, like *Tyrannosaurus* and *Allosaurus*, were put into one evolutionary clan and their smaller-bodied cousins, like *Velociraptor*, into another. However, this arrangement proved to be far too simplistic. Paleontologists now recognize several groups of theropods, most of which included a substantial range of body sizes. So it appears that gigantism evolved multiple times within the group, raising the twin issues of how bigness evolved and why it conferred an evolutionary advantage. We'll return to these problems shortly.

Today, the detailed arrangement of branches on the theropod family tree remains controversial. Each year at the annual meeting of the Society of Vertebrate Paleontology, experts lay out new hypotheses and cladograms, arguing for some revised set of relationships. Clearly, then, we still have some work to do to nail down the history of this famous family. Nevertheless, based on current knowledge, the major branches and their nested relationships are summarized in table 1. To highlight one example, birds (Aves) are members of Paraves, which also includes *Velociraptor* and its sickle-clawed kin. Paraves is one of several groups within Maniraptora, which in turn is part of Coelurosauria, a subset of Tetanurae. For comparison, think of the famous Russian matryoshka dolls nested one inside another.

TABLE I.   The Theropod Family

Theropoda
    Coelophysoidea (e.g., *Coelophysis, Dilophosaurus*)
    Ceratosauria
        Ceratosaurus
        Abelisauroidea (e.g., *Carnotaurus, Masiakasaurus*)
    Tetanurae
        Spinosauroidea (e.g., *Spinosaurus, Megalosaurus*)
        Allosauroidea (e.g., *Allosaurus, Carcharodontosaurus*)
        Coelurosauria
            Tyrannosauroidea (e.g., *Tyrannosaurus, Albertosaurus*)
            Ornithomimosauria (e.g., *Ornithomimus, Gallimimus*)
            Maniraptora
                Oviraptorosauria (e.g., *Oviraptor*)
                Therizinosauroidea (e.g., *Therizinosaurus, Falcarius*)
                Paraves
                    Deinonychosauria
                        Troodontidae (e.g., *Troodon*)
                        Dromaeosauridae (e.g., *Velociraptor, Deinonychus*)
                    Aves (e.g., *Archaeopteryx, Confuciusornis,* other birds)

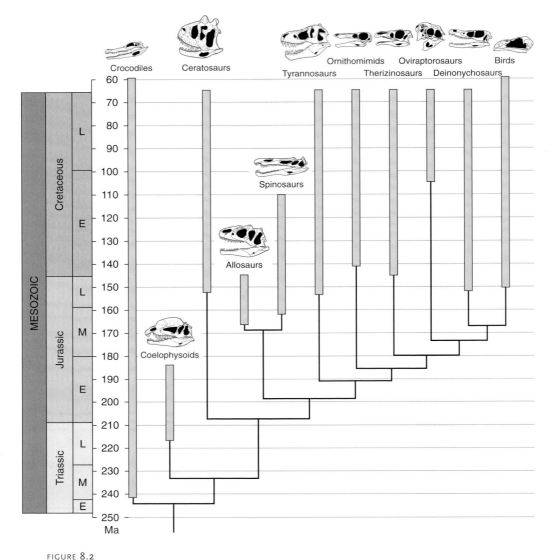

FIGURE 8.2

The family tree of theropod dinosaurs depicted as a branching diagram, or cladogram, showing major groups (clades) and their occurrences in time. The shaded bars represent time intervals for which particular groups are known. A geologic time scale (depicting millions of years ago, abbreviated "Ma") is shown on the left. Although crocodiles are not dinosaurs, the group is shown on the left to depict the branching point with nondinosaurian reptiles.

Theropods share a suite of features, such as thin-walled bones, that separate them from virtually all other dinosaurs. The first theropods to branch off the dinosaur stem were the coelophysoids, a group that includes relatively smaller-bodied forms like *Coelophysis* from North America and *Syntarsus* from Africa, as well as a few midsized examples like *Dilophosaurus,* the putative poison-spitting menace from *Jurassic Park* (which, by the way, almost certainly did not have venom glands and thus did not engage in this unseemly behavior). As expected of the most primitive members of a group, they are also among the oldest theropods, living in the Late Triassic and Early Jurassic.

The next branch up the tree is Ceratosauria, an odd assemblage that first appeared in the Jurassic with forms like *Ceratosaurus* but persisted under the evolutionary radar until the Early Cretaceous, when the group radiated in the form of large and small abelisaurs such as *Carnotaurus*, *Majungasaurus*, and *Masiakasaurus*. While *Tyrannosaurus* was terrorizing *Triceratops* and *Edmontosaurus* up in North America, smaller-bodied abelisaurs were preying on titanosaur sauropods on various southern continents.

In contrast to the ceratosaurs, the succeeding branch of theropods, TETANURAE, had a strong presence from the Middle Jurassic through to the end of the Cretaceous. Virtually all theropods other than coelophysoids and ceratosaurs fall within this diverse group. The name *Tetanurae* means "stiff tail," in reference to the interlocking bony processes that connect adjacent tail vertebrae. Relative to more primitive forms, most tetanurans had larger brains. Many were specialized runners, and a few possessed well-developed binocular vision. Near the base of the tetanuran part of the tree are the crocodile-snouted spinosaurs like *Spinosaurus* and *Suchomimus*, part of a larger group we call spinosauroids. Other spinosauroids include megalosaurs such as *Torvosaurus* and *Megalosaurus*. The next branch is the allosaurs, including the Late Jurassic *Allosaurus* as well as the giant Cretaceous CARCHARODONTOSAURS such as *Giganotosaurus* and *Carcharodontosaurus*. Gigantism is particularly prevalent among spinosaurs, megalosaurs, and allosaurs.

Nested well within tetanurans is the most diverse group of theropods, the coelurosaurs, which includes the largest (at least in terms of body mass) and smallest predatory dinosaurs. At the big end of the spectrum are tyrannosaurs, a group of predators that emerged in the Northern Hemisphere during the Middle Jurassic and became the standard top theropods in Asian and North American ecosystems by the Late Cretaceous. Also found within coelurosaurs are ornithomimosaurs, enigmatic ostrich-like forms that spawned the fastest dinosaurs ever to grace the planet. Finally, the most derived of all the major theropod groups are maniraptors, a diverse branch on its own that encompasses the sickle-clawed DROMAEOSAURS and TROODONTS, as well as the birdlike oviraptorosaurs, the bizarre therizinosaurs, and birds. By far the smallest dinosaurs known are also found among the maniraptors.

Given this great diversity, it's perhaps surprising that most theropods were evolutionarily conservative, retaining the general conformation present in distant nondinosaur ancestors. This is not to say that theropods were homogenous. Various subgroupings of carnivorous dinosaurs evolved interesting specializations. And, to be fair, there were a handful of true oddities, among them the long-snouted, crocodile-like spinosaurs. Most skull variations in this group are attributed to differing hunting and feeding strategies, although many species possessed bizarre structures, such as horns or crests, which apparently had little to do with ingesting sufficient calories.

Theropods were also true "airheads," not in terms of intelligence (they were actually the biggest-brained dinosaurs) but in the sense that many of their skull bones were literally

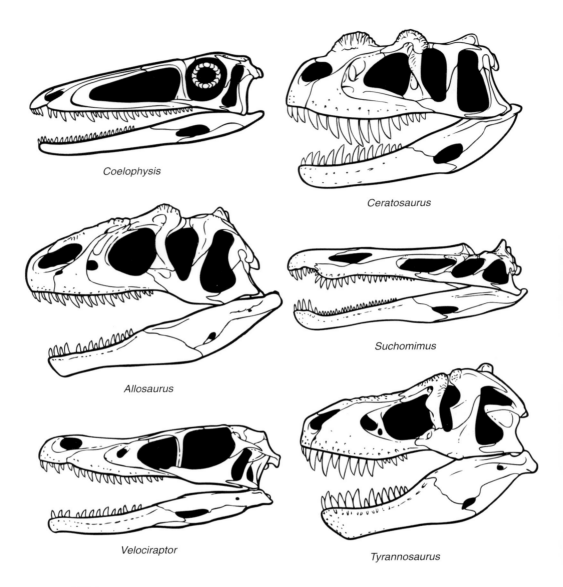

Coelophysis

Ceratosaurus

Allosaurus

Suchomimus

Velociraptor

Tyrannosaurus

FIGURE 8.3
Variation in the skulls of theropod dinosaurs: the Late Triassic coelophysoid *Coelophysis*; the Early Jurassic ceratosaur *Ceratosaurus*; the Late Jurassic allosaur *Allosaurus*; the middle Cretaceous spinosaur *Suchomimus*; the Late Cretaceous dromaeosaur *Velociraptor*; the Late Cretaceous tyrannosaur *Tyrannosaurus*.

filled with air, or "pneumatic." AIR SACS lined with specialized, bone-eating cells carved out cavities of varying size within the nose, ear, and throat regions of theropod skulls. The most obvious example of this PNEUMATICITY is the large opening between the bony holes that housed the nose and eye—the ANTORBITAL OPENING—formed by an air sac that branched off the nasal cavity. Other air sacs penetrated the ear region and consumed parts of the bones around the brain.[2] In a recent study, Larry Witmer and Ryan Ridgely (Ohio University) discovered that big theropods like *Tyrannosaurus* and *Majungasaurus* were some of the biggest airheads of all, with more space within the skull devoted to air sinuses than to brain tissues. Despite this airy and rather fragile appearance, evolution honed the skulls of predatory dinosaurs into highly efficient weapons. Indeed, although they may frequently have used their hands and feet to help subdue prey, if modern carnivores are any guide, the most lethal piece of armament in the theropod arsenal was the head.

Turning to the remainder of the skeleton, we know that all theropods were bipeds, a characteristic inherited from the earliest dinosaurs. Rather than holding the body upright and dragging the tail, as long thought, we're now confident that the body was held almost horizontally and balanced directly over the hindlimbs, with the long tail acting as a counterbalance for the trunk and head. In contrast to the broad-bellied herbivores, with enlarged guts packed with microbes, theropods tended to have narrow hips and gut regions. In keeping with a rapacious lifestyle, theropod claws tended to be sharp and pointed instead of the blunt, rounded hooves typical of herbivores. The trackway evidence shows that theropods (together with the bipedal herbivorous dinosaurs) held their hind legs close to the body during locomotion so that, like us, the left and right feet fell almost in front of each other. The skeletal and trackway evidence also tells us that they walked on their toes, with the ball of the foot held high off the ground. This tip-toe condition, referred to as DIGITIGRADE, evolved independently in a variety of nondinosaur vertebrate lineages, including various groups of mammals (e.g., horses, deer, and antelope). In contrast, we human mammals walk not only on our toes but on our foot bones as well, a condition called PLANTIGRADE.

Many theropod dinosaurs, particularly smaller-bodied forms and juveniles of larger species, were slender and long legged, with relatively elongate shin and foot bones. This arrangement, also present in many mammals, enabled these animals to move their limbs faster and take longer steps, resulting in higher speeds. Another characteristic of theropod limb bones is that they're hollow—or, more accurately, thin walled—a distinction inherited and elaborated by birds. Dinosaur paleontologists in the field picking up fragments of fossilized limb bones are frequently able to distinguish theropods from nontheropods simply by examining bone thickness.

Life poses major challenges for any animal. Perhaps the three most important are eating sufficient amounts of nutritious food, locating and attracting mates, and avoiding a premature demise in another animal's gullet. Many of the features we observe in animals can be related directly to addressing one or more of these fundamental challenges. For example, herbivorous dinosaurs evolved a range of strategies to cope with the formidable defenses of plants. Clearly, however, of equal concern to plant-eating dinosaurs would have been the task of thwarting the consumptive efforts of theropods.

Imagine for a moment that you are a giant plant-eating dinosaur living in the Jurassic or the Cretaceous. Take your pick—maybe you like the idea of being a crested hadrosaur or a horned ceratopsid. Perhaps you have a penchant for armor and decide that a stegosaur or an ankylosaur strikes your fancy. Or you might opt instead for taking on the guise of a long-necked sauropod, giant among giants. Whatever your choice, for the reasons cited earlier, life in your Mesozoic world is tough. First, in order to keep your internal fires burning, you must continually locate and choke down massive amounts of fibrous vegetation. Second, once you reach sexual maturity, there is the strong drive to reproduce, which may bring you into direct, even violent, competition with others of your kind. And, of course, survival in this world means dealing with various theropods that survive by feasting on meaty beasts like you. Although managing the last of these tasks might be easier as a giant, you must first survive the growth years from hatchling to adult. What's a poor dinosaur to do?

Fortunately, whatever your chosen prehistoric alter ego, you have been ably prepared by a remarkable ally—evolution. As a dinosaur megaherbivore (or any animal, for that matter), you are the product of millions of years of evolution that has weeded out unsuccessful forms and perpetuated those best adapted for feeding, avoiding predation, and passing on genes to subsequent generations. What are your options for dealing with theropods? The most obvious strategy would be to remain hidden, a tactic that most varieties of herbivorous dinosaurs likely used on occasion. CRYPSIS, the ability to remain out of sight or blend in with the background, may have been particularly important for juvenile dinosaurs, which could have been much slower moving than adults and less able to defend themselves.

Surprisingly, even adult megaherbivore dinosaurs may have used this technique to avoid detection. You might raise a skeptical brow, conjuring up an image of a sauropod attempting to conceal its vast bulk behind a single sequoia. Yet even multiton animals can be difficult to discern in dense vegetation. On many occasions while driving in African game parks, the vehicle I was in came up directly alongside a giraffe or an elephant before someone noted its presence. And many large animals are simply passed by, undetected by big-game watchers. It's remarkable how silent even huge animals like elephants can be and how their coloration can blend completely with the background. Depending on the density of the Mesozoic vegetation (which varied greatly through time and around the globe), crypsis may have been an excellent survival strategy, particularly for herbivorous dinosaurs that were solitary or lived in small groups. Like mammals, some dinosaurs may have employed camouflage coloring in order to enhance this ability. Coloration, however, tends to be less pronounced among large-bodied animals today, suggesting that elephant-sized dinosaurs may have been quite dull looking. As yet we have no direct evidence of skin or feather coloring among dinosaurs, and thus the concept of camouflage (and coloring more generally) is founded only on speculation informed by possible modern analogues.

Let's assume for the moment that you (i.e., you as a large herbivorous dinosaur) have been spotted by a big theropod, perhaps a lone allosaur or tyrannosaur, or maybe several raptorlike dromaeosaurs. You have two obvious options: fight or flight. The latter might seem to be the most sensible given your opponent. In modern ecosystems, running escape is the most frequent and successful response of large plant-eating mammals confronted by one or more predators. Even if the herbivore's top speed can't match that of the predator—say, in the case of a gazelle and a cheetah—running away is often very effective, in part because predatory mammals are typically unable to sustain top speeds for more than a few seconds.

But wait a minute. As a megaherbivorous dinosaur, you have a serious problem. You are significantly slower than the carnivore now sizing you up for a meal, too slow in most instances to execute a running escape. In contrast to present-day mammals, it appears that dinosaurian plant eaters were much less adept runners than theropods, an inference based on a number of skeletal features. Most telling are the leg bones. In sauropods and ornithischians, the thigh bone (FEMUR) tends to be relatively long compared to the remainder of the limb. For reasons noted earlier, this also would have greatly limited running ability. The same pattern holds for front limbs of four-legged stegosaurs, ankylosaurs, and ceratopsids, which tend to have a large and elongate upper arm bone (HUMERUS). A study of dinosaur locomotory abilities by Matt Carrano of the Smithsonian Institution indicates that none of the big, bulky herbivores—including sauropods, hadrosaurs, ceratopsians, ankylosaurs, and stegosaurs—were capable of true running, which entails a brief suspended phase during each stride when all limbs are off the ground. Rather, all appear to have been slowpokes compared to the hunters of their world.

So, if you can't run and you can't hide, fighting for your life seems to be the sole alternative. Here again we enter the realm of speculation. However, a number of statements can be made based on a mixture of functional analyses and observations of living large-bodied animals.

It was long thought that many bizarre structures of dinosaurs—for example, the horns of ceratopsids and the domes of pachycephalosaurs, as well as the tail spikes and tail clubs of stegosaurs and ankylosaurs, respectively—functioned as antipredator weapons. However, for several reasons that we will explore in chapter 10, many paleontologists now think that few, if any, of these bony appendages were used primarily for stabbing or whacking theropods. In addition, some groups, like ornithopods and sauropods, did not possess any weapon-like appendages.[3]

Unable to flee the scene and (at least in many groups) lacking overt weaponry, how did herbivorous dinosaurs manage to fend off theropods? Certainly it would be foolish to regard these oversized plant munchers as defenseless, because they survived in great diversity for about 160 million years, living alongside some of the greatest terrors Earth has known. As I see it, the two most likely antipredator "strategies" were giant size and communal living.

Fast or slow, horns or no horns, dinosaur megaherbivores must have been formidable opponents simply by virtue of their big bodies. With increased mass comes greater inertia and higher impact forces. A mere bump from an adult sauropod could have broken the limb of a large theropod and ended its days. Healthy adults of large-bodied species are usually the least vulnerable to predators. Today, for example, adult elephants are all but immune to predation, though rare instances have been recorded of lion attacks during periods of starvation. It's been argued that one of the reasons so many different plant-eating dinosaurs grew to gigantic proportions was the need to avoid carnivores. In other words, large species may have had a distinct evolutionary advantage over their smaller relatives, resulting in the bigger forms persisting longer and spawning even larger descendant species. Multiplied over millions of years, this trend could have been a major factor pushing dinosaur lineages toward gigantism. Of course, the predators would have followed along, evolving bigger bodies as well.

The previously mentioned king coprolite attributed to *Tyrannosaurus rex* contained numerous bone fragments from one or more juvenile ornithischian dinosaurs, prompting speculation that large theropods preferentially hunted young prey rather than risking life and limb attacking adults. Although a sample size of one certainly cannot form the basis of any firm conclusions, it's probable that theropods frequently sought out younger, smaller prey. Dinosaurs were egg layers, producing large clutches of eggs each year. This reproductive strategy meant that a substantial portion of the total species biomass (the cumulative mass of all individuals alive at any one time) was composed of nonadult animals.[4] If all of these little dinosaurs survived, Mesozoic ecosystems would soon have been overrun with hungry mouths devastating vegetative landscapes. Therefore, growing dinosaurs must have experienced high rates of mortality, and it seems entirely reasonable that large numbers of these youngsters suffered a premature demise in the guts of theropods.

Sociality was another antipredator strategy apparently used by some plant-eating dinosaurs. As the saying goes, there's strength in numbers. Many big-bodied mammals, particularly those that live in open habitats, form large herds. It's difficult to hide out in the open, so herding provides a measure of protection, with many pairs of eyes continually on the lookout for predators. Moreover, if the group includes adult males and females as well as numerous juveniles, the larger adults may be able to fend off the advances of carnivores. A variety of herbivorous dinosaurs have been found in mass death assemblages called bonebeds, which preserve tens to hundreds of individuals of one variety. Typically, these sites preserve the jumbled remains of juveniles through adults. Examples include horned dinosaurs like *Centrosaurus*, *Pachyrhinosaurus*, and *Einiosaurus*, and ornithopods like *Iguanodon*, *Hypacrosaurus*, *Edmontosaurus*, and *Maiasaura*. Although these sites confirm only that the animals were entombed in the same place, the bonebed evidence is highly suggestive that these animals did form large herds at least some of the time. This conclusion is supported by numerous examples of dinosaur trackways showing animals of varying sizes (and presumably ages) traveling in the same direction.

In addition to giant size and social groups, herbivorous dinosaurs may have adopted another strategy to avoid predators. In the previous chapter, I described some of my experiences in the Masai Mara ecosystem of Kenya. Although it was astounding to bear witness to such vast numbers of animals, sometimes stretching to the horizon in all directions, it was easy to find the opposite situation. Many times, after feasting our eyes on a sea of snorting, munching ungulates, we would continue driving through the park and find ourselves in a nearly identical setting that differed in one crucial aspect—not a single animal in sight. Only the wind could be heard whistling through the tall grass. Such patchy distributions of herbivores result from their tendency to gather in vast, mixed-species aggregations. While these multispecies groupings may be related in part to the previously described sequence of plant-eating species consuming grassland forage, such congregating behavior is thought by many biologists to be a key antipredator strategy. More species translates into still more eyes scanning the horizon for danger. In modern ecosystems, the alarm call of one species often alerts others to the presence of predators, thereby enhancing the survival potential of all (except the predators, of course). It's interesting to speculate—and speculation it is, because we have no direct evidence—that herbivorous dinosaurs gathered in mixed-species groupings for the same reason.

Having addressed life from the herbivore's perspective, let's shift our focus from hunted to hunter. If large, plant-eating dinosaurs were pretty much sluggards across the board, what about the idea of sprinting theropods? In other words, if their prey were so slow, why would the predators have evolved the capacity for fast running? Some small-bodied theropods, such as ornithomimosaurs and troodonts, appear to have clocked relatively high speeds, on the order of 40–60 kilometers per hour (25–40 miles per hour). Evidence for such locomotory prowess can be found in bones of the hindlimb, as well as in trackway evidence. Unlike sauropods, hadrosaurs, ceratopsids, and other big plant-eating dinosaurs, these theropods have elongate shinbones and feet, evolutionary hallmarks for running. However, as will be discussed later, some of the more diminutive theropods may have been omnivorous or even herbivorous. In contrast, recent work by John Hutchinson of the Royal Veterinary College, London, provides strong evidence that giant theropods, though faster than their prey, did not approach the top speeds of carnivorous mammals. *Tyrannosaurus*, for example, likely topped out somewhere in the range of 16–40 kilometers per hour (10–25 miles per hour), respectable but still rather slow compared to many modern animals. This estimate is also much slower than many previous estimates for the Tyrant King. The findings make sense, however, once we consider not only the incredible mass of the predator but the poor locomotory abilities of the large herbivorous dinosaurs.

With pursuit speeds apparently under control, the next concern for theropods was weaponry. As noted, subduing any large, healthy dinosaur would have been a formidable task. Unsurprisingly, then, theropods evolved a diverse range of weapons well suited for dispatching and dismembering their prey. These structures were concentrated in three regions of the body—hindlimbs, forelimbs, and head—though evolution often emphasized one or two of these within a particular group.

We'll start with the forelimbs. One of the great advantages of walking on two legs instead of four is that it frees up the arms and hands for other functions. Most theropods took full advantage of this potential, possessing large and flexible hands attached to strong, mobile arms. Long fingers, present in the earliest theropods and many of their descendants, further enhanced grasping abilities. This trend was taken to the extreme in maniraptors, with subgroupings like oviraptorosaurs and dromaeosaurs possessing particularly elongate hands. The thumb of most theropods was semiopposable—that is, it was flexible enough to rotate and touch the other digits. As described in chapter 3, this "pivotal" adaptation enabled the hand to be used for grasping. The claws of theropods, especially those on the hands, are typically curved and elongate, tapering to sharp points. The long fingers and large, sharpened claws are thought to have made fearsome slashing weapons.

Whereas more primitive theropods had four fingers, many later groups reduced this number to three (allosaurs, spinosaurs, and others), two (tyrannosaurs), or even just one (a strange, birdlike group of maniraptors called alvarezsaurids). Evolutionary reduction in the number of digits tended to occur from the outside in, such that the fifth finger was the first to go, then the fourth, and so on. The reduction in digit number in tyrannosaurs was accompanied by a dramatic decrease in forelimb size. Abelisaurs such as *Majungasaurus* retained four fingers but also possessed puny arms and hands.

What could the little forelimbs of tyrannosaurs and abelisaurs have been used for? So short are the arms and hands that these animals could not have reached up to their mouths. Paleontologists have speculated on a number of possible functions, including such unlikely alternatives as pushing the body off the ground from a sleeping position. Another possibility is that the tiny forelimbs were vestigial structures, akin to the pelvic bones of whales and the nonfunctioning eyes of cave-dwelling fishes. Vestigial structures, the flotsam and jetsam of evolution, tell us more about an organism's ancestry than its lifestyle. Yet the arms and hands of tyrannosaurs and abelisaurs, although small, appear to be fully functional and well muscled, with unusual specializations that would seem unlikely if the forelimbs were not somehow functional. Perhaps they were used like grappling hooks to hold onto carcasses during feeding. Paleontologists are still scratching their heads over this one.

Turning to the feet, we know that theropods tended to have three functional toes. In this case, it was the first and fifth digits that were reduced or lost, such that the three working toes were digits two, three, and four. Many birds retain this same familiar, three-toed form, easily identified in modern footprints. In contrast to their curved, sometimes scythe-like hand claws, the toes of most theropods were tipped with claws that, although sharp, tended to be straight and more like hooves. The well-known sickle-clawed exceptions, celebrated in major motion pictures, occur on the second toe of maniraptors like *Velociraptor*. In these birdlike forms, including dromaeosaurs and troodonts, the so-called killing claw was carried off the ground during walking and running, presumably to protect it. The hyperextendable second digit gave the claw a wide arc of movement that, combined with a rapid kicking motion, may have made this toe a lethal stabbing or slashing tool. Alternatively, it may have been used to secure struggling prey, open a carcass, or stabilize the predator's body during feeding.[5]

Looking up front at the true business end, it appears that theropods had good eyesight. A few, like *Tyrannosaurus*, possessed forward-facing eyes indicative of well-developed stereoscopic vision, though not to the degree seen in primates like us. *T. rex* and some other theropods possessed an acute sense of smell, evidenced by a pair of expanded bony depressions immediately in front of the brain cavity. At least among the sickle-clawed troodonts, a pneumatic connection between the ears suggests a heightened ability to accurately assess distance with their ears. It isn't difficult to imagine how all of these modifications—bigger brains, acute senses of smell and hearing, and stereoscopic vision—would have been useful for dinosaurian predators.

For the most part, theropod skulls are relatively long, narrow, and lightly built, with four sizable openings on each side. From front to back, these are the nose opening (which housed the fleshy nostril), the antorbital opening (dominated by a large air sac), the orbit or eye socket (which housed the eye), and a jaw muscle opening (filled in life by jaw-closing muscles). One of the most obvious differences between mammalian and dinosaurian carnivores is the relative lack of dental variation in the latter. Mammals show numerous specializations in their teeth and jaws that are reflected in varying hunting and feeding behaviors. For example, the omnivorous diet of bears is reflected in their dentition—canines up front for piercing flesh, molars at the rear of the mouth for processing plant foods. Then there are bone-cracking specialists such as hyenas, with teeth and jaws capable of snapping bones so as to access the highly nutritious marrow and mineral salts within. Among giant theropods at least, there is no clear evidence of omnivory or specialized bone cracking. Rather, the sharp, recurved, and serrated tooth crowns typical of theropods indicate a singular devotion to consuming flesh, what ecologists aptly refer to as "hypercarnivory."

An oft-mentioned theropod predation strategy, one that is frequently portrayed in popular books and movies, is pack hunting. In particular, maniraptors such as *Velociraptor* are often depicted living and hunting in social groups or "packs" analogous to those of wolves. This interpretation goes back at least to John Ostrom, father of the dinosaur renaissance and discoverer of the raptorlike dinosaur, *Deinonychus*. The first remains of this small (3.5-meter, or 10-foot) sickle-clawed maniraptor included bones from several individuals scattered around the skeleton of a midsized (7-meter, or 21-foot) ornithopod, *Tenontosaurus*. The dramatic disparity in size between carnivore and herbivore led Ostrom to speculate that *Deinonychus* hunted in packs in order to take down bigger-bodied animals. Since then, various workers have claimed that at least some giant theropods also engaged in pack behavior. What's the evidence?

Just as the remains of various herbivorous dinosaurs—including prosauropods, sauropods, ornithopods, and ceratopsians—have been found in bonebeds containing remains of many individuals, the same is true for a handful of theropods. As with the herbivores, these localities have been used to bolster arguments of social living and gregarious behavior. One of the most famous examples is the *Coelophysis* quarry at Ghost Ranch, New Mexico, which preserves hundreds of skeletons of this primitive little carnivore. Bonebeds dominated by the remains of big theropods, such as allosaurs, tyrannosaurs, and carcharodontosaurs, have also been excavated. Renowned dinosaur researcher Philip Currie of the University of Alberta studied a site in Dry Island Buffalo Jump Provincial Park, Alberta, that has yielded remains of at least twenty-four tyrannosaur individuals referable to *Albertosaurus*. The collection includes representative animals from several age groups, including juveniles, subadults, and adults.

Currie thinks that the Dry Island locality is the best evidence to date of gregarious behavior in large theropods, noting that such bonebeds coincide with those of duck-billed and horned dinosaurs. If the big herbivores were living in large social groups, this may have been an antipredator strategy that evolved in part to counter pack-hunting tyrannosaurs. Currie goes further to speculate about hunting strategies. Citing bony evidence that the juveniles would have been faster than the adults, he speculates that tyrannosaur predation sometimes involved a division of labor, with the fast-running, more agile youngsters driving potential prey toward the larger, more powerful adults. He adds, however, that enhanced locomotor prowess may also have enabled juvenile albertosaurs to feed on different, smaller-bodied prey than the adults.

Dinosaur pack-hunting scenarios, particularly those that involve complex, cooperative behavior within a group of social predators, are fun to ponder. Yet they are exceedingly difficult to demonstrate, given the limitations of the fossil record. The fact that many individuals of a single species are preserved together does not necessarily mean that they lived together, let alone worked in cooperative packs. There are a number of circumstances, such as drought, that could congregate predators (say, around a waterhole), which, of course, would tend to deter herbivores. So an essential part of this kind of work is a reconstruction of the circumstances surrounding death. Although the "crime scene" is millions of years old, with only a tiny fraction of the original clues preserved, paleontologists study fossil sites to determine not only cause of death but mode of life. At present, the evidence for theropod pack hunting can best be summed up as equivocal.

All known dinosaur carnivores were theropods. But the reverse was not true. If we consider birds to be dinosaurs, this is a trivial statement. Hundreds of living bird species subsist on seeds, fruit, or plants. Yet, although descended from meat-eating ancestors, it appears that at least one group of nonavian theropods became devoted vegetarians, and one or two others appear to have shifted at least to omnivory, if not all-out herbivory. I conclude this discussion of theropods with a closer look at these three oddities.

First up are ornithomimosaurs, the ostrich-mimic dinosaurs. Although generally considered to be close relatives of tyrannosaurs, members of this group evolved in a novel direction. Other than their long, bony tails, these animals truly do resemble ostriches, bearing long necks, elongate limbs, and gracile, long-snouted skulls. At least some species, and perhaps all, even had feathers. Various features of ornithomimosaur limbs and trunks demonstrate that they were among the fastest of dinosaurs. And, whereas the most primitive ornithomimosaurs retain a few teeth, most forms apparently possessed jaws lined only with a horny beak. Notably, with one possible exception from Asia, all members of the ostrich-like dinosaurs are midsized, weighing much less than 1,000 kilograms (2,200 pounds).

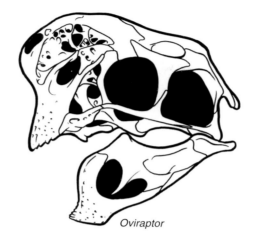

Oviraptor

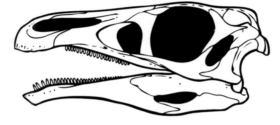

Erlikosaurus

FIGURE 8.4

Skulls of possible omnivorous and/or herbivorous coelurosaurs: the Late Cretaceous oviraptorosaur *Oviraptor*; the Late Cretaceous therizinosaur *Erlikosaurus*; and the Late Cretaceous ornithomimosaur *Gallimimus*. *Oviraptor* and *Erlikosaurus* are part of the subgrouping of coelurosaurs known as maniraptors, which also includes "raptor" dinosaurs like *Velociraptor*, as well as birds.

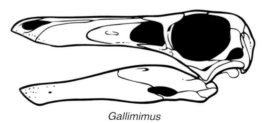

Gallimimus

The toothless skulls of ornithomimosaurs have long intrigued paleontologists, prompting speculations about unorthodox feeding habits. Over the years, these poor creatures have been lumped into just about every imaginable dietary slot, from carnivore to herbivore to OMNIVORE. One recent idea is that ornithomimosaurs were the flamingos of the Mesozoic, with a sieve-like beak adapted for straining everything from plankton to shrimp. But the evidence for filter feeding is poor, and other investigators argue convincingly for an herbivorous diet. Among the clues are stomach stones, or GASTROLITHS, suggestive of a gastric mill better suited to processing high-fiber plants.

The next enigmatic group is oviraptorosaurs—or, more informally, oviraptors—an assemblage of maniraptors that, with few exceptions, also lack teeth. These theropods are also extremely birdlike in many respects, including the presence of feathers and a well-developed, toothless beak. Some of the more specialized members have bizarre, boxy, air-filled skulls with deep jaws that, at first glance, are difficult to orient front to back. The word *oviraptor* means "egg thief," in reference to the original presumed diet of these animals. A single oviraptor specimen was found in Mongolia during the original American Museum expeditions of the 1920s. Much more common in these Cretaceous-aged, wind-blown sediments were clutches of dinosaur eggs preserved in nests. And most common of all were skeletons of the small ceratopsian dinosaur *Protoceratops*. Roy Chapman Andrews and his team assumed that the numerous egg clutches belonged to the nearly ubiquitous *Protoceratops*. They then hypothesized that the swift, toothless oviraptors survived in large part by feeding on these eggs, and they named these theropods accordingly.

In the 1990s, however, this interpretation was turned on its head. Several oviraptor specimens discovered by crews from the American Museum of Natural History were found preserved on top of nests. The animals appear to have been brooding the eggs at the time of death, although other alternatives, such as shielding the eggs from the sun, are conceivable. Just as in birds, oviraptors placed their bodies directly over the eggs, with their feathered arms surrounding the nest, presumably to regulate the interior temperature of the eggs. One embryo was recovered inside an egg, revealing the identity of the egg makers as oviraptors rather than ceratopsians. Overnight, the putative egg thief was transformed into a nurturing parent, and another link was forged between dinosaurs and birds. But if not eggs, what did oviraptors eat? Once again, the honest answer is that we don't know. Yet few paleontologists argue that these animals pursued a strictly carnivorous diet. Most lean toward herbivory or regard this group of small-bodied theropods as gustatory switch-hitters, eating a combination of plants and animals.

In this brief overview of theropods, I have reserved the strangest for last. Therizinosaurs are the only theropod group that most investigators agree evolved undoubtedly herbivorous forms. Like their close cousins the oviraptors, therizinosaurs were maniraptors that diversified during the Cretaceous. The most specialized members of this group, which lived toward the end of that period, tended to be bulky, short-legged beasts with small heads, long necks, pot bellies, and humongous scythe-like claws. Imagine a nightmarish amalgam of ostrich, ground sloth, and Edward Scissorhands, and you begin to get the idea.[6] So bizarre are these animals that, for a long time, paleontologists did not know what to do with them. Over the past few decades, as additional fossilized bits and pieces were unearthed, therizinosaurs have been placed into almost every major group of dinosaurs and even linked to turtles! Finally, the discovery of nearly complete skeletons, some with feathers, demonstrated clearly that therizinosaurs were maniraptor theropods.

Until recently, the origin of therizinosaurs from carnivorous ancestors was shrouded in mystery. This all changed with the discovery of a new, primitive therizinosaur, *Falcarius utahensis*, from Utah's Early Cretaceous, dating to about 125 million years ago. The lead authors on this study were Jim Kirkland of the Utah Geologic Survey and Lindsay Zanno, one of my graduate students at the University of Utah. *Falcarius* resembles *Velociraptor* in a number of respects, including its relatively short, sharp claws. Yet, although we can't be certain whether this 4-meter- (12-foot-) long beast was omnivorous or strictly herbivorous, the long neck, broad belly region, and leaf-shaped teeth indicate that it had already embarked on the path toward herbivory. *Falcarius* is an exceedingly rare example, perhaps the only one among dinosaurs, of evolution caught in the act of a major dietary shift. Some of the media reports got it right when they ran the headline "Killer Dinos Turned Vegetarian."

All three of these dietary misfits—ornithomimosaurs, oviraptorosaurs, and therizinosaurs—are members of the Coelurosauria, the most advanced major grouping of theropods that also includes tyrannosaurs and birds. As far as we can tell, all other theropods followed closely in the carnivorous footsteps of the very first theropod. Why didn't the other groups of carnivorous dinosaurs mentioned in this chapter give rise to omnivores or herbivores? Keep that question in the back of your mind. It may just be a clue toward solving a long-standing mystery about the physiology of dinosaurs.

Having examined the flow of solar energy from plants to dinosaurian herbivores and then to carnivores, let's turn our attention now to the movement of energy and nutrients through some of the smaller Mesozoic life-forms. We shall find that, although tiny, these organisms had critical roles to play in the odyssey of dinosaurs.

Microcosm in a semiarid coastal forest, Early Cretaceous Araripe Basin, Crato Formation, Brazil. From top to bottom, depicted invertebrates are wasp, dragonfly, cricket, centipede, whip scorpion, cockroach, and earthworm. Araucarian (monkey puzzle) conifers tower above, and gnetales like *Ephedra* and *Welwitschia* make up much of the understory groundcover of plants.

9

HIDDEN STRANDS

IT'S HUMBLING TO STAND in the belly of an ancient redwood, to be enveloped by one of the largest living things on Earth. The thick trunk surrounding me on this particular day was heavily charred, its base hollowed out by some long-ago fire, leaving behind an expansive interior cavity. Fires don't ravage this forest often, at least not in the human sense. But measured in redwood lifetimes, sometimes well in excess of 1,000 years, fire is as inevitable as death. Fortunately, evolution has endowed these majestic trees with flame-retardant bark and other survival strategies, even allowing them to benefit from the periodic cataclysms.

Exiting the tree, I shifted my gaze skyward, feeling vertigo as my eyes followed the trunk to dizzying heights. Although the uppermost branches were obscured by a sea of green foliage, I knew that the topmost needles were processing sunlight about 90 meters (300 feet) above me. This was a coast redwood, *Sequoia sempevirens*, a species that includes the tallest organisms on Earth. The largest known examples achieve heights of about 115 meters (375 feet), taller than a 35-story building. Today, only three kinds of sequoia remain, but their legacy, like their trunks and life spans, is long. Sequoias not so very different from these graced the world of dinosaurs, so a visit to a redwood forest is like stepping back into deep time. As such, these forests provide another excellent place to contemplate the dinosaur odyssey.

This particular hollowed-out tree lives in Muir Woods National Monument, a stand of old-growth redwoods named after the famous tree lover and environmentalist John Muir. It is also the closest patch of old-growth forest to the city of San Francisco. As a result, the Monument receives almost a million visitors each year, a fact that would have pleased Muir greatly. Although I had visited Muir Woods many times, on this day I had come not to dwell on the majestic trees but to contemplate some of the less obvious, hidden aspects of the forest.

Muir Woods is home to several varieties of nonredwood trees. Meandering farther down the loamy path, my attention was drawn to a large overhanging leaf from a California bay laurel. Several holes were visible within it, and a few chunks were missing from the leaf margin. Still closer inspection revealed that these imperfections were edged by characteristic patterns left by the jaws of feeding insects. Though I could not see or hear them, I knew that the world around me was full of tiny herbivores.

A butterfly flitted by chaotically, looking for a meal of flower nectar, totally unaware that its activities were part of the plant's sexual dance. My next encounter was with a massive fallen redwood paralleling the path. After inhaling carbon dioxide and exhaling oxygen for more than a millennium, and after providing sustenance, shade, and safe haven to countless shrubs, microbes, fungi, and animals, this monstrous plant fell to earth, hammering several smaller trees on the way out. Yet the shattering event was not so much an end as a beginning.

When a tree dies, it is occupied by a slow parade of decomposer organisms, all of which, in one way or another, break down and consume its carbon, returning its nutrients from whence they came. The process often begins with beetles, which tunnel into the bark in order to generate larval nurseries. The battery of holes in the tree bark before me revealed that beetle activity was already well underway, providing access points for water and a host of other decomposer organisms—bacteria, mites and worms, among others. Beetles carry fungi on their bodies, which transfer to the wood and begin to consume and further decompose the tree. When the beetle larvae emerge in the tunnels, they feed on the newly arrived fungi before boring still deeper into the fallen tree. About the same time, more kinds of beetles arrive, as well as ants; together these insects penetrate the heart of the dead log. Wasps enter many of the tunnels, locate the beetle larvae, and lay eggs nearby, so that the newly hatched wasps have a plentiful food supply. After a year or so, different kinds of fungi capable of processing the highly fibrous redwood interior follow the insect highways, spreading throughout the log's interior. They are followed closely by termites and their gut bacteria, which together process the extremely coarse cellulose through a fermentation process. Bacteria in the termite guts also capture nitrogen and return it to the environment as their hosts die.

This fallen giant may take three or four centuries to fully decompose. Throughout the lengthy process, the redwood will provide a constant and vital source of nutrients to the environment. In fact, the tree will become a complex microecosystem unto itself, home to a succession of tiny life-forms. Without this complex organic cavalcade, the tree would take much longer to decompose and the valuable chemicals locked within its body would be unavailable for use by the larger system. Paradoxically, this redwood will become more alive in death than it was in life. During its lifetime, a tree is composed of about

5 percent living tissue; other than a thin layer near the surface that transmits water and nutrients up and down, the massive trunk is made up of dead matter. (This is why a tree with a fire-hollowed trunk can still thrive for many years.) A fallen tree in an advanced state of decay may be 20 percent living tissue by weight, thanks largely to occupation by decomposers. In a sense, then, trees have two lives, the first dominated by the capture of solar energy and the second by breakdown of that energy into a bounty of nutrients. The rotting log also underlines an important ecological truism: every species has a unique role in maintaining the larger ecosystem of which it is a part.

Stooping down, I picked up a handful of loose earth. There in my hand was the heart and soul of decomposition, capable of transforming any fallen leaf, tree, or animal into the chemical building blocks of future generations. Running the rich, soft substance through my fingers revealed a moisture-laden mixture of plant detritus, fungal strands, and tiny organisms. I reminded myself for the umpteenth time that microbes make the living world go round, as they always have. A random fistful of forest soil like this one contains billions of microscopic life-forms, mostly bacteria, exceeding the number of humans currently living on the planet—a staggering thought.

Of course, microbes are not restricted to the soil. They exist virtually everywhere in every ecosystem, so small and so firmly embedded in all aspects of life that we barely notice them. To give an example close to home, the human mouth contains more than seven hundred different kinds of symbiotic bacteria, each of which has carved out a unique existence amid a complex topography of tongue, teeth, and gums. Before you run off and brush every square millimeter of your mouth raw, remember that, by combating disease-causing bacteria, this phalanx of microbes is vital to your health. The same is true for many of the bacteria in your gut, on your eyelashes, in your nose, and indeed throughout your body. What you regard as your physical self includes on the order of 10 trillion cells, yet nine out of ten of these are not human cells. Put another way, your body is home to many more life-forms than there are people on Earth or stars in the Milky Way galaxy. Like it or not, you are a walking colony—or, better still, an ecosystem—living in unwitting harmony with these smidgens of life. This phenomenon held true for dinosaurs as well; it is awe inspiring to contemplate the long-necked sauropods, largest animals ever to walk the Earth, as gargantuan, four-legged bacterial colonies.

The previous pair of chapters followed the flow of energy from plants to herbivores to carnivores. Yet the changing web of life includes additional paths of energy flow involving the microscopic extremes of life's spectrum. Although human attention is naturally diverted toward dinosaurs and other creatures at the macro end of the spectrum, this bias should in no way lead us to conclude that absolute size is somehow proportional to ecological importance. Indeed, most ecosystems would likely persist in some form if the large vertebrate contingent were suddenly removed. But so critical are the strands of energy flow maintained by insects and microbes that expulsion of either would lead to immediate systemic collapse. Although *Tyrannosaurus* never stalked a fly or a bacterium, the complex interlinking of ecosystem subparts means that these Lilliputians of the

Mesozoic were every bit as important to the Tyrant King's survival as was *Triceratops*. Therefore, any discussion of the dinosaurian web of life would be grossly incomplete without some consideration of the Mesozoic microworld.

Unfortunately, the fossil record of Mesozoic insects, fungi, and bacteria is grossly incomplete, not surprising given their small sizes and lack of hard parts. By necessity, then, much of the following discussion is derived from the living realm. Yet even our understanding of modern microbes is woefully inadequate. Despite sending manned and unmanned vehicles to other worlds, we have the barest acquaintance with the surface of our own planet. Our gross ignorance of Earth's biota is revealed by the fact that we can only make the roughest of guesses as to the total number of species currently living on Earth. Estimates vary wildly, ranging from about 4 million to greater than 100 million, with 15 million an oft-cited ballpark figure. Of that astounding total, less than 2 million kinds of life-forms have been recognized as distinct and given Latinized genus and species names. Of those formally named species, only a tiny fraction is known much beyond this label. For most of nature's burgeoning diversity, then, we lack even basic information regarding life span, diet, and reproduction, let alone details of interactions within and between species.

On the plus side, the fossil record is far from silent with regard to these hidden strands. The past two decades have witnessed an abundance of insights relating to Mesozoic plants and insects in particular, and several exciting discoveries even point to a key role for bacteria in the process of fossilization. The scientific strategies used to find and unravel these fossil strands closely resemble those portrayed nightly in the highly successful crime scene investigation (CSI) genre of television dramas. A diverse range of forensic-style tools and techniques tease out bits of evidence preserved at prehistoric death scenes. In the current media-driven lingo, these investigations into ancient processes amount to the ultimate in "cold cases."

Our record of Mesozoic plants comes from fossilized tree parts—particularly trunks, branches, leaves, and flowers—as well as from microscopic pollen grains. Unlike dinosaurs and other vertebrates, for which mostly complete skeletons are not infrequently recovered, complete trees are virtually absent from the fossil record. Like several blind people touching different parts of an elephant, paleobotanists do their best to connect the pieces and describe entire plant species. Despite this uncertainty, we can be sure that Mesozoic plants faced the same kinds of challenges as their modern counterparts. And, as discussed in chapter 7, they also developed various kinds of mechanical, chemical, and biological defenses to ward off herbivores. Many defenses observed in living plants appear to have arisen during the Mesozoic, when both insects and dinosaurs were proliferating. Like the herbivorous dinosaurs, insects undoubtedly responded by evolving specialized jaw mechanisms and digestive strategies.

Given that insects represent a major, even dominant herbivorous impact on plant communities, it would be very useful to know about the insects that coexisted and coevolved with dinosaurs. Despite their relative lack of hard parts, we have an extensive fossil record of insects. Much of this evidence comes in the form of insects trapped in tree sap and subsequently turned to amber. The vast majority of the insect groups alive today lived alongside dinosaurs from the Jurassic onward. Insect species tend to be long-lived, persisting on average about 10 million years, and many Mesozoic insects appear closely related to forms alive today. This remarkable longevity stands in stark contrast to mammals and other vertebrates, whose species typically persist for about one million years. Most mammal groups, not to mention the species within them, evolved well after the major dinosaur extinction 65.5 million years ago.

Although Mesozoic insects are plentiful at a growing number of amber localities around the world, much more abundant are fossil leaves that preserve distinctive insect-feeding traces. Go outside to the nearest tree, and chances are that many leaves will show insect feeding traces akin to those that I saw in Muir Woods. Spend some time examining

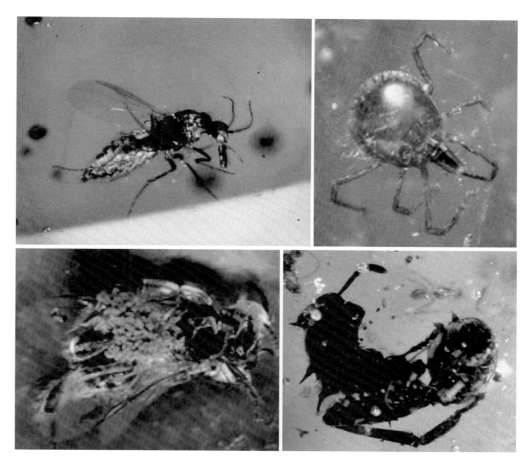

FIGURE 9.1
Photographs of fossilized insects preserved in amber. Clockwise from top right: a tick, a leafcutter ant, a bee, and an unidentified insect.

those traces under a microscope, and you will soon find that there are several distinct types in any particular place. Studies of living insects demonstrate that these variations in feeding traces correspond to specialized mouth parts and feeding behaviors that represent alternative strategies to counter plant defenses. As evolution adjusts plant composition or leaf shape, it also counters with insect-feeding innovations.

In an exceptional piece of detective work, paleontologists Conrad Labandeira and Peter Wilf of the Smithsonian Institution discovered that many distinctive feeding traces left by insects on modern plants can be found on fossil leaves. This deep time correlation has provided an amazing diagnostic tool. Just as human footprints provide conclusive evidence of people, insect-feeding traces from the Mesozoic can be matched to modern groups to demonstrate the presence of those insects (or near relatives) in the fossil record. One example involves a group of leaf beetles known as hispines, which today feed on ginger and leave behind long, narrow feeding marks about 2.5–3.0 millimeters (about 0.1 inch) wide. The researchers found the same traces on fossil ginger leaves dating to the end of the Cretaceous, pushing back the known fossil record of leaf beetles more than 20 million years. The presence of these specialized beetles feeding on Mesozoic gingers implies that this relationship is ancient, dating back at least to the Cretaceous. Together, the evidence of insect body fossils and feeding traces provides a surprisingly detailed picture of ancient insect diversity for certain places and times, an understanding that is certain to grow in the coming years.

How important were insect pollinators during the Mesozoic? Some Cretaceous insects, particularly flies, have elongate noses, or probosces, that likely served as straws to

FIGURE 9.2
Photograph of Late Cretaceous angiosperm leaf fossil (*Leepierceia perartocarpoides*) showing two kinds of insect damage.

PLATE 1

The Berivotra field area, northwestern Madagascar.

PLATE 2

Site 93-18 in the Berivotra field area, Madagascar, which has produced numerous fossils of Late Cretaceous dinosaurs.

PLATE 3

Wildebeest herd crossing the Mara River, Kenya.

Mark Loewen

PLATE 4

Three *Diplodocus* sauropods—two adults and one juvenile—saunter across a Late Jurassic floodplain.

Michael Skrepnick

Michael Skrepnick

PLATE 5

The Late Jurassic plated wonder *Stegosaurus.*

Michael Skrepnick

PLATE 6

While a Late Cretaceous tyrannosaur, *Albertosaurus,* feeds on a rotting carcass, the small maniraptor theropod *Bambiraptor* must be content with scraps.

Mark Loewen

PLATE 7
Exposures of the Late Triassic Chinle Formation, Utah.

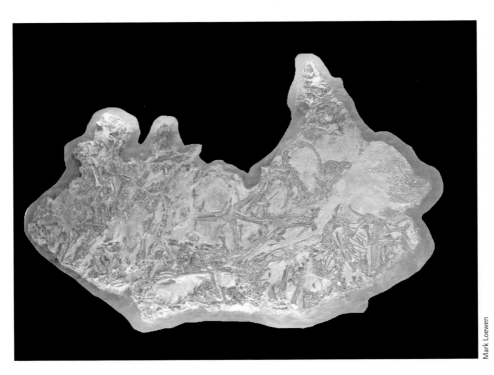

Mark Loewen

PLATE 8
Photograph of a *Coelophysis* fossil block collected at Ghost Ranch, showing a jumble of skeletons of this Late Triassic coelophysoid theropod.

Michael Skrepnick

PLATE 9

A Late Jurassic Morrison scene. A large theropod, *Allosaurus*, attempts to chase down a small ornithopod, *Dryosaurus*.

Jelle Wiersma

PLATE 10

Exposures of the Late Jurassic Morrison Formation, Utah.

PLATE 11

Achelousaurus, a short-frilled (centrosaurine) horned dinosaur from the Late Cretaceous of Montana that was named by the author.

PLATE 12

Corythosaurus, a crested (lambeosaurine) hadrosaur from the Late Cretaceous of Alberta.

PLATE 13

Predator and prey in Late Cretaceous Montana. A large tyrannosaur theropod, *Daspletosaurus,* attacks a herd of horned dinosaurs, *Einiosaurus.*

ReBecca Hunt-Foster

PLATE 14
Exposures of the Late Cretaceous Kaiparowits Formation, Grand Staircase–Escalante National Monument, Utah.

sip flower nectar. Direct (and remarkable) evidence of nectar feeding includes nectar pre-served within the guts of Mesozoic insects. Bees, the most important group of modern pollinators, appear to have evolved and spread during the Cretaceous alongside many other insect groups. In all, considering evidence for both herbivory and pollination, it's likely that insects, perhaps much more so than dinosaurs, underwent extensive coevolu-tion with plants (particularly flowering plants) during the latter half of the Cretaceous.

So insects clearly played important roles guiding energy flow within Mesozoic eco-systems. Through their efforts as pollinators, some participated in plant reproduction and energy production. As major plant consumers, others participated in the flow of energy through food webs. And just as plant toxins kept insect populations in check, insects must have returned the favor, limiting the spread of plants and helping maintain ecosys-tem balance. As we shall discover, insects were also important decomposers, converting organic materials into nutrients for future generations of plants.

Waste is an empty concept as applied to ecosystems. Imagine a forest in which dead plant and animal matter—branches, leaves, bones, trunks, and desiccated flesh—collected on the forest floor rather than decomposing. In addition to being a treacherous and smelly place to walk, such an ecosystem would not long survive, because organisms of successive generations are made up largely from the reconstituted remains of previ-ous generations.

Ecology and evolution have worked in tandem millions upon millions of years to all but perfect the recycling of organic matter, whatever its source. Just as for plants, the death of any animal—be it a beetle or a buffalo—triggers a succession of decomposing organisms. For up to a century after a whale dies and sinks to the bottom of the sea, its body becomes the focal point—indeed, the foundation—of a richly diverse ecosystem. These "whale falls," which often descend to depths in excess of 2,000 meters (6,500 feet), inject a tremendous amount of food—up to 145 tons (320,000 pounds) of organic material—into an otherwise resource-poor setting. The animals that congregate around cetacean car-casses include scavengers such as sharks, hagfish, and crabs, which devour the whale's soft tissues. Others organisms exploit the sulfide-rich habitat generated by bacterial decomposition of blubber. Many of the greater than four hundred species that inhabit these deep sea menageries appear to have evolved specifically to take advantage of these transient islands of resources. Perhaps most bizarre are bone-burrowing worms lacking eyes, mouth, or stomach yet sporting feathery plumes that behave like gills. The worms also carry symbiotic bacteria specialized for digesting oils and fats within whale bone.

It is interesting to contemplate a parallel situation involving "dinosaur falls." A dead sauropod would also have yielded a multiton island of resources, attracting a spectrum of terrestrial decomposers to the feast. And, as we shall see, some of these Mesozoic waste- recyclers were also specialized to exploit bone.

In the case of large, land-living vertebrates, the distinctly disgusting sequence of car-cass decay has been well studied and can be summarized as follows. During the first two days following death, the carcass remains fresh looking, though it is beset upon by a variety of bacteria, protists, and nematode worms. Various vertebrate carnivores are also most likely to devour their share of the remains during the first day or so. Days 2 to 12 comprise the period of putrefaction. Blowflies and wasps arrive, laying their larvae on the carcass. Within the body, the activities of gut-living bacteria have continued despite the death of their host, causing the buildup of gases. Unable to escape the GI tract, these internal gases lead to bloating and distortion. From days 12 to 20, the degrading flesh changes texture, becoming "creamy," and the greatly distended gut finally ruptures, generating a foul odor. The collapsed carcass then becomes the target of still more wasps. From about days 20 to 40, the carcass dries out and dermestid beetles arrive in great numbers, rapidly devouring any remaining flesh. Mites and fungi also get in on the act, now able to infiltrate almost any region of the carcass. For about a year following day 40, a more gradual transformation occurs, marked by further drying and decay. Dermestids, mites, and fungi continue their work unabated until only the bones remain. In some instances, the bones are targeted by boring beetles or other insects, which often leave larvae within the excavated tubes. In some ecosystems, specialized vertebrates like porcupines or hyenas assist in the bone decomposition. Ultimately, left exposed to the elements, the bones break down and disintegrate, disappearing within a few years.

In recent decades, paleontologists have begun to look more closely at dinosaur fossils, with an eye toward identifying other life-forms that interacted with the decaying carcasses. On the macro side are such modifications as trampling breaks and tooth marks, some left behind by dinosaurian carnivores that scavenged the carcass or perhaps made the kill. On the micro side are bone modifications made by other groups of organisms, providing clues to broader ecological interactions. Just as with leaves, ancient insects left distinctive traces on bones. Most consist either of borings that penetrate the bone's interior or meandering trails on the surface. The former have been interpreted as pupal chambers excavated by carrion beetles, whereas the latter are thought to be feeding traces. In some instances, the same osteophilic (bone-loving) beetles, or at least close relatives, may have been responsible for both kinds of traces, the first associated with survival (surface traces), and the second with reproduction (borings).

Occasionally, residues found around dinosaur skeletons appear to be the fossilized remnants of bacteria and perhaps fungi. Several recent studies indicate that bacteria are integral to fossilization, particularly in the rare instances of soft tissue preservation. This might seem counterintuitive, because many bacteria are excellent decomposers, notorious for breaking down and recycling organic matter rather than preserving it. Yet decay and fossilization may sometimes go hand in hand. Particularly under watery conditions, bacterial colonies can generate gel-like secretions that generate a biofilm halo around a carcass. Through a process called BIOMINERALIZATION, these bacteria can also secrete minerals, either within their cell walls or directly into the surrounding environment, and thereby play a critical role in fossilization. This topic is likely to see a lot of attention in the coming years, and my suspicion is that bacteria will turn out to be critical in the preservation of fossils generally.

I once saw a sewer-and-drain plumbing van emblazoned with the cartoon of a proud, uniform-clad man bearing a broad smile and a Roto-Rooter-style cleaning hose. Beneath was the slogan "Your waste is my bread and butter." Although unappetizing, this motto is apt for many poop-loving organisms. After all, dead animals are not the only organic debris in need of recycling within ecosystems. Throughout their lifetimes, animals generate excrement, cumulatively depositing many tons of it each day in almost every ecosystem. And the "no waste" rule imposed for carcasses applies equally to fecal matter. Dung provides a rich source of nutrients for organisms that have evolved to consume it. Regardless of their metabolic strategies, giant dinosaurs undoubtedly produced whopping masses of dung. Surely we would expect important clues to be hidden inside ancient poop, were we to find any.

Coprolite is the term scientists use to refer to fossilized dung, and Karen Chin of the University of Colorado, Boulder, is a world expert. Fortunately, her chosen subject matter is matched with a wry sense of humor. So it's not surprising that Chin has variously been dubbed "Queen of Coprolites," "Duchess of Dung," and "Princess of Poop," among other unmentionable titles. She has identified a variety of different kinds of fossilized scat ascribed to dinosaurs, including both herbivores and carnivores. She was lead author of the study that described the king-sized *T. rex* coprolite mentioned in the previous chapter. Perhaps contrary to your first squeamish impressions, "paleoscatology" (as the field is sometimes labeled) is neither dirty nor disgusting. Feces that survive the rigors of fossilization are fully lithified and thus bereft of squishiness and foul smells. Like their modern counterparts, fossilized dung comes in a variety of shapes and sizes. Some appear classically scatlike, complete with pinched ends and striations left behind by the anal sphincter, whereas others are nondescript masses whose identification requires an expert eye.

To tease out secrets hidden within coprolites, Chin uses many of the same technologies employed by dinosaur histologists. A diamond-studded saw blade cuts thin slices from rock-hard dung specimens. These sections are then ground down to thicknesses on the order of a few thousandths of an inch before being mounted on glass slides. Viewed through a high-powered microscope, the slices reveal a complex amalgam of organic flotsam and jetsam. Carnivorous dinosaur scat may preserve bits of bone, teeth, and even muscle, whereas herbivore coprolites often include a mash of leaf parts, stems, branches, seeds, pollen, and/or spores. Interestingly, Chin thinks that the preservation of dinosaur dung is also due, at least in part, to the mineralization-promoting metabolic activities of bacteria.

Poop spotting, at least when it involves the lithified scat of dinosaurs, is as much art as science. One has to get the right search image. To complicate matters, coprolites, like other kinds of fossils, vary in color and texture, as well as shape. I remember a cool summer morning in the late 1980s, walking with Chin through scenic Late Cretaceous aged badlands in northern Montana. The region is known as Egg Mountain because of the spectacular discoveries of dinosaur nests and eggs found there by Jack Horner and his crews. Egg Mountain had also yielded several large mass death assemblages of a hadrosaur that

Horner named *Maiasaura*, the "good mother lizard," because its fossils were associated with nests of juveniles that appear to have benefited from parental care. Chin told me that she had identified several large coprolites from the same area that were attributable to a large herbivore, perhaps even *Maiasaura*. I was skeptical, as I had walked this small chunk of terrain daily for several weeks and seen no sign of any fossilized dung. However, Chin walked purposefully to several irregular broken masses of rock, the largest being more than a foot across. Kneeling down, she pointed out numerous plant fragments within the rock. She told me the woody fibers, mostly derived from conifers, had traveled through the intestinal tract of an animal (a big one, given the magnitude of the dung blocks). She also noted what appeared to be small tunnels through the specimen and conveyed her suspicion that these might be fossilized insect burrows, perhaps left behind by dung beetles.

This suspicion was later confirmed when Chin contacted dung beetle specialist Bruce Gill in Ontario, Canada. Although the ancient scat contained no remains of the dung beetles themselves, many of the burrows were backfilled with organic matter from the dung pat. Plenty of invertebrates can dig through a pile of dung, but only dung beetles (formally known as paracoprids) backfill their tunnels. The beetles burrow through fresh dung in order to create nutrient-filled brood chambers for their young. Chin's subsequent studies of these same dung masses revealed two additional hidden web insights. First, the coprolites contain dozens of snails of four different species; the large, moist, microbe-laden maiasaur pies apparently offered excellent food sources and roosting sites for the snails. Second, the chunks of conifer wood in the dung are not mixed with leaves, stems, or branches, suggesting that the maiasaurs deliberately ate wood from the core of the tree. Living wood is typically not nutritious, but the partially digested chunks in the dung were degraded by fungus. Chin's idea is that the hadrosaurs fed on trees that were not only dead but in an advanced state of rot, loaded with fungus and other detritus feeders. If so, the giant dinosaurs may have, at least on occasion, consumed the rotten wood not so much for the wood itself but for the calories and carbon in the decomposers.

Chin's discoveries, made with a series of colleagues, are significant on several fronts. They provide direct evidence of an odd food item, rotting wood, in the diet of a Cretaceous dinosaur. They push back the known fossil record of dung beetles millions of years into the Mesozoic, demonstrating that these poop-loving insects evolved alongside dinosaurs instead of mammals, as previously thought. And, perhaps most intriguing, this research illuminates multiple ecological links in a Cretaceous food web—from conifer producers to dinosaur consumers to insect, snail, and fungal decomposers—a rare and marvelous look at a 75-million-year-old ecosystem.

Another example comes from India, where researchers studying a collection of Late Cretaceous coprolites recovered not only plant remains within the fossilized dung but several kinds of fungi as well. Remarkably, these prehistoric fungi appear to belong to forms still alive today. Some of the fungi are plant parasites that attach to leaves. One of these, *Colletotrichum*, causes leaf spot and red rot diseases, whereas others, exemplified by *Ersiphe* and *Uncinula*, produce powdery mildew on leaves (something I see in my own garden). The study also revealed examples of a mycorrhizal soil fungus named *Glomus*.

Here, then, we glimpse another ancient food web. The herbivorous dinosaurs (perhaps the titanosaur sauropods recovered from the same deposits) consumed leaves inhabited by parasitic fungi; after passing through the mouths and GI tracts of the dinosaurs, undigested portions of these meals were defecated and later inhabited by a different kind of fungus, which removed nutrients from the fecal matter and perhaps distributed them via symbiotic alliances to other plants.

Discoveries of Mesozoic fungi provide an appropriate segue from aboveground habitats to subterranean realms, where the real heavy lifting of ecosystem recycling takes place. Fungi also propel us back to the present day.

Soil is Mother Nature's mistreated stepchild, either overlooked or treated with disdain as mere "dirt." Walking through field, forest, or desert, we tend to think of the substrate beneath (if we think of it at all) as merely a physical substance. Yet this view is entirely wrongheaded. Yes, soil contains plenty of minerals, air, and water. But, as noted earlier, it is also brimming with life-forms, living and dead—plants, fungi, earthworms, insects, mites, millipedes, centipedes, nematodes, and sow bugs, to name a few. While we are being entranced by soaring trees, colorful birds, and majestic predators, a greater mass of life-forms is living within the soil than above it.

Many of the animal denizens inhabiting soils are smaller than the period at the end of this sentence. One entomologist calculated that, with every step taken in a temperate forest, your foot is supported on the backs of 16,000 invertebrates, in turn held up by an average total of 120,000 legs![1] To focus on just two groups, the total weight of all ants in the world is roughly equal to the cumulative mass of all 6.8 billion humans, and the tiny nematode worms, virtually ubiquitous in soils worldwide, are estimated to comprise four-fifths of all animals on Earth. A single cubic inch (16 cubic centimeters) of soil can contain more than a mile of interwoven mycelia, the fungal mats that pervade the first few inches of soil. As if these figures weren't sufficiently mind-boggling, a single gram of fertile soil held between your fingers will contain on the order of ten billion bacterial individuals and perhaps six thousand bacterial kinds, or species. To date, biologists have formally named just over six thousand bacterial species, about the same number found in that gram of soil.[2]

This means that every handful of forest soil is a burgeoning microecosystem—complete with a stunning array of plants (or at least plant parts), herbivores, carnivores, and decomposers—about which we have only the faintest inkling. Beneath our feet are tiny wildernesses in which the dramas continually playing out are every bit as compelling as lions stalking zebra or tyrannosaurs preying on horned dinosaurs. In this alien microrealm, in the darkness amid tangled roots of trees and shrubs, vast interlinked networks of diaphanous fungal strands transfer nutrients between plants, ants are frequently the dominant predators and scavengers, and a succession of ever-smaller organisms continually grind up and recycle the dead.

Soil, then, is the epitome of the hidden web. We know that soils are essential to the health of terrestrial ecosystems, that they provide, to use a recent phrase, vital "ecosystem services." Primary among these is the recycling of several essential elements, including nitrogen, phosphorus, and sulfur. Yet our ignorance of this fertile, belowground realm is such that we don't really understand why there are so many species packed into such small spaces. Larger, burrowing animals, including many mammals, mix air, water, minerals, and organic materials, a churning task taken to another level by smaller organisms such as ants and earthworms. Still-smaller life-forms break down and recycle organic remains, the feces of one becoming the food of another. As one researcher noted, soil is perhaps best thought of as "bug poop." The activities of these myriad composters, all supported by specialized bacteria within their guts, replenish available nutrients and enable plant growth, thereby allowing the system to continue capturing solar energy. In addition to sustaining the terrestrial food supply, soil is also the primary water filter, cleansing and storing water for ecosystem use. Moreover, together with climate, soil determines which life-forms can live in a given area.

Although numerous examples exist of aboveground symbioses (e.g., the aforementioned ants and bull's-horn acacia), such partnerships are even more plentiful belowground. Some of the most important are the multispecies unions that partner plants with bacteria. All the nitrogen used by plants is derived from the decomposition of previous generations of plants and animals. However, because plants are unable to absorb nitrogen directly from the soil, an intermediary is required. This is where bacteria comes in, offering their unique talents to the relationship. Plant roots are generally accompanied by stringy bacteria, seen as swellings on the roots. The bacterial side of this partnership takes up nitrogen from the soil and transforms it into amino acids that can be used by the plant for growth. The bacteria in turn receive nourishment from the plant roots.

Mycorrhizae also come in fungal varieties, which absorb phosphorus for plants and enable them to deal with harsh environmental conditions such as drought, high temperatures, and toxins. It turns out that fungi are remarkable organisms, far more critical to the health and survival of terrestrial ecosystems than ever imagined just a few decades ago. Together with bacteria, fungi are the most important organic recyclers. A mushroom is merely the small fruiting body of a fungus, the proverbial tip of the iceberg. The great bulk of fungal biomass is hidden belowground, forming dense mats of interwoven mycelia. Virtually ubiquitous in the first few inches of soil all over the planet, mycelia rank among the most important soil builders. Today, and likely for most of the past half billion years, the largest organisms on Earth have not been dinosaurs or sequoias but fungi, in the guise of these mycelial mats. The largest example known, a 2,400-year-old honey mushroom (Armillaria ostoyae), runs only 1 meter (3 feet) deep but spans about 9 square kilometers (3.5 square miles, or 2,200 acres) and weighs an estimated 550,000 kilograms (1,200,000 pounds)!

Like all parts of an ecosystem, soil is always changing. Forming millimeter by millimeter, soil is constructed from a mixture of rock below and organic debris above, a process dependent on rock weathering. Physical or mechanical weathering—driven by wind, rain, and freezing—causes rocks to break into smaller fragments. Water percolating into cracks may freeze and thaw, leading to expansion, contraction, and, ultimately, splitting of rocks. Chemical weathering, often the result of rain and water flow, alters the chemical structure of rock. Over large spans of time, water with dissolved carbon dioxide literally erodes rock surfaces. A third factor, biological weathering, was overlooked until recently. Roots penetrate rocks in search of nutrients and lead to fracturing. Fungi in lichen produce more acids that corrode rock faces. And many kinds of bacteria secrete compounds that cause swelling inside fissures and cracks within the rock, leading to further fragmentation. In all, it is estimated that rates of rock weathering and erosion are a thousand times greater than they would be on a lifeless planet. Over geologic time, entire mountain ranges are flattened by this cadre of forces.

But what about Mesozoic soils? Can we be sure that soils comparable to modern examples, with such a bewildering array of tiny and microscopic diversity, were present beneath the feet of dinosaurs? After all, it's not like we can sample the diversity of, say, a Late Cretaceous soil horizon. The truth is, we can actually dig into ancient soil horizons and find evidence of life, often including abundant root traces, burrows, and even signs of bacterial activity. Beyond these clues, however, scientists are extremely confident that Mesozoic terrestrial ecosystems, like their modern counterparts, were nourished and dependent on a highly diverse, organic-rich interface between earth, water, and air.

The early Earth lacked soils altogether. Prior to life's migration onto land, the unconsolidated material derived from rock was a barren substance that we refer to as "regolith," the kind of stuff we find on the moon's surface. The spread of life into the terrestrial realm was initiated by bacteria, whose metabolic activities began to dissolve exposed rocks. It's likely that fungi were the next major group to follow, ultimately forming their dense mats of mycelia. Later came the plants, establishing a terrestrial beachhead more than 350 million years ago. Through their inputs into rock weathering, countless generations of these varied life-forms generated the first true soils, disintegrating boulders while alive and contributing their own remains to the organic mix postmortem. This earthy matrix provided a rich substrate for land-based evolution, including the invertebrates and vertebrates that subsequently migrated inland from the water's edge. Over millions of years, an increasing diversity of organisms occupied more and more land, with additional incursions likely dependent in large part on the generation of soils. This thin veneer of organic matter began to envelope the continents, carrying with it a suite of complex molecules that enabled plants to flourish, together with the consumers and decomposers dependent on them. Dinosaurs, latecomers on the terrestrial scene, arrived into a world of complex soils replete with mycelial mats, mycorrhizal symbionts, and numerous other specialized soil organisms. Importantly, without this previous, lengthy occupation of land by soil-dependent life-forms, there would have been no dinosaurs—and no us, for that matter.

So life has depended on soils virtually from the time it first ventured beyond its aquatic confines. Whether or not the numbers of Mesozoic invertebrate and bacterial soil species approximated the astonishing modern levels of diversity is an open question. But, without doubt, that all-important brown, crumbly, poop-filled matter was there, providing the substrate for the dinosaur odyssey. The take-home message here is that, even in a world graced by such biological splendors as dinosaurs, redwoods, and humans, we are forced to concede that microbes rule.

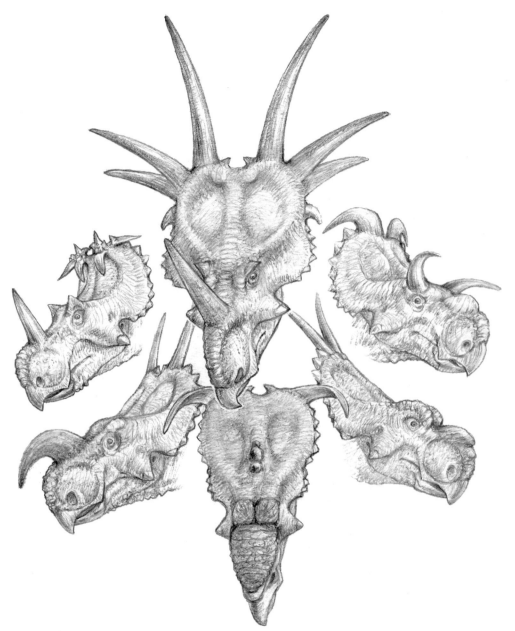

Bizarre headgear in short-frilled (centrosaurine) horned dinosaurs. Whereas *Styracosaurus albertensis* (top middle) displays one of the more flamboyant expressions of frill ornamentation, others possess equally unusual skulls: (from left to right) *Centrosaurus brinkmani, Einiosaurus procurvicornis, Pachyrhinosaurus lakustai, Achelousaurus horneri,* and *Albertoceratops nesmoi.*

10

OF HORN-HEADS
AND DUCK-BILLS

IN THE SUMMER OF 1987, I was a young, eager, and naive graduate student at the University of Toronto trolling for a great research project. I was already well along the path to becoming a paleontologist and had pretty much settled on dinosaurs as my target group. In particular, I liked the idea of working either on theropods or the horned ceratopsids like *Triceratops*. But I still needed access to some unstudied fossils. While returning east after digging up ichthyosaurs (those dolphin-like, seagoing reptiles from the Mesozoic) in my home province of British Columbia, my coworkers and I stopped by the home of dinosaur expert Philip Currie in Drumheller, Alberta. Late in the evening, I told Currie about my search for a graduate project, and he suggested that I speak with another dinosaur paleontologist, Jack Horner, at the Museum of the Rockies in Bozeman, Montana. Word was out that the Montana group had recently dug up a new horned dinosaur and the bones might be available for study. Following this advice, I traveled to Montana a few days later to meet with Horner.

By the mid-1980s, Jack Horner was already well established as a leader in the field, following his discoveries of baby dinosaurs, eggs, and nests. I was soon to learn, however, that Museum of the Rockies field crews had excavated many adult dinosaurs as well, particularly from the Two Medicine Formation, a rock unit that dates to near the end of the Cretaceous. I first met Horner at one of his field camps in the spectacular Big Sky country of northern Montana. Sitting amid Blackfeet teepees (the accommodation of choice at the time), with the Rocky Mountains looming in the darkening sky, I excitedly explained my interests in horned dinosaurs and inquired about the new animal. Horner briefly described this new beast, and I soaked up every word.

The adult skulls were about 1.8 meters (6 feet) in length, with a single pronounced horn over the nose that was narrow side to side and hooked forward like some prehistoric can opener. Additional adornments included a small horn over each eye and a pair of long spikes pointing rearward from the back of the bony neck frill. The spiked frill suggested a close affinity with *Styracosaurus*, a horned dinosaur from about the same time period recovered from Dinosaur Provincial Park, Alberta. The new Montana dinosaur had been recovered from a rich pocket of badlands called Landslide Butte, tucked up against the border with Canada. Importantly, many specimens of this new animal had been excavated from two bonebeds, including the fossilized remains of juveniles, subadults, and adults. So there would be an opportunity to examine multiple growth stages, a rarity for dinosaurs (at least at that time). I had heard enough, and apparently so had Horner. Minutes later we had struck a deal. My PhD research would entail naming and describing the can-opener ceratopsid, assessing its relationships with other horned dinosaurs, and studying its growth patterns.

I could barely contain myself. How could a grad student interested in dinosaurs ask for a better project? I thanked Horner profusely and returned to Toronto determined to read everything I could get my hands on about ceratopsid dinosaurs. I knew they were a diverse group of giant herbivores known only from the Late Cretaceous of North America. I was surprised to learn that the many species were distinguished from one another almost entirely on the basis of differences in a few bones on the top of the skull. In particular, diagnostic features were concentrated in ornamentations over the nose and eyes, as well as on the frill. Armed with the brashness of youth, I convinced myself that other researchers must have missed some evidence. Surely if I were to look carefully enough, *I* would identify bony features elsewhere on the skull, as well as in the skeleton, that could distinguish closely related species.

A few months later I found myself sitting in the basement of the Museum of the Rockies poring over bones of the can-opener ceratopsid. It was then that Horner asked whether I would like to add a second newly discovered horned dinosaur to the project. It, too, had been discovered at Landslide Butte, but from a higher layer of rocks, indicative of a somewhat younger time interval. This second animal was closely similar to the first, also possessing a skull about 1.8 meters (6 feet) long, with an intimidating pair of backward-projecting spikes on the frill. However, instead of horns over the eyes and nose, there were thick platforms of bone covered with a roughened texture reminiscent of blowtorched metal. Here was the classic example of a face only a mother could love. (To be fair, this face would also have been attractive to certain other horned dinosaurs as well, a point driven home later.) Similarly lumpy platforms of bone were known from another horned dinosaur found in Alberta called *Pachyrhinosaurus*. The additional Montana fossils made the story all the more interesting, and I gladly accepted Horner's offer.

With detailed notes and photographs in hand, I began studying horned dinosaurs housed in other North American museums like the Royal Tyrrell Museum, the Canadian Museum of Nature, the Royal Ontario Museum, the American Museum of Natural History, and the Academy of Natural Sciences. It soon became apparent that previous paleontologists working on this group had got it right after all. Key differences were pretty much limited to variations in ornaments on the skull roof. I could lay out multiple examples of, say, the maxilla (the toothed upper jawbone) on a table and find as much or more variation within a given species as I did between two or more of them. The same was true of vertebrae and limb bones. In other words, even if you were to find a complete skeleton, without the top of the skull it would be difficult to identify exactly which dinosaur species the bones belonged to. Although subtle differences might be revealed through measurements, the small number of bones available for most species limits the application of quantitative approaches.

So, like others before me, I focused my attention on the skull and, for the first time, began to contemplate the implications of this recurrent pattern. Why were horned dinosaurs distinguished predominantly on the basis of odd bony growths? Granted, bones tell only part of the story of any vertebrate animal, yet the lack of obvious differences in teeth, jaws, and limbs across closely related species suggested that all these dinosaurs were doing pretty much the same thing in order to make a living. This conclusion seemed at odds with what I knew about ecology, because every animal in an ecosystem tends to have its own particular niche. A second implication was that evolutionary change within this ancient group had primarily targeted horns, frills and spikes. So it suddenly became important to understand the function of this strange headgear.

I am regularly asked why dinosaurs continue to be so popular year in and year out. What could possibly fuel such an exotic passion? Surely part of the answer is the sheer size of many dinosaurs. It's awe inspiring to contemplate gargantuan carnivores prowling ancient landscapes in search of even larger prey. Yet I submit that generation after generation of humans finds these long-dead beasts so magnificent because they were not only big but bizarre. As noted in earlier chapters, dinosaurs sported an amazing array of bony appendages. In addition to the horns and frills of ceratopsids, there are the crests of hadrosaurs, the plates and spikes of stegosaurs, the domes of pachycephalosaurs, and the body armor and tail clubs of ankylosaurs.

Over the years, many ideas have been put forth to account for these odd features. For instance, it was suggested that the long, hollow crest of *Parasaurolophus* functioned as a snorkel, allowing the animals to feed underwater while maintaining continual breathing. A fatal flaw with this idea, however, is that there is no opening at the tip of the crest to allow the passage of air. Similarly, it was long argued that the elongate frills of ceratopsids were merely enlarged surfaces for attaching very large jaw-closing muscles, offering these mega-herbivores a great advantage in processing tough plant fodder. But muscle strength derives more from cross-sectional thickness than length, and various bony

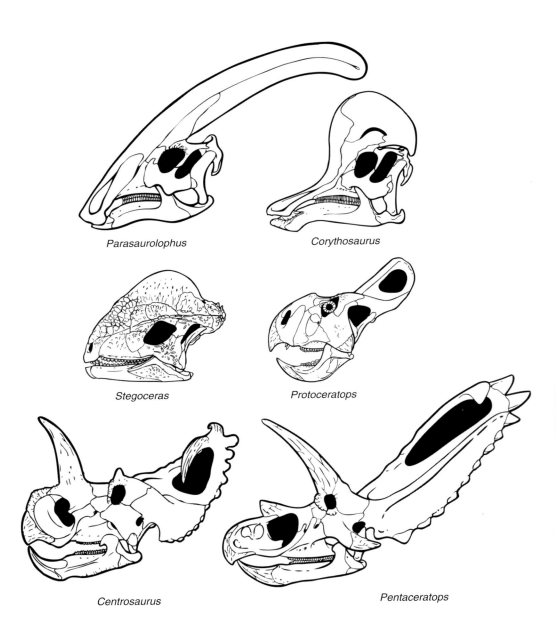

Parasaurolophus

Corythosaurus

Stegoceras

Protoceratops

Centrosaurus

Pentaceratops

FIGURE 10.1

Bizarre structures in ornithischian dinosaurs: top row, the lambeosaurine hadrosaurs *Parasaurolophus* and *Corythosaurus*, bearing hollow crests; middle row, the dome-headed pachycephalosaur *Stegoceras* and the frilled ceratopsian *Protoceratops*; bottom row, the centrosaurine ceratopsid *Centrosaurus* and the chasmosaurine ceratopsid *Pentaceratops*, both of which display bony frills and horns above the nose and eyes.

indicators suggest that the jaw muscles did not extend significantly onto the frill. Perhaps, thought some investigators, the diversity of strange bony features on dinosaurs required a more general explanation that encompassed many groups.

Traditionally, the most commonly accepted hypothesis is that these structures functioned to ward off predators. Dinosaur reconstructions during most of the twentieth century often depicted prehistoric standoffs or all-out battles between carnivore-herbivore pairs like *Tyrannosaurus* and *Triceratops*, or *Allosaurus* and *Stegosaurus*. At first blush, the antipredator hypothesis seemed entirely reasonable, particularly for slow-moving vegetarians sharing a world with toothy, multiton meat eaters. But closer examination revealed various problems. First, most of these odd structures would have made lousy weapons. The frills of most ceratopsids are extremely thin in parts, as little as a millimeter in some regions. Second, there is also evidence that this lengthy rearward extension of the skull had a rich blood supply. Therefore, one solid bite from a large theropod would have resulted in extensive blood loss and perhaps death. The same is true for the plates of stegosaurs and the crests of hadrosaurs, which also have extremely thin regions.

What about other structures, such as the horns of ceratopsids? Wouldn't they have made good weapons? Here, too, are nagging issues. For starters, it's difficult to imagine how some of these horns—like those of the new can-opener beast, with its downward-directed tip—could have been wielded effectively as weapons. More fundamentally, remember that every species tends to have its own uniquely shaped horns. It turns out that this trend applies not just to ceratopsids but to the marvelous smorgasbord of bizarre dinosaurian structures. If fighting off predators was the primary function, we would expect to see evidence that evolution driven by natural selection settled on one or a few effective designs and then modified them very little. Instead, evidence points to the opposite phenomenon: rampant variation and evolutionary malleability. Then there is the fact that the carnivorous theropods also got into the act with their own, sometimes delicate, bizarre structures. For example, *Dilophosaurus* has a pair of thin crests emanating from the top of the face, whereas the Asian theropod *Monolophosaurus* has a single crest. Others, like *Allosaurus* and *Carnotaurus*, possessed well-developed horns over their eyes. Once again, in all of these instances we find that no two species are exactly alike in their ornamentations, although they tend to be quite consistent within species.

A second explanation put forth to account for at least some bizarre structures is control of body temperature, or THERMOREGULATION. As we shall see in the next chapter, for reasons associated with volume and surface area, big animals tend to retain heat much more than small animals. So evolution sometimes increases the overall surface area of larger animals as a means of dumping excess heat and preventing overheating. The large ears of elephants, for example, shunt heat from inside the body to the outside environment. Perhaps stegosaur plates and some other bizarre structures of dinosaurs also reflect heat-dumping strategies. Yet some shapes are much more effective radiators of heat than others. So a prediction of this hypothesis is that dinosaurian ornaments will tend toward designs that maximize this ability to shed excess heat. Once again, the many

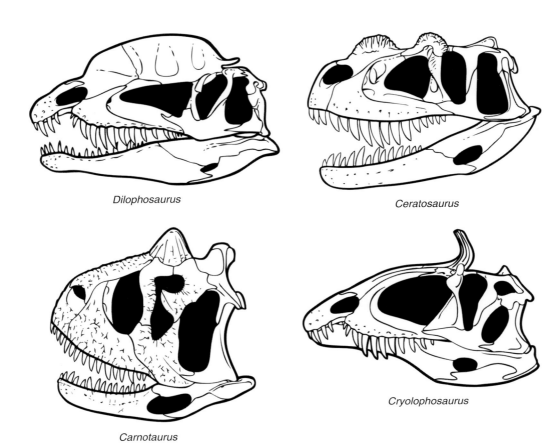

Dilophosaurus

Ceratosaurus

Carnotaurus

Cryolophosaurus

FIGURE 10.2

Bizarre structures in theropod dinosaurs: the coelophysoid *Dilophosaurus*, bearing an elongate pair of crests; the ceratosaur *Ceratosaurus*, with horns above the nose and eyes; the abelisaur *Carnotaurus*, with a horn above each eye; the allosauroid *Cryolophosaurus*, with a thin, curved crest atop the head.

species-specific variations in these structures suggest that we must look to an alternative explanation.[1]

A third idea, recently promoted by Kevin Padian (University of California, Berkeley) and Jack Horner, is that many of the odd bony features of dinosaurs were signaling structures that allowed animals to recognize members of one's own species. Species recognition is important if two or more closely related species share the same habitat, because animals do not want to waste valuable energies attempting to mate with individuals from these other forms. Therefore, evolution may generate features that facilitate assessments of species membership and prevent such inadvertent interspecies philandering. One of the problems with this idea is that there is minimal evidence of multiple, closely related animals sharing the same ecosystem.

Following publication of several landmark papers—in particular, a 1975 offering from Jim Farlow and Peter Dodson—increasing numbers of paleontologists began to support yet another alternative, one that regards these elaborate structures primarily as objects for courtship and combat with members of the same species. Evidence for this shift in scientific thinking can be seen in the recent history of dinosaur art. Images of the legendary confrontation between *Triceratops* and *Tyrannosaurus* have largely been replaced by wrestling ceratopsians, honking hadrosaurs, and head-butting pachycephalosaurs. Such depictions often include striking, even riotous coloration patterns, particularly on the bizarre structures. Is there solid evidence to support this latest hypothesis, or will future generations of paleontologists look back on this period as a time of unchecked, misguided speculation? Although undeniably biased on the matter, I'm confident that we now have a good handle on this issue. Just look at animals alive today.

The bulk of animal behavior can be loosely described as facilitating either survival or reproduction. The previous several chapters have focused on aspects of survival, particularly feeding and avoiding becoming food. But the drive to reproduce is equally powerful in evolution. Animals invest considerable energy in reproduction, in terms of both finding mates and raising young.

As briefly described in chapter 6, this survival/reproduction dichotomy was first understood and outlined by none other than Charles Darwin. Darwin knew that natural selection tends to propagate features that enhance the chances of survival. Yet he also noted that many animal characteristics seemingly did not provide any advantage in the fight for survival. Indeed, some, like peacock tails and deer antlers, actually appeared to increase chances of a premature demise. Nevertheless, it was evident that these same features also increased the probability of reproductive success. In particular, males with elaborate characteristics were frequently more successful in the competition for mates. To account for the occurrence of ornate structures like peacock plumes, Darwin devised the concept of sexual selection and devoted his second-longest book to the subject. He reasoned that, because animals must reproduce as well as survive, any features that increased the chances of reproducing would also tend to be passed on to future generations, even if they were detrimental to survival. In other words, sexual selection involves variation in reproductive success arising from differences in the ability to acquire mates.

Sexual selection can take various forms, involving males competing against males, females competing against females, males choosing females, and females choosing males. Of these four alternatives, male-male competition and female choice are the two that have been the most widely documented and discussed. Thus, for example, if male birds with more elaborate plumage are preferentially selected by females, plumage may become increasingly elaborate over time. Or if male deer with larger, more ornate antlers are able to outcompete and dominate other males through DISPLAY and physical combat, antlers in that lineage may tend toward further increases in size and elaboration over time because of differences in reproductive success. The most obvious examples of mating signals are visual ones, involving unique coloration and/or physical structures such as horns, crests, and elaborate tails. But sexual selection can and does exploit every other sensory modality: acoustic, olfactory, tactile, and chemical. Most familiar are the varied calls of animals, particularly birds but also frogs, insects, and mammals.

Over the past two decades, sexual selection has migrated from biology's backwater to its roaring mainstream. Hundreds of published studies and numerous books now detail the theory of mate competition and document its occurrence in wild populations.[2] Importantly, the female preference and male signaling structure can form a positive, or reinforcing, feedback loop in which the females' unwavering predilection results in a trend toward increasingly elaborate—that is, bizarre—signals.

It turns out that the modern realm includes thousands of examples of bizarre structures. In the vast majority of instances where these features have been studied in detail, they function predominantly either to attract mates or to repel rivals. In other words, they are mating signals whose role is to increase the odds of successful reproduction. This finding also applies to the vast majority of animals with horns or hornlike structures—from ants, beetles, and chameleons to cassowaries, antelope, and deer. The primary function in virtually all cases involves the competition for mates. Such an overwhelming pattern from this living realm generates confidence that the bizarre structures of dinosaurs served a similar role. Yet there are other lines of evidence, too.

Further study of the two new horned dinosaurs from Montana, which I named *Einiosaurus* and *Achelousaurus*,[3] reaffirmed initial observations that these animals and their close relatives are distinguishable from one another only by unique and decorative arrays of hooks, horns, spikes, and/or lumpy growths. This finding is unsurprising if we think of these bony features as mating signals. The reason that paleontologists can identify dinosaur species based on these weird features is that the dinosaurs themselves probably used the very same cues to recognize fellow species members and compete for mates.

As often happens in science, my investigation into the growth patterns of *Einiosaurus* and its kin was revealing in unexpected ways. As noted earlier, the two *Einiosaurus* bonebeds preserved multiple growth stages, showing how skulls changed shape as individuals grew larger. Working with two colleagues—Michael Ryan, now of the Cleveland Museum of Natural History, and Darren Tanke of the Royal Tyrrell Museum of Palaeontology—I compared growth changes in *Einiosaurus* with those of related animals like *Centrosaurus* and *Styracosaurus*. The same growth pattern was present in all of these forms. Juveniles began life with short and simple horns over the eyes and nose, as well as relatively abbreviated and unadorned frills. It was not until the animals approached adult size that these features sprouted into the varied, species-specific "bells and whistles" that we use to tell them apart. In other words, the juvenile and subadult ceratopsids of closely related species look very much alike, whereas the adults can usually be readily differentiated.

Such delayed growth, commonly seen in living animals as well, makes little sense if the primary function of these bizarre structures was to fend off predators. After all, immature herbivores might need to defend themselves, too. Yet, if these features were used as mating signals, we would predict delayed growth. Large bony structures can be energetically costly to build and maintain, so it makes sense that animals delay full expression until sexual maturity, when they begin to compete for mates.

Delayed growth of mating signals among living animals is often associated with dominance hierarchies, or "pecking orders." Showy features like horns serve as obvious indicators of an animal's age and level of maturity. Delayed and progressive expression of such signals enable individuals living in social, mixed-sex groups to determine their rank within the group without risking life and limb in physical combat. Based on the growth evidence, as well as multiple occurrences of vast bonebeds, it is certainly plausible, and perhaps likely, that some dinosaur species—particularly among such groups as ceratopsids and hadrosaurs—lived in complex, hierarchically organized herds for at least a portion of the year.

An important implication of this finding is that some horned dinosaurs known only from juvenile and subadult fossils may actually belong to other known species. In other words, it is difficult to assess species membership unless one has adult materials because species-specific features in the horns and frills were not fully expressed in preadult individuals. Had these immature animals survived to adulthood, they might have developed the horn and frill characteristics that we associate with some other species. This same argument was made earlier in classic studies of growth and variation in duck-billed dinosaurs.

If you've been thinking like a scientist, a possible prediction of the mating signal hypothesis may have occurred to you. Males of many living species are often larger or more elaborately colored than their female counterparts, and males frequently possess social structures such as horns or crests that are reduced or absent in females. Sex-based variations within species are referred to as SEXUAL DIMORPHISM. So if the primary function of dinosaurs' bizarre structures was to compete for mates, wouldn't we expect to see evidence of sexual dimorphism, likely with males larger and more elaborately ornamented than females? The answer appears to be yes and no.

Sexual dimorphism in skull shape has been suggested for many kinds of dinosaurs. Given that no two animals are exactly alike, the key becomes finding a way to partition adults into two groups that could plausibly represent males and females. Of course, one must be cautious and consider whether the variations might be the consequence of other factors such as age, nonsexual differences among adults, or even species-level differences. The problem is that, despite the boon of information offered by bonebeds, there are surprisingly few examples of species known from sample sizes that are sufficiently large to allow statistical testing.

However, not all animals with bizarre structures are sexually dimorphic. Among modern large-bodied, terrestrial mammals, for example, sexual dimorphism in both size and weaponry tends to be least in small-bodied forms (body mass less than 20 kilograms, or 44 pounds), greatest in medium-sized forms, and reduced in the largest-bodied species (body mass greater than 300 kilograms, or 660 pounds), particularly those inhabiting open environments. It appears that in the largest forms, females tend to mimic males in bizarre structures, although subtle sexual differences in horns are common.[4] If this pattern observed in modern animals applied to dinosaurs, we would predict maximal sexual dimorphism in midsized dinosaurs, and minimal dimorphism in body size and weaponry among large to giant-sized dinosaurs, many of which greatly exceeded 300 kilograms (660 pounds). Current evidence supports this prediction, at least for ceratopsid dinosaurs like *Triceratops* and *Einiosaurus* where dimorphism, if present at all, is limited to horns and frills.

The best we can say at the moment is that sexual dimorphism in dinosaurs, assuming it was present at all, was likely concentrated in bizarre structures, because this is where most of the adult variation occurs.[5] So, given that dinosaur species are distinguished largely on the basis of bizarre structures, and given that, at least in some cases, they tend to show delayed growth or expression, I firmly believe that many of the remarkable features present in dinosaurs did indeed function first and foremost as mating signals. This does not mean that they did not have additional functions in some instances. Among modern animals, horns and similar structures sometimes seem akin to Swiss army knives, aiding the animal not only in competing for mates but in recognizing mates, warding off predators, and controlling body temperature, among other functions. To give one quirky example, I have heard reports from park rangers that the large African antelope known as kudu sometimes use their long, corkscrew-like horns to reach up into trees and break off branches bearing tasty buds. Nevertheless, to my mind at least, it's likely that the primary role of such features involved competition for mates.

It's one thing to argue that a particular dinosaur structure was used to compete for mates. It is another thing altogether to say exactly how. For example, was the tubelike crest of *Parasaurolophus* a striking (perhaps brightly colored) visual signal, a trombone-like resonator used to generate vocalizations, or both? The real problem comes with testing alternative hypotheses. It's often highly problematic to determine the function of a particular structure in living, breathing, behaving animals, let alone extinct forms represented by a bunch of bones. Indeed, in most cases, when it comes to the behavior of extinct organisms, it is simply not possible to eliminate all but one hypothesis, and one is left with the unhappy circumstance of several plausible explanations.

Nevertheless, an important first step in this process is to reconstruct the anatomy of nonbony or "soft" tissues as accurately as possible. Were ceratopsid horns covered in a sheath made of keratin, as in antelope, or some other tissue? Were the hollow crests of duck-billed dinosaurs lined with tissues that would have enhanced the resonating capability of this convoluted chamber? Attempting to answer questions like these runs us headlong into the ever-present problem faced by dinosaur paleontologists; the material basis of our science is restricted largely to bones, which are but one of many tissue types present in vertebrates. To confound matters further, various soft tissues (blood vessels, nerves, muscle, cartilage, skin, etc.) are often more informative with regard to anatomical function. Fortunately, soft tissues frequently interact with bones and leave telltale signatures, from muscle attachment scars to holes that once transmitted nerves and blood vessels. The problem comes with interpreting these features, underlining the need to study living vertebrates, both close relatives (e.g., birds and crocodiles) and more distantly related animals possessing features that might be analogous to those of dinosaurs. Led by such anatomically minded folks as Larry Witmer, there has been a recent dramatic increase in anatomical sleuthing aimed at reconstructing soft tissues, and this kind of detailed comparative work has only just begun. We can be certain that many important insights into dinosaur behavior are still to come as we increase our literacy in reading the subtle messages preserved in bones.

Another useful approach is to conduct biomechanical tests to ascertain whether a given structure is well designed for a putative function. Thus, while it's difficult to assess whether or not a particular bizarre structure was used in display, it may be possible to test the proposition of combat with other members of a given species. For example, the skull and neck of *Triceratops* and its close ceratopsid relatives appear well suited for confrontations in which opponents locked horns and wrestled to establish dominance. Possible adaptations for fighting include a pair of large horns above the eyes, a thick double-skull roof over the brain, reinforced bony eye sockets, and fusion of the first three neck vertebrae. This interpretation is further supported by the fact that living horned animals engage in similar contests. In contrast, the frills of many species are more delicately built, making them poorly suited for combat but exceptional devices for visual display. One can easily envision frontal displays in which the head was held vertical and perhaps moved from side to side to enhance the appearance of size. If this show was not sufficient to ward off a rival, the horns could have been brought into play in pushing-and-shoving contests.

A final source of evidence worthy of mention is paleopathology—that is, evidence of trauma endured by a dinosaur during its lifetime. For example, as documented in recent work by Andrew Farke (Raymond Alf Museum) and others, the frills of some horned dinosaurs exhibit injuries that may well have resulted from the horn thrust of a same-species opponent.

Turning to the dome-headed pachycephalosaurs, Mark Goodwin of the University of California, Berkeley, and Jack Horner have argued on the basis of histological evidence that the commonly reconstructed head-butting behavior in this group is highly improbable. Head butting also seems questionable from a functional standpoint. Among living horned mammals such as bighorn sheep that engage in dramatic head-smashing contests, the horns tend to be broad up front, creating a wide contact platform that reduces the chance of spinal cord injuries resulting from neck twisting. The rounded skull domes of pachycephalosaurs would seem to be exactly the opposite of what one would predict in animals that ram heads. Think of two people standing at opposite ends of a billiard table, each one holding a single ball. If both simultaneously roll their ball toward the center of the table, it's unlikely that the collision will send the balls directly backward. More often than not, the rounded shape will result in glancing blows, sending the balls

FIGURE 10.3
Flank butting as a possible alternative function for pachycephalosaur domes.

careening to either side. If pachycephalosaurs rammed heads in a manner anything like bighorn sheep, the result might well have been lethal injuries. Of course, it's possible that pachycephalosaurs engaged in less forceful head-butting contests, but once again rounded domes seem ill suited for this type of behavior, because the heads could not be locked together in the manner of horned animals. Proposed alternatives to head butting include flank butting, species recognition, sexual display, or some combination of these.[6]

Two humbling points should be kept in mind when considering the social behavior of dinosaurs. First, although we know that mating signals can exploit the entire spectrum of sensory channels, fossils generally provide direct evidence for only one of these: visual signals on bones. So paleontologists face strict limits on the amount and quality of evidence potentially available.[7] Second, the Mesozoic was a very different time, and dinosaurs diverged in some fundamental ways from any other group of animals before or since. Considering dinosaurs as a whole, we can be certain that we're missing a large part of the mating signal picture. Barring the invention of time travel technology, this situation is unlikely to alter substantially, and the study of dinosaurian social lives is destined to remain a highly speculative endeavor.

On the flip side, of the entire potpourri of animal mating signals, only a handful involve bone and are therefore prone to fossilization. In this sense, the bizarre structures of dinosaurs are highly significant, because they provide a glimpse into ancient behaviors that are typically invisible to paleontologists. If the recent past is any indication, future studies employing techniques such as soft tissue reconstruction and biomechanical analysis will find innovative ways to assess alternatives and answer questions that today seem beyond our reach.

Sex and species are inextricably interwoven. The boundaries that separate biological species are built on the potential for reproduction. If two organisms breed and produce viable offspring (i.e., young capable of breeding), they are considered members of the same species. If not, they belong to two distinct species.[8] At a minimum, then, the origin of a new species entails some change in the mating system. One example would be alterations in mating behaviors. Sexual reproduction typically involves a complex chain of behaviors. Whereas partners of the same species are able to complete the appropriate sequence of signals and responses and thereby mate successfully, the sequence typically collapses between partners of different species. An evolutionary modification to a single link of this chain that spreads in one population but is absent in another can be enough to prevent interbreeding and erect a new species boundary (see chapter 6). So if two populations become isolated and diverge in their mating signals, they may no longer recognize each other as members of the same kind if subsequently reunited. Viewed in this way, it's not surprising that mating signals frequently play a major role in the origin of new species or that species are often distinguished mostly or solely on the basis of such signals.

Sexual selection has been linked to the evolution of diversity in many groups of animals. One of the greatest known evolutionary radiations of vertebrates involves a group of fishes called cichlids that swim the waters of several African Great Lakes. More than 1,500 species of cichlids have evolved in these lakes, and recent evidence indicates that this great piscine blossoming occurred over a relatively short period, perhaps 12,000 years. Many of the Great Lakes cichlids are distinguished at the species level on the basis of mating signals, and it's been hypothesized that sexual selection played a key role in their rapid diversification. As lake levels fluctuated up and down, populations of fishes frequently became isolated from one another for extended periods, perhaps long enough for mating signals to diverge to the point of new species formation.[9]

Additional examples are known from the fossil record. During the mid- to late Pleistocene, a group of deer in the Old World diversified into a wide array of forms distinguished largely on the basis of variations in body size and antler shapes and sizes. The most famous member of this family is *Megaloceros*, the Irish elk (a misnamed beast, as it was neither strictly Irish nor an elk). Like other deer, *Megaloceros* males annually grew and then dropped their antlers. Unlike other deer, the antlers of the Irish elk reached an astounding 3.1 meters (10 feet) in breadth and weighed approximately 36 kilograms (80 pounds)! We know from observations of living deer that antlers function chiefly in competing for mates, particularly male contests, and it's almost certain that this was also true of the Irish elk. So it seems probable that forces such as sexual selection exerted a strong influence, causing antlers to morph into a broad array of forms over time. Interestingly, this great radiation of deer took place against a backdrop of environmental change, including advancing and retreating ice sheets.

Spanning a range of recent and extinct vertebrate groups, current evidence suggests that bizarre structures played a key role in the evolution dinosaurs. Specifically, given the close relationship shared between mating systems and the origin of species, these same features may have been pivotal in the formation of new kinds of dinosaurs. For example, the Late Cretaceous radiations of ceratopsids and hadrosaurs may owe much of their species diversity to the divergence of mating signals driven by processes such as sexual selection. If sex and the competition for mates did have an exaggerated influence on the origin of new dinosaur species, the very same features that make dinosaurs so strange and wonderful to us may also have been vital to their evolution and overall success. And, like the Old World deer example, these dinosaur radiations may also have been closely tied to episodes of environmental change. We will return to this topic in chapter 14.

Within the broad sphere of animal reproduction, the competition for mates is only one component. Others include nesting and parental care, as well as the actual act of sexual reproduction. With regard to the act itself, we know surprisingly little about how dinosaurs "did it." Presumably, males approached females from behind in order to bring their sexual organs into contact with those of the female and transmit sperm. Among quadrupedal forms, this sequence of events must have involved males rearing up on their hind legs, no mean feat for gargantuan animals such as sauropods. And it's intriguing to contemplate two stegosaurs, with their respective repertoires of back plates and tail spikes, engaged in sex. Fortunately, we know somewhat more about the postmating reproductive behaviors of dinosaurs.

Biologists recognize two divergent strategies toward reproduction and parental care that represent end points on a spectrum. At one extreme is the scattershot approach—maximize the number of offspring and minimize parental investment. A single oyster produces thousands of eggs, but very few of these survive, let alone reproduce. All reproductive energies are directed toward the numbers of eggs, with no investment in parenting. At the other extreme are organisms like humans, who typically give birth to a single offspring that cannot survive without abundant parental care, including nourishment, protection, and socialization. Here the approach is to place all of one's eggs in one or a few baskets, maximizing parental care of a small number of offspring. Whereas the first strategy emphasizes quantity, the second focuses on quality. All organisms fall somewhere on this spectrum, which represents a trade-off between producing large numbers of offspring and providing care to help those offspring survive and contribute to the next generation. Where were dinosaurs positioned on this continuum? Current evidence suggests that different groups occupied varying positions in the middle range between the two extremes.

All dinosaurs were egg layers, not surprising given that the closest living dinosaur relatives, birds and crocodiles, also lay eggs. The first recognized dinosaur eggs were discovered during the Mongolian expeditions led by Roy Chapman Andrews in the 1920s. In recent years, many additional discoveries of eggs, nests, and even nesting grounds have greatly increased our understanding of dinosaur reproduction. Perhaps the most important of these discoveries is an amazing assemblage of nests, eggs, and embryos in the Patagonia region of Argentina that covers on the order of 5 square kilometers (2 square miles)! This Late Cretaceous locality, dubbed Auca Maheuvo, preserves tens of thousands of large eggs and appears to have been a breeding ground for titanosaur sauropods. Amazingly, many of the eggs contain fossilized skeletons of embryos, some with preserved skin. Layers of densely packed nests suggest that the animals returned to the same breeding ground year after year.

Now you might imagine that the biggest dinosaurs generated proportionally gigantic eggs, but this is not the case. In order to preserve the integrity of the egg, maximum egg size is limited by shell thickness. Shells become thicker as egg volume increases. Yet embryos must receive air from the outside environment through tiny pores in the egg shell. If egg sizes were to become too large, the increased shell thickness would close up the pores and prohibit air from reaching the developing embryo. So even dinosaurs that grew to 50,000 kilograms (110,000 pounds) or more hatched from eggs much smaller than a basketball. For example, the sauropods at Auca Mahuevo hatched from spherical eggs approximately 13–15 centimeters (5–6 inches) in diameter, significantly smaller than the eggs of some recent ground birds.

The total number of eggs per nest, referred to as "clutch size," for the Auca Mahuevo sauropods ranges from fifteen to thirty-five. Current evidence suggests that other dinosaurs had similar clutch sizes. For example, hadrosaurs, which are much smaller than sauropods, averaged about twenty to thirty eggs per nest, and similar clutch sizes have been found for the still-smaller oviraptor theropods. The Mongolian oviraptors are among the best-known dinosaurs in terms of their reproductive behavior. Females within this theropod group apparently laid two eggs at a time, adjusting position slightly after placing each pair until the nest was filled with an egg spiral. To an extent, egg-laying behavior places dinosaurs a little closer to the oyster end of the reproductive spectrum, because this strategy enables females to produce many more young than could be achieved through live birth.

Yet, in contrast to the oyster example, growing evidence indicates that many dinosaurs engaged in some form of parental care. The Auca Mahuevo locality includes hundreds of nests in a single nesting horizon, suggesting that these animals may have formed annual breeding colonies. Incubation of the eggs was likely accomplished in part through placement of vegetation on the nests; rotting of the plant material would have produced heat that in turn kept the nest somewhat warmer than the surrounding environment. Moreover, with a number of sizeable predators sharing the same habitat, it seems unlikely that adults would simply have abandoned the nests, thereby providing a rich and abundant food resource for local carnivores. Rather, it has been suggested that the adults (males? females? both?) likely engaged in nest-guarding behavior in an attempt to protect the developing embryos. With potentially hundreds of nests in a single nesting horizon, those numbers translate into many mouths to feed. Imagine for a moment one of these vast sauropod breeding colonies; the commotion, the din, the smell, and the sight of so many animals would have been truly remarkable to behold.

Even more extensive parental care has been postulated for duck-billed dinosaurs, or hadrosaurs. Many duck-billed dinosaurs also appear to have formed breeding colonies and used the same nesting sites year after year, perhaps also utilizing rotting vegetation as a means of incubating the eggs. In addition, Jack Horner has argued that some species cared for their young for a time following hatching. Evidence for extended parental care includes fragmented egg shells within the nests, suggesting that the young remained nest bound for a period after hatching. The bones of hadrosaur hatchlings also tend to have poorly formed ends, indicating minimal walking abilities. It's likely that the young were helpless for weeks or months after hatching and fully dependent on one or both parents. In contrast, hatchling skeletons of smaller ornithopods like hypsilophodonts are better ossified, suggesting that freshly hatched young were able to leave the nest. Hadrosaur adults may have supplied food to their young through regurgitation, as many birds do. Large hadrosaur bonebeds provide further evidence, though circumstantial, that some species traveled in large herds that included animals ranging from juvenile through adult. If so, young may have received protection from adults long after departing the nest.

Some maniraptor theropods brooded their nests like modern birds. Specimens of both oviraptors and troodonts indicate that an adult (likely the mother, father, or both—we can't be certain) sat on top of the nest to incubate the eggs with its own body heat. This behavior was facilitated by feathers, which acted as insulators to retain heat within the nest. Skeptics have suggested that the adults were not brooding but rather simply died atop the eggs. But this scenario is highly unlikely, because fully one-quarter of all known oviraptor skeletons are found in the stereotypical nest-brooding posture, with the limbs arranged symmetrically on either side and spread out so as to cover the eggs. Although many crocodiles engage in some form of parental behavior, both pre- and posthatching, they do not brood their nests in this characteristic pose. Conversely, many modern birds use the same brooding posture, indicating that some of the reproductive behaviors we associate with avians evolved in a prebird theropod ancestor.

My final example represents one of the latest and most exciting discoveries in dinosaur parental care. In 2007, David Varricchio (Montana State University) and coauthors presented amazing evidence of the first known burrowing dinosaur, a small hypsilophodont-like ornithopod they called *Oryctodromeus*, which means "digging runner." *Oryctodromeus*, recovered from rocks of Late Cretaceous age in southwest Montana, possesses several bony features of the snout, shoulder, and PELVIS associated with diggers. Remains of an adult and two juveniles were found at the expanded end of a burrow, providing perhaps the best evidence to date of extended parental care in dinosaurs. Varricchio proposed that burrowing was likely an important parenting strategy for this animal, further noting that burrows may have allowed some dinosaurs to inhabit environments with seasonal extremes, such as deserts, alpine regions, and polar latitudes.

Although most aspects of dinosaur lives remain shadowy, a picture of dinosaur reproductive behavior is slowly emerging. It's now clear that numerous groups devoted considerable energies to elaborate mating signals and, by correlation, the competition for mates. All dinosaurs were egg layers, allowing every generation to produce large numbers of young. This strategy allowed population numbers to recover quickly when decimated by unexpected crises. Yet many, perhaps most, dinosaurs also engaged in some form of parental care, helping ensure that the young survived to adulthood. This nurturing apparently took a variety of forms. Prehatching care likely included nest guarding (some sauropods) and brooding (some maniraptoran theropods), whereas posthatching attention included feeding and protection of nest- (and burrow-) bound hatchlings (some ornithopods) and protection of juveniles and subadults within social groupings (hadrosaurs and ceratopsids). Although we have only begun to unravel this aspect of the dinosaur odyssey, it's becoming clear that advanced reproductive behaviors were a critical facet of their lives.

The topic of parental care, particularly among feathered dinosaurs, brings us to one of the most hotly debated topics in dinosaur paleontology—physiology.

A feathered maniraptor theropod, *Oviraptor*, broods a nest of eggs in the windblown sands of a Late Cretaceous Asian desert.

11

THE GOLDILOCKS
HYPOTHESIS

DINOSAUR PHYSIOLOGY IS ONE of the great paleontological mysteries. Were dinosaurs warm-blooded, like birds and mammals, or cold-blooded, like lizards, crocodiles and most other animals? Among vertebrates, these divergent strategies are often associated with active or sluggish lifestyles, respectively. The recognition of birds as direct descendants of small, carnivorous dinosaurs makes this enigmatic question all the more intriguing. Did the warm-blooded metabolisms of birds arise after the split with dinosaurs, or was it inherited from a common ancestor from within the dinosaurs or perhaps some other still more primitive group? Finding a satisfactory solution to this metabolic mystery is essential in our endeavor to understand the dinosaur odyssey, because the answer will, to a great extent, determine how we conceive of dinosaurs.

As long as we're ranking dinosaur mysteries, gigantism must also fall somewhere near the top of the list. Dinosaurs pushed—and, in the case of sauropods, exploded—the size limits of land-dwelling animals. How and why did so many groups of dinosaurs become giants, in many cases far exceeding the sizes of all other land-living animals before or since? Supersizing brings with it a host of physiological and ecological implications, not least of which is the need to consume vast amounts of generally poor-quality food. What was the evolutionary pathway that enabled this repetitive birthing of giants during the Mesozoic?

Over the past few decades, great strides have been made in addressing both dinosaur physiology and gigantism, yet much discord persists. I think that we are close to resolving both of these mysteries and that the two may, at least in part, have a single solution.

This chapter explores dinosaur growth and metabolism in an attempt to present the key issues and outline a possible resolution.

Before delving further into dinosaur physiology, we need to review a few metabolic fundamentals. METABOLISM refers to the cumulative chemical activity of cells necessary for maintaining life. In all organisms, energy is obtained from the breakdown of a remarkable molecule called ATP (adenosine triphosphate), the universal fuel of life. ATP is generated and stored through the process of cellular respiration, involving the breakdown of carbohydrates in a series of chemical reactions that consume oxygen and generate heat. As the energy output of an organism increases, the need for ATP increases apace, leading to further oxygen consumption and heat production. That's why breathing rate and heart rate both increase during exercise. The chemical processes of metabolism are speeded up at higher internal body temperatures, so increased metabolic rates, in addition to consuming more oxygen, enable increased activity levels. Eventually, however, organisms reach a limit beyond which they can no longer generate ATP fuel. Once this point is exceeded, lactic acid begins to form, severely restricting an organism's ability to remain active. This threshold occurs at the boundary separating aerobic from anaerobic activity—that is, the point at which oxygen reserves are fully depleted. So, for example, whereas long-distance runners work almost entirely within the aerobic range of activity (with muscles receiving a constant supply of oxygen), sprinters *exceed* this limit and are in the anaerobic range (muscles lacking oxygen) when they finish a race.

Unfortunately, "warm-blooded" and "cold-blooded" are imprecise and inadequate terms. We must replace them with four words that better serve our needs. A key aspect of metabolism is temperature regulation, something practiced by all organisms. Those that regulate body temperature through internal heat sources are called "endothermic" (*endo* = "inside"; *thermic* = "heat"), and those that do so via external sources are referred to as "ectothermic" (*ecto* = "outside"). Conversely, organisms that maintain a relatively constant internal body temperature are called "homeotherms" (*homeo* = "same"), whereas those that undergo considerable fluctuations in their internal temperatures are dubbed "poikilotherms" (*poikilo* = "varying"). So, whereas ENDOTHERMY and ECTOTHERMY refer to how body temperature is regulated (inside or outside the body), HOMEOTHERMY and POIKILOTHERMY pertain to whether body temperature is held relatively constant. So far so good.

Most mammals, including us, are endothermic homeotherms. That is, we closely regulate our body temperature internally through high metabolic rates coupled with a precise balance of heat production and heat loss. If your core body temperature fluctuates by more than a few degrees it means that you're sick. Conversely, lizards and snakes are classic examples of ectothermic poikilotherms. Their lower metabolic rates mean that they are dependent on external heat sources, typically the sun, to regulate internal temperatures. However, many of these animals are able to be functionally homeothermic—that is, they can maintain constant internal body temperatures for extended periods through various kinds of behaviors. For example, reptiles move in and out of the sun and become active as a means of generating heat internally.

All would be simple if life-forms could be subdivided into these two tidy physiological camps: endothermic homeothermy (internally generated constant body temperature) and ectothermic poikilothermy (externally driven variable body temperature). Nature, however, is much more messy and interesting, with a diverse mixture of metabolic strategies. For example, bats are endothermic poikilotherms; they regulate their temperature internally but vary their body temperature considerably. When stationary, these flying mammals are capable of slowing their metabolic rate and entering a state of torpor with a much lower body temperature. Bears do the same seasonally, dropping their internal temperature and metabolic rate (and thus their food requirements) during hibernation. It turns out that most birds are also endothermic poikilotherms, with body temperatures that fluctuate in accordance with different activities. Moreover, mammals and birds have not cornered the market on endothermy. Despite being members of the "cold-blooded" fishes, tuna are able to raise their internal body temperatures through a system of blood flow that shunts heat from warm blood back to the body core, thereby minimizing heat loss to the water. A variety of insects—particularly among flying forms like dragonflies, bees, wasps, butterflies, and moths—are functionally endothermic as well.

On the flip side are the ectothermic homeotherms, "cold-blooded" animals capable of maintaining relatively constant internal temperatures. Galápagos tortoises, for example, weigh about 200 kilograms (440 pounds) and can maintain body temperatures several degrees higher than ambient even on chilly nights. This metabolic feat is made possible by a simple relationship relating to volume, surface area, and heat flow. As a sphere increases in size, its volume (the space on the inside) increases faster than the area of its surface. More specifically, the surface area increases with the square of the radius, whereas the volume increases with the cube of the radius. To give a concrete example, a beach ball has a smaller outer surface area relative to its internal volume than, say, a Ping-Pong ball, which has a relatively large surface area and small volume. Although animals are not spherical, the same relationship applies: small animals have relatively more surface area and less volume than do large animals, and vice versa. Heat loss is directly proportional to surface area: more area, faster heat loss; less area, slower heat loss. As intuitive proof, think about how we curl up into a ball to conserve heat when we're cold. Galápagos tortoises, being giant and compact animals, tend to hold onto body heat because of their great volume and relatively low surface area. The tendency to maintain internal temperatures by virtue of size rather than greater metabolic rate is sometimes referred to as GIGANTOTHERMY.

The surface area–volume relationship has dire consequences for endotherms, particularly at either extreme of the body size spectrum. At the small end, mice and hamsters have a large surface area for their size and consequently tend to lose heat very fast. To compensate, they require an insulating fur coat and tend to shiver to generate additional body heat. Elongate, tubular body shapes that dramatically increase surface area are relatively common in ectotherms (e.g., many lizards and snakes). Yet they are extremely rare (and energetically costly) among birds and mammals because of the tremendous increase in heat loss. At the giant end of the spectrum, animals such as elephants have relatively greater volume and less area devoted to surface, so their problem is not the retention of body heat but the dumping of it. It is for this reason that elephants are prone to overheating. Their large ears, long trunks, and wrinkly skin all help to increase surface area and dissipate heat. They also tend to be most active during the cooler nighttime hours and take frequent baths during the day. For obvious reasons, modern elephants tend to have a minimal coating of hair (although their ice age relatives like mammoths and mastodons grew thick, hairy coats as a means of coping with the frigid Pleistocene temperatures).

Because Mesozoic animals could not have avoided the same laws of physics that affect modern animals, the surface area–volume relationship has obvious implications for dinosaurs. If dinosaurs possessed mammal- or birdlike endothermic metabolisms, their high volume/low surface area bodies would have faced the same heat retention problems experienced by elephants. Shedding excess body heat would have been a particularly acute issue for the largest dinosaurs. It's been suggested that the long necks and tails of sauropods may have functioned in part to increase surface area and radiate body heat. Many investigators have argued that giant dinosaurs could not have been endotherms, that instead it makes more sense to regard them as gigantotherms, maintaining steady internal body temperatures (and thus higher activity levels) by virtue of their huge sizes rather than heightened metabolic rates.

As mammals, we tend to think of endothermy as superior to its cold-blooded counterpart. The benefits of endothermy include a greater capacity for sustained activity, potentially a major advantage when hunting prey or escaping capture.[1] Endotherms can also be fully active at night, and many have adapted to life in extremely cold climates. Thus, we find polar bears rather than crocodiles in the high arctic. In short, endotherms are more independent of their environments than are ectotherms. Yet this behavioral flexibility comes at great cost. Comparing similarly sized animals, endotherms typically require ten to thirty times the energy of endotherms. A Komodo dragon must eat its own body weight in prey every 90 days, whereas a lion of the same size must do so every nine days. If ectotherms are the fuel-efficient hybrids of the animal world, endotherms are the gas-guzzling SUVs. Birds and mammals effectively lack an idle setting; with the engine constantly revving, they (and we) must fuel up more often.

Making matters worse is the fact that endothermy becomes increasingly costly at small body sizes. The relative increase in surface area at smaller sizes translates into increased heat loss and thus an inability to regulate body temperature internally. It is largely for this reason that the smallest birds and mammals weigh about 3 grams (0.1 ounce), whereas many ectotherms are about one-tenth of this size (0.3 gram; 0.01 ounce). Birds and mammals attempt to thwart this heat loss with feathers and fur, respectively, which cover the body in a layer of insulation. Increasing the thickness of this insulation layer by raising feathers or fur is an efficient means of trapping even more heat.[2]

The extremely high costs of endothermy should caution us from regarding ectothermy as some sort of evolutionary backwater. Ectothermic animals have evolved several solutions to compensate for their lack of endurance. Although they can't sustain high levels of energy output, many are capable of short bursts of activity lasting several minutes. This strategy, more like that of a sprinter than an endurance runner, is very useful in a range of behaviors, including catching prey, guarding nests, and avoiding predators. Ectotherms have also adapted to extreme climates, inhabiting blistering hot deserts and high altitudes where temperatures drop below freezing every night. Ectothermy, then, is a highly successful strategy. In the end, ectothermy and endothermy are best regarded as fundamentally different approaches to life, each with distinct advantages and disadvantages. Whereas ectothermy is the low-cost, economical option, concentrating energies into short bursts of activity, endothermy is the relatively rare, high-cost alternative that enables sustained activity but requires much more energy intake. Of the two, ectothermy is the primitive condition for vertebrates and has remained the predominant physiologic strategy of animals for more than 500 million years.

So what about the physiology of dinosaurs? Evolutionary evidence from the closest living relatives of dinosaurs is ambiguous; crocodiles are "cold-blooded" ectotherms, whereas birds are card-carrying endotherms. Because we can't measure the body temperatures or metabolic rates of extinct animals directly, paleontologists search for circumstantial evidence. John Ostrom ignited the dinosaur renaissance in the late 1960s when he described *Deinonychus*, the first known "raptor" dinosaur and argued that many aspects of its anatomy indicated a highly active, endothermic metabolism. In the intervening years, numerous lines of evidence have been cited for and against the notion of endothermic dinosaurs. Investigators in favor of the dinosaur endothermy hypothesis have sought key anatomical features shared by dinosaurs, birds, and mammals yet absent in other vertebrates. In contrast, those advocating for ectothermic dinosaurs have attempted to identify features that link birds and mammals to the exclusion of dinosaurs.

The hot-blooded camp has marshaled numerous, often ingenious, lines of evidence to bolster the idea of endothermic dinosaurs. Examples include relative brain size, predator-prey ratios, and high-latitude dinosaurs. Based on measurements of the large bony cavity in the skull, the brains of dinosaurs, especially among the small maniraptor theropods, approached bird and mammal proportions. Yet relative brain size in dinosaurs was extremely variable, and some researchers question the accuracy of such estimates. Similarly, censuses of fossils collected from particular areas suggest that dinosaur herbivores were approximately ten times more common than carnivores, a result that closely resembles predator-prey ratios seen in living endotherms (mammals and birds) and stands in contrast to assemblages of ectotherms, which tend to be closer to 1:1. However, such censuses are prone to many biases, and we don't even know who ate whom, raising major concerns about the utility of such inferences. Finally, the occurrence of dinosaurs in both northern and southern polar regions is cited as evidence of endothermic metabolisms because the latter would be necessary to survive frigid temperatures (the aforementioned polar bear–crocodile contrast). But recall that the Mesozoic was a hothouse world lacking polar ice caps, and, at least in some areas, dinosaurs may have been able to migrate to lower latitudes during the cold, dark winter periods.

Other arguments put forth in support of endothermic dinosaurs relate to stance and internal bone structure. Dinosaurs held their limbs upright, as birds and mammals do, whereas ectothermic vertebrates such as lizards tend to be sprawled. Yet even if upright posture is a prerequisite for endothermy—perhaps associated with higher speeds and improved lung ventilation—it's always possible that the posture came first and the elevated metabolic rates much later within a subset of the descendants. Similarly, in contrast to most ectotherms, the bones of birds and mammals tend to be fast growing and riddled with blood vessels. When thin sections of these blood-rich bones are viewed in cross section under a microscope, the numerous blood vessel channels result in a repetitive bull's-eye pattern known as HAVERSIAN CANALS. Haversian canals, typically associated with mammals and birds, are essentially narrow tubes that run parallel to the long axis of the bone carrying both nerves and blood vessels. Another kind of bone associated with both mammals and rapid growth is called FIBROLAMELLAR BONE, which contrasts with the slower-growing variety, LAMELLAR-ZONAL BONE, in which distinct growth rings are typically present. However, occasionally we find living ectotherms, including certain crocodiles and turtles, with Haversian canals and/or fast-growing fibrolamellar bone, whereas these bone types are absent in some small-mammal endotherms. To muddy the waters further, many dinosaurs show distinct growth rings, generally associated with slower-growing ectotherms. So the correlation between bone type and metabolism is far from perfect. In response, investigators have refined their hypotheses to account for body size and assess the density of Haversian canals and fibrolamellar-type bone rather than merely their presence or absence. This debate is closely related to the issue of growth rates, a topic I will return to shortly.

Equivalent kinds of claims have been made by those who contend that dinosaurs, particularly the giants among them, could not have been endotherms but rather must have had much lower, lizardlike metabolic rates. Arguments from the ectothermy camp have been equally innovative, ranging in scope from the physics of heat flow to ecology on the scale of continents. In terms of physics, it's difficult to imagine how dinosaurian giants could have possessed metabolic rates on a par with birds and mammals, because their relatively high-volume/low-surface-area bodies would have been prone to overheating. Instead, many paleontologists claim that dinosaurs, or at least the giants among them, are best regarded as gigantotherms, enjoying the benefits of elevated and constant body temperatures simply by virtue of their body mass, but without the added strain of supporting endothermic-grade metabolic rates.

In a similar vein, the high-cost lifestyle embodied by endothermy depends on prodigious amounts of fuel. Larger animals must consume more food than smaller animals, so giant endotherms face a double whammy, forced to take in massive amounts of energy in order to sustain both higher metabolic rates and more substantive bulks. Elephants spend up to 18 hours per day feeding, their herds moving across the landscape like gangs of oversized Hoovers. A single adult elephant weighing between 4,000 and 7,000 kilograms

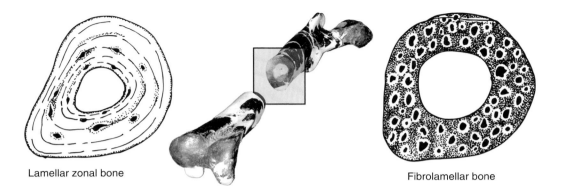

Lamellar zonal bone

Fibrolamellar bone

FIGURE 11.1

Variation in bone microstructure. Idealized section at left depicts lamellar zonal bone, typically associated with slow bone growth and characterized by the presence of growth rings and minimal evidence of blood supply within the bone. Section at right shows fibrolamellar bone, typically associated with rapid bone growth and characterized by a rich blood supply and minimal evidence of growth rings. The *Allosaurus* limb bone (femur) in the center shows a typical cross section where slices are typically taken for histological study.

(8,800 and 15,500 pounds) must consume about 150 kilograms (330 pounds) of food (about 6–8 percent of its body weight) every day, and requires on the order of 8,000 acres (32 square kilometers, or 12.5 square miles) of productive land. Of course, many kinds of dinosaurs were larger than elephants, with the biggest sauropods about ten times more massive. Trackway and bonebed evidence suggests that, like elephants, many dinosaurs traveled in large groups, exacerbating the problem of locating sufficient food.

The body size problem is still more acute for giant carnivores because they occupy the highest link in the food chain, the peak of the trophic pyramid. Recall from chapters 5 and 8 that, as energy flows from plants to herbivores, and then from herbivores to carnivores, the bulk is lost to heat. As a result, top carnivores are able to access only a tiny fraction of the total energy initially captured by plants. This phenomenon explains why the largest mammalian carnivores within any given ecosystem tend to be smaller and far fewer in number than the biggest plant eaters. Lions, for example, must exist at much lower population densities than wildebeest, requiring continent-sized geographic ranges to maintain sustainable population numbers.[3] Now consider *Tyrannosaurus rex*. At a whopping 5,000 kilograms (11,000 pounds), one of these giants would have equaled the mass of about twenty-five African lions. If we assume that continent-scale ranges are necessary to support a single species of lion in the long term, clearly we cannot extrapolate to the conclusion that *Tyrannosaurus* required an area equivalent to twenty-five continents, because such a supercontinent would have exceeded even the size of Pangaea. So what gives? Possible explanations include lower (nonendothermic) metabolic rates, a greater supply of plants (supporting more herbivores and carnivores), or a combination of the two.

Given all of these arguments, pro and con, what can we conclude from this academic feud between the cold-blooded Hatfields and warm-blooded McCoys? Some claim that this debate has generated far more heat than light. I disagree, and think that the two camps are not so widely separated. Endothermy and ectothermy do not represent a simple dichotomy but rather end points on a metabolic spectrum, with a broad range of possible intermediate strategies. Instead of construing the evidence as being in favor either of high or low metabolic rates, a middle ground alternative is called for.

Although the data supporting elevated metabolic rates in dinosaurs is currently insufficient to demonstrate that they were high-cost, bird- or mammal-like endotherms, this evidence does provide strong support for the notion that dinosaurian metabolic rates were elevated well above those of lizards. Similarly, the dinosaurs-as-ectotherms advocates cannot provide convincing evidence of lower, lizardlike metabolic rates, yet they do make a firm case that most dinosaurs did not possess high-cost avian-/mammalian-style physiologies. Present evidence points instead to dinosaurs possessing metabolic rates intermediate between those of lizards and birds. This conclusion, argued previously by several authors such as R. E. H. Reid (Queen's University of Belfast), is a striking one. With a few exceptions like those noted above, animals alive today can be subdivided into endotherms (birds and mammals) and ectotherms (everything else). Yet an intermediate, Goldilocks metabolism—"not too hot, not too cold, but just right"—may have led dinosaurs (or at least the bulk of them) into a physiological "no man's land" of sorts, enabling them to explore intermediate physiologies with great success. Let's pursue this avenue further and see whether it helps break the logjam separating the two academic camps.

Science proceeds by identifying patterns—combinations of features that show a particular arrangement—and comparing them with other patterns. Patterns, then, are the clues used to solve scientific mysteries. If a newly observed pattern is most consistent with a particular hypothesis, that hypothesis receives support. If a well-supported pattern runs counter to the predictions of a hypothesis, the latter must be modified or discarded. Over the past several years, a number of important patterns have been established for dinosaurs, several of which pertain to the metabolic debate.

The first set of patterns relates to body size and the issue of gigantism. If we define the term *giant* as all species exceeding 1,000 kilograms (1 metric ton, or 2,200 pounds) in mass, then we can say that gigantism evolved numerous times within dinosaurs and indeed occurred in all major groups. Within these giant forms, "ultragiants" exceeding 3,000 kilograms (6,600 pounds) evolved independently at least eight times. So it's fair to say that dinosaurs frequently trended toward giant sizes. Focusing on the meat-eating theropods, we find giants among all major subgroups but one: the maniraptors, the diverse clan that gave rise to *Velociraptor*, *Oviraptor*, and birds. (A couple of possible exceptions to the previous statement have been announced in recent years, and I will address these shortly.) Even if we expand the scope to include all birds that have lived in the 65.5 million years since the major extinction of the dinosaurs, no known species has approached body masses of 1,000 kilograms (2,200 pounds). Interestingly, however, small body sizes—say, less than 5 kilograms (11 pounds)—apparently did not occur in dinosaurs, once again with the exception of maniraptor theropods. The smallest known example is the aforementioned *Microraptor*, at about 1 kilogram (2 pounds). So, for reasons that have remained unclear, the rogue maniraptors carved their own path, evolving plenty of little dinosaurs but few giants.

Additional patterns relate to coelurosaurs, the broader theropod group that includes ornithomimosaurs, tyrannosaurs, and maniraptors. Based on several indicators such as teeth, it appears that all theropods were devoted meat eaters except for a handful of oddities among coelurosaurs. Tyrannosaurs are the definitive prehistoric carnivores, but, as described in chapter 8, the closely related ornithomimosaurs may have been herbivorous or at least omnivorous. And among the maniraptors, oviraptorosaurs and therizinosaurs are now commonly thought either to have been omnivorous or to have included a combination of omnivores and herbivores. Moreover, feathers and their precursors (so-called protofeathers) are known only in coelurosaurs. Current wisdom is that all coelurosaurs either possessed feathers or evolved from feathered ancestors. Feather-like structures have even been documented in an obviously nonflying, small-bodied tyrannosaur found in China. If protofeathers initially functioned as insulating structures to conserve body heat, the implication is that these animals possessed metabolic rates elevated into the high-cost, endothermic range.

Finally, a wealth of exciting patterns has recently emerged from studies of dinosaur growth, once again based on study of bone microstructure in dinosaurs. In order for us to understand growth in any species, information on life span is a primary requirement. There are a couple of ways to estimate longevity from fossils. The best-established technique involves counting annual growth lines. Like trees, bones tend to grow at uneven rates over the course of a year, adding a lot of tissue in the good times and slowing down in the tough times. Tree rings, deposited as concentric circles radiating out from the tree's core, are familiar examples of annual growth fluctuations. The bones of many animals, including most dinosaurs, exhibit a parallel phenomenon, showing lines of arrested growth (LAGs) in their cross sections that represent intervals of minimal bone deposition. Numerous studies show that animals, like trees, typically add one LAG per year. So a count of these lines can reveal the age at death of the organism. If this method is applied to many individuals, it's even possible to estimate the average life span of animals within a given species.

A new generation of paleontologists, including Greg Erickson of Florida State University, has applied the LAG count method to many different kinds of dinosaurs; as a result, these researchers have been able to tackle a whole host of previously intractable questions. The most obvious of these is, How long did dinosaurs live? Prior to the application of quantitative techniques like counts of growth lines, it was suggested that the largest dinosaurs may have topped well over a century in age. After all, how else could they have achieved such gargantuan proportions? It turns out, however, that the largest sauropods lived on the order of 50 years. Giant theropods lived no more than about

30 years, whereas life spans for small to midsized dinosaurs were 7–15 years and those of the tiniest dinosaurs played out in a mere 3–4 years.

The next question to consider is, How fast did they grow? Were dinosaurian growth rates relatively slow, like those of lizards, crocodiles, and other reptiles, or more rapid, as in birds and mammals? The intuitive answer is that giant sizes plus brief life spans equals fast growth rates. But paleontologists have gone much further than this common-sense inference, using a simple but elegant method to discern many aspects of dinosaur life histories. Imagine you have access to a sample of, say, twenty-five dinosaur specimens of varying ages from a single species. For each specimen, you estimate two key parameters. The first is age at death, accomplished by counting annually deposited LAGs. The second estimate is total body mass, assessed using a method based on measurements of the thigh bone. Next, make a simple graph that plots the estimates of longevity against body size for every individual. The end result is an age-versus-size growth curve. This straightforward tool enables you to see how large individuals were at a particular age and thereby estimate rates of growth. Additionally, rather than simply calculating an average growth rate for a given species, these plots can be used to assess rates of growth at particular periods during a dinosaur lifetime.

Erickson and his colleagues applied this technique to assess growth in a diverse range of dinosaurs. Mammals and birds have characteristic, three-step growth trajectories. Growth starts out relatively slow, shifts into high gear for some period, and then

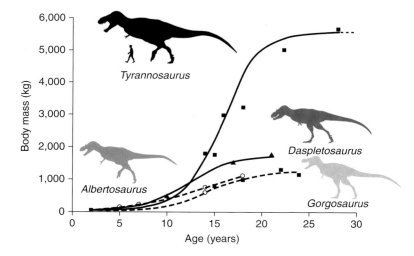

FIGURE 11.2

Growth patterns in four North American tyrannosaurs—*Gorgosaurus, Albertosaurus, Daspletosaurus,* and *Tyrannosaurus*—estimated by comparing body mass with age. All four animals show a similar, three-step S-curve pattern in which growth is slow in young animals, undergoes a spurt in juveniles, and then slows again in adults. The largest example, *Tyrannosaurus rex,* evolved by increasing the rate of growth in the fast-growing juvenile phase. Based on the work of Greg Erickson and colleagues. Human added for scale. See text for discussion.

tapers off or ceases for the remainder of life. When viewed in an age-versus-size growth curve, the life span of an individual appears roughly S-shaped. Erickson and others have found that dinosaurs grew much the same way. *Tyrannosaurus rex*, for example, experienced a remarkable growth spurt that started at about age 10 and peaked between about 15 to 19 years of age. Interestingly, the larger the dinosaur, the faster it grew during the accelerated juvenile phase. The smallest dinosaurs, weighing less than 1 kilogram (about 2 pounds), added about 0.3 gram (0.01 ounce) of body mass per day during this interval, which may seem slow but represents twice the rate seen in living reptiles of equivalent size. Somewhat larger dinosaurs, between 1 and 20 kilograms (2 and 44 pounds), added about 1–20 grams (0.04–0.7 ounce) daily, a rate similar to that of marsupial mammals. Midsized dinosaurs, with adults weighing between 100 and 1,000 kilograms (220 and 2,200 pounds), had correspondingly faster growth, packing on between 100 and 800 grams (3.5 ounces to 2 pounds) per day. During four ungainly years of maximal growth, *T. rex*, with an adult mass of 5,000 kilograms (11,000 pounds), added about 2 kilograms (4 pounds) of body mass per day, a rate on a par with living birds and mammals. Gigantic sauropods like *Apatosaurus*, with an adult mass of about 25,000 kilograms (55,000 pounds), tipped the scales during their growth spurt by adding as much as 20 kilograms (44 pounds) per day! The astounding growth rate of sauropods is well within the range of living whales, the fastest-growing animals known. It's safe to conclude, then, that dinosaurs grew very fast indeed and in this respect were more like endotherms than ectotherms.

Erickson and his colleagues have taken the process one step further, comparing the growth curves of closely related dinosaur species in an attempt to unravel the evolution of giants. Their first major study of this type compared growth curves in tyrannosaurs, including *Tyrannosaurus* and several smaller-bodied (1,000–2,000 kilograms, or 2,200–4,400 pounds) tyrannosaurs such as *Daspletosaurus* and *Albertosaurus* that lived earlier in the Cretaceous. The smaller tyrannosaurs showed the same stages of growth as their Godzilla-like cousin, but with much lower growth rates (less than 0.5 kilogram, or about 1 pound, per day) during the period of maximal size increase. This finding led the team to conclude that *Tyrannosaurus rex* evolved to such gargantuan proportions through a dramatic evolutionary increase in growth rate during the juvenile phase. In contrast, other studies have found that at least three extinct reptiles (two crocodiles and one lizard) achieved gigantic body sizes not by increasing their growth rate but by delaying maturity and thereby prolonging the period of maximal size increase.

Keeping in mind the diverse range of patterns described so far, next I outline an alternative model that I've been developing with Jim Farlow of Indiana University–Purdue University. I refer to this model informally as the GOLDILOCKS HYPOTHESIS because it suggests that the metabolisms of most dinosaurs were intermediate between those of ectotherms and endotherms (not too hot, not to cold, but just right). This idea examines physiology and growth through the lens of energy flow, and it argues that evolution endowed dinosaurs with a unique metabolic strategy, one that enabled them to achieve gigantic body sizes and likely contributed fundamentally to their overall success.

The energy budget of an animal—that is, the sum total of ATP generated by metabolic processes—can be subdivided into three primary activities, which we can think of metaphorically as bins. First, a portion of this energy is lost to waste products—urine and feces. Second, a portion goes toward body MAINTENANCE, including such diverse activities as renewing cells, generating heat, and finding the next meal. The third and final portion of energy is devoted to PRODUCTION—that is, building or adding something new. Production activities include growth, fat storage, and reproduction. Of these three energy bins—waste, maintenance, and production—waste can be taken as a relative constant and need not concern us further. That leaves the energy available for maintenance and production.

It turns out the ectotherms and endotherms differ greatly in the relative sizes of their maintenance and production energy bins. Recall that ectothermy can be thought of as an efficient, low-cost lifestyle and endothermy as its opposite. Studies of living animals have shown that ectothermic reptiles and amphibians are able to devote about 40 percent of their assimilated energy supply to production (growth, fat storage, and reproduction). Conversely, because of the exceptionally high maintenance costs of endothermy, birds and mammals typically allocate a meager 3 percent of available energy to this bin. Constant, internally driven body temperatures and high activity levels require endotherms to burn large amounts of energy at low efficiency, with the bulk of it lost as body heat.

Modern biological studies suggest further that an increase in metabolic rate among ectotherms results in a corresponding increase in maintenance but, surprisingly, also an increase in production.[4] That is, although a lineage of ectothermic reptiles undergoing an evolutionary increase in metabolic rate would have to devote more energy to maintenance (energy devoted to such activities as feeding and making body heat), there would be an overall increase in the total available energy as well, allowing more of it to be devoted to growth, reproduction, and fat storage. On the basis of preliminary calculations, Farlow and I hypothesize that this trend would continue until a tipping point was reached, one at which the cost of maintenance became so high that the amount of energy available for production would drop precipitously. After surpassing this tipping point, greater and greater proportions of the energy budget would be channeled into the generation of body heat. The animals that completed this transition would be fully endothermic, with a huge bin for maintenance energy and a relatively small bin for production.

The core idea of the Goldilocks hypothesis is that most dinosaurs possessed metabolic rates intermediate between those of low-cost ectotherms and high-cost endotherms. Let's refer to them informally as "mesotherms" (*meso* = "middle"). If so, it's likely that the earliest dinosaurs inherited somewhat elevated metabolic rates from some as-yet-unidentified nondinosaur ancestor among archosaurs. Further increases in metabolic rate would have heightened the cost of maintenance. Yet, as discussed in the previous paragraph, faster metabolisms would also have given these dinosaurs access to a larger total pool of energy, a substantial portion of which could have been directed toward the building

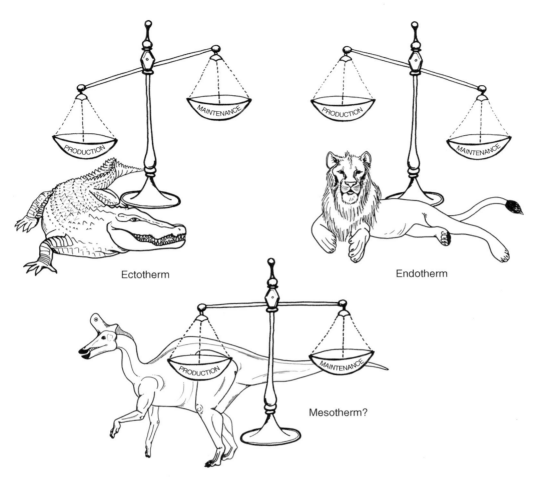

FIGURE 11.3

Variation in the energy budgets of low-cost ectothermic reptiles, high-cost endothermic mammals, and possibly intermediate, "mesothermic" dinosaurs. The scales represent the relative amounts of energy devoted to production (growth, reproduction, and fat storage) and maintenance (sustaining bodily functions). See text for discussion.

activities of production. In other words, bumps in metabolic rate would have increased the pool of energy for activities such as reproduction and growth. Among other things, the greater energy available for reproduction could have been used for parental care. Alternatively, the bounty of reproductive energy may help explain the recurrent evolution of elaborate and bizarre structures in dinosaurs such as horns, crests, plates, and spikes. Perhaps more important, a greater supply of energy in the production bin could have fueled rapid growth rates, likely a necessity for animals attaining giant sizes in a world full of carnivorous monsters. Harnessed over geologic time, with progressive increases carried through numerous speciation events, this extra growth potential may well have been used to generate progressively larger descendants, ultimately resulting in the evolution of giants.

This conclusion is supported by the finding that at least some giant dinosaurs achieved such humongous proportions through dramatic increases in growth rates.

The Goldilocks hypothesis addresses the *how* of dinosaur gigantism. The evolutionary and ecological reasons behind dinosaur gigantism (the *why* of giants) remain unclear. Perhaps bigger bodies opened up new food niches, allowing herbivores to eat more readily available, poorer-quality fodder. If so, the carnivores would have been forced to follow suit, increasing in size alongside their prey. Alternatively, in some instances larger bodies may have offered an advantage in the competition for mates. A further possibility is that the increased body heat associated with giant sizes (resulting from the reduction in relative surface area) offered a cheap means of maintaining elevated internal temperatures, enabling dinosaurs to remain active for a greater proportion of the day. Whatever the impetus (and there may well have been more than one, as these explanations are not mutually exclusive), the recurrent evolution of gigantism among dinosaurs indicates that larger body sizes frequently conferred a strong selective advantage.

Another way of considering the problem of dinosaurian gigantism is that neither ectothermy nor endothermy is conducive to gargantuan body sizes for land-dwelling animals. On the one hand, ectotherms are highly efficient, with about half of their assimilated energy available for production (including growth), yet the total pool of available energy is much less than for endotherms. On the other hand, endotherms are able to access a much larger supply of energy, but their inefficient metabolisms demand that the bulk of this fuel be devoted to maintenance, with little left over for the evolution of giants. The middle road of MESOTHERMY may have offered the best of both strategies—relatively low cost and high efficiency combined with an expanded pool of energy for growth and other production needs. The issue of heat dissipation at enormous sizes would have been ameliorated to a great extent by this metabolic strategy, which would have minimized the costly generation of heat while maximizing production and growth rates. Similarly, the food requirements of these putative mesotherms would be substantially reduced relative to those of high-cost, avian- or mammalian-grade endotherms, allowing dinosaurs to maintain smaller home ranges and thus higher population densities.

As outlined here, mesothermy is not merely the otherwise undefined middle ground between ecto- and endothermy. Rather, it represents a distinct metabolic strategy, distinguished by the relative amount of energy available for production. As far as I know, no mesotherms as defined here are alive today. Yet ancestral forms of both birds and mammals must have passed through intermediate stages on the way to acquiring their high-cost metabolisms. Ectotherms generate less energy for production because their overall energy budget is smaller. Endotherms have access to a larger pool of energy, but their jacked-up metabolic rates demand that most of this fuel be allocated to maintenance. Mesotherms, in contrast, would have access to a relatively large store of energy (engendered by heightened metabolic rates) and, thanks to a relatively low-cost lifestyle, be able to devote much of it to production activities. Perhaps mesothermic dinosaurs regularly took advantage of this extra production energy to fuel the evolution of gigantic sizes.

If so, the uniqueness of dinosaur gigantism among terrestrial animals may be due at least in part to the fact that no other group of mesotherms has diversified in the history of life. The corollary is that metabolic evolution within dinosaurs may have occurred at least in part as a response to strong selection pressures for larger body sizes.

The Goldilocks hypothesis suggests further that one group of dinosaurs, coelurosaur theropods, evolved still-higher metabolic rates, entering into the fully endothermic range. Supporting evidence comes from several sources, but most telling of all is the presence of feathers. Serving as efficient insulators, feathers created the potential to retain a substantial portion of the body heat otherwise radiated to the outside world. (Feather comforters are an excellent example of this capacity.) Once captured, this additional body heat could be put to use maintaining a stable body temperature. A less conclusive but nevertheless intriguing line of evidence comes from the observation that only a handful of theropod dinosaurs appear to have deviated from carnivory, and all of these exceptions (ornithomimosaurs, oviraptorosaurs, and therizinosaurs) are to be found among coelurosaurs. What could possibly induce a devoted carnivore to alter its diet to become omnivorous or even herbivorous? Perhaps this repeated move to a lower trophic level relates to amount of available energy. If coelurosaurs possessed higher metabolic rates than other theropods, their increased energy needs would have become increasingly difficult to satisfy at the top of the trophic pyramid, because considerably less energy is available to top carnivores than to plant eaters. One possible strategy would have been a broadening of the diet to include plants. This is exactly the approach that many mammalian meat eaters, including bears, have adopted in order to make a living.[5]

Finally, mounting evidence indicates that a subset of coelurosaurs, the maniraptors, completed the metabolic journey, becoming full-blown endotherms within the range of living birds. Many functions, such as increased temperature regulation and activity levels, might feasibly have been involved in such a metabolic shift. Yet one particularly important function, reproduction, is backed by a solid fossil record. As described in chapter 8, *Oviraptor* and several of its maniraptor kin brooded their egg clutches (and perhaps hatchlings as well), tapping into the excess body heat generated by higher metabolic rates to care for their young. In effect, then, some feathered dinosaurs (including birds) managed to rescue a portion of the energy allocated to the maintenance bin (body heat) and move it over into the production bin (for reproduction). Natural selection may have favored theropods with higher metabolic rates, because the excess body heat could now be put to use among these feathered dinosaurs to increase the level of parental care.

Yet this high-cost physiological strategy would also have placed new limitations on maniraptors, including a lower ceiling for maximum body sizes. Because endothermy is so demanding energetically, putative giant maniraptors would likely have been unable to find sufficient food and space to sustain their populations. In addition, because larger body sizes typically come with decreased surface area and increased volume, giant maniraptors with high-cost metabolisms would have been prone to overheating. This reasoning may explain why maniraptors, uniquely among major groups of theropods, did not spawn any giant forms (greater than 1,000 kilograms, or 2,200 pounds), despite their great diversity, longevity, and broad geographic span. The only possible exceptions to this rule seem to be found among therizinosaurs and oviraptors that are thought to have broadened their diets to include plants. The biggest known maniraptor may be *Gigantoraptor*, a Late Cretaceous beast described by Chinese paleontologist Xu Xing and his colleagues. Xu estimates that *Gigantoraptor* may have reached an astounding 1,400 kilograms (over 3,000 pounds). My prediction is that *Gigantoraptor* and any other giant maniraptors will turn out to be herbivorous, or at least omnivorous, because high metabolic rates would have precluded meat as the sole food source.

Then we must consider the opposite realm—small body sizes. Why is there a lack of small dinosaurs (say, less than 5 kilograms, or 11 pounds) outside of maniraptor theropods? After all, the vertebrate family tree includes numerous examples of small-bodied forms among both ectotherms—from fishes and frogs to lizards and snakes—and endotherms—for example, hummingbirds and mice. This particular pattern has received little attention, the standard explanation being that endothermic mammals and a variety of reptilian ectotherms packed the small-bodied niches, blocking dinosaurs from filling these ecological roles. Another possibility is that many kinds of dinosaurs did evolve to small sizes, but we simply haven't found them yet. Yet both of these explanations seem inadequate, given the 160-million-year duration of dinosaurs and their overwhelming success in dominating medium to giant body size categories. Perhaps most dinosaurs possessed some inherited constraint that prohibited the evolution of little forms.

The Goldilocks hypothesis suggests an explanation for this pattern, one founded on internal physiology rather than external forces. If dinosaurs other than coelurosaur theropods possessed intermediate metabolic rates, small body sizes may have been too costly to maintain. At smaller sizes, an intermediate-grade metabolism would not generate sufficient heat to maintain a stable core temperature. In addition, lacking any kind of heat-trapping insulation such as feathers or fur, the relative increase in surface area associated with small sizes would have led to rapid loss of internal heat, further prohibiting the maintenance of constant body temperatures. However, with the evolution of feathers, in combination with high-cost endothermy, maniraptors were able to attain small body sizes, perhaps for the first time among dinosaurs. Maniraptors, including birds, made full use of this potential, evolving a range of smaller species.

To take this line of speculation one step further (perhaps to the breaking point), intermediate metabolic rates may even account for the glaring absence of marine dinosaurs. As noted previously, many groups of land-living vertebrates gave rise to lineages that became fully aquatic. For example, during the late Paleozoic, early amniotes gave rise to a group of small, aquatic predators called mesosaurs. During the Mesozoic, multiple groups of reptiles (sauropsids) produced fully aquatic descendants such as ichthyosaurs, mosasaurs, and plesiosaurs. Finally, during the Cenozoic, mammals spawned several marine clans, including seals and cetaceans (whales and dolphins). Even the archosaur line that led to the evolution of dinosaurs spawned freshwater and marine crocodilians. So why no aquatic dinosaurs? Perhaps because of their intermediate metabolic rates. Water is an excellent conductor of heat, so animals tend to lose body heat much faster in water than in air. For a given body size, the midrange metabolisms postulated here for dinosaurs would have required the retention of significantly more body heat than is necessary in ectotherms. Yet animals with that same intermediate Goldilocks physiology likely would not have been able to generate sufficient heat in an aquatic setting. Only after the evolution of high-cost endothermy in maniraptor theropods was there sufficient production of body heat to enable the evolution of aquatic-adapted forms like penguins.

In sum, the Goldilocks hypothesis proposes that the great majority of dinosaurs possessed intermediate, "mesothermic" metabolic rates, enjoying many of the low-cost advantages of ectothermy without the extremely high costs of endothermy. With more energy available for production, dinosaurs achieved rapid growth rates, which in turn offered an evolutionary pathway for generating giants. Other factors facilitating the evolution of giants on this metabolic middle road would have been reduced energy needs and body temperatures relative to endotherms. The profound physiological transition from ectothermy to endothermy has been difficult to envision because we lack living examples with intermediate metabolisms. A single group of dinosaurs, the coelurosaur theropods, appears to have pushed the metabolic trend further, with the feathered maniraptors becoming fully endothermic. Once the evolutionary ratchet had turned, establishing metabolic rates at this higher rate, maniraptors embarked on an independent evolutionary path, one that prohibited the evolution of giant sizes (with a few plant-eating exceptions), yet facilitated the emergence of small-bodied forms. Sometime during the Jurassic, a subset of these dinosaurs made full use of their small size, feathered insulation, and endothermic metabolisms to take to the air as birds and become the sole strand within the dinosaur thread to persist through a major extinction to the present day.

What is the unique contribution of the Goldilocks hypothesis? The real strength of any scientific hypothesis is its ability to make testable predictions and account for a range of observations. The Goldilocks hypothesis not only provides an explanation for the occurrence of presumed intermediate metabolic rates in dinosaurs. Perhaps for the first time, it also proposes a concrete link between the evolution of higher metabolic rates and the evolution of giant body sizes, postulating that both were intimately tied to greater amounts of energy available for production, particularly for higher growth rates. This hypothesis further accounts for the lack of very small-bodied dinosaurs outside of maniraptors, as well as the virtual lack of giants and the presence of tiny dinosaurs within maniraptors. If this idea is supported by future discoveries and research, it's fair to say that dinosaurs owed much of their tremendous success to a unique, intermediate-grade metabolism that enabled them to diversify into an array of otherwise-unavailable ecological niches.

Archosaur reptiles in a Chinle Forest, Late Triassic of western North America. The crushing grip of a rauisuchian, *Postosuchus*, on a nearly lifeless aetosaur, *Desmatosuchus*, captures the attention of a small, lightly built theropod dinosaur, *Coelophysis*.

12

CINDERELLASAURUS

IN HIS 1997 PULITZER PRIZE–winning book, *Guns, Germs and Steel*, Jared Diamond asked this provocative question: Why was it that Europeans became world conquerors, invading distant lands and establishing colonies in the name of Spain, Portugal, Germany, England, and France? Why wasn't it, say, Africans or South Americans or Australians spreading around the globe to invade Europe instead? Many previous explanations incorporated either explicit or implicit notions of racism, assuming that Caucasians were somehow superior to other races. In contrast, Diamond's powerful answer pointed to a combination of external factors, including geography (the major east-west trending axis of Eurasia enabling broad diffusion of technologies), domestication (Eurasia having had more than its fair share of plants and animals with domestication potential), and population size (the larger Eurasian population providing more opportunities for invention and diffusion). Among other things, these three factors resulted in the spread of microbes carrying infectious diseases; whereas Europeans had slowly developed a high degree of immunity to these diseases, indigenous populations on other landmasses exposed for the first time had virtually no defenses. Together, these various elements "conspired" to make it far more likely that Europeans would become the first to harness key technologies and spread around the world. Caucasians, in other words, were in the right place at the right time, opportunists rather than superior beings.

An analogous discussion has revolved around the great success of dinosaurs. Why did they, rather than some other group of animals, diversify on a grand scale and become dominant, maintaining this reign for the majority of the Mesozoic, about 150 million years? After all, several other candidate groups were present at the time that might have radiated their way to supremacy. The mammals were already there, originating about the same time as the dinosaurs. So, too, were therapsids, an ancient group that gave rise not only to mammals but also to a horde of related forms, many of them large bodied. Also present were a variety of other archosaurs, the "ruling reptiles," which gave rise not only to dinosaurs and crocodiles but to several other dinosaur-like lineages as well. The Triassic even saw a score of small- to large-bodied amphibians, many of them land dwelling, with voracious predators among their ranks.

So were the first dinosaurs in some way superior to their contemporaries and thereby able to outcompete all rivals? Or was their success a matter of dumb luck, with the small-brained reptiles merely opportunistic winners of an ancient evolutionary lottery? In this chapter, we explore the early days of dinosaurs in an attempt to answer this question.

The content of this book has built up in a stepwise fashion. First, I established the foundations of the dinosaurian web, including the history of life and the diversity of dinosaurs, as well as the physical, ecological, and evolutionary processes that shaped the Mesozoic world. Then I examined and wove together various threads, focusing in particular on the flow of energy from plants to dinosaurs and other animal consumers, ultimately closing the circle with decomposing insects, bacteria, and fungi. Also considered were the pivotal roles played by reproduction and metabolism. We now embark on the third and final phase of our journey, in which I attempt to convey the changing world of dinosaurs and spin some prehistoric webs. Along the way, we will consider a few of the many interacting factors that made the Mesozoic a unique and pivotal chapter in our planet's history.

How, you might wonder, can I hope to summarize the lives of a diverse, globally distributed group of animals that persisted more than 160 million years? Surely such bewildering variation in both time and space is too much for a single treatment. Perhaps, but by concentrating on a few key "place-based windows in time," I'll address a number of themes relevant to the Mesozoic world. Of the entire 160-million-year Mesozoic tenure of dinosaurs, I have elected to focus on three multimillion-year "snippets"—one each for the Triassic, Jurassic, and Cretaceous—exploring a key mystery related to each. To help synthesize these windows into a single narrative, all three derive from the Four Corners region of the southwestern United States, the right-angled confluence of Utah, Colorado, Arizona, and New Mexico. (Actually, I will freely transcend the boundaries of this region, but it serves as a useful focal point.) Why Four Corners? Because we know more about dinosaurs from western North America than from anywhere else in the world. And within the North American West, only the Southwest preserves an abundant fossil record that spans the Triassic through the Cretaceous.

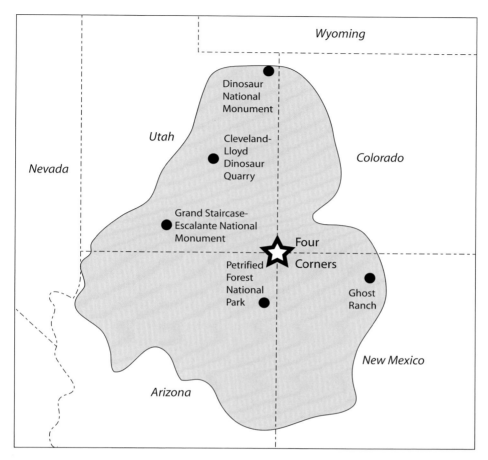

FIGURE 12.1

Map of the Colorado Plateau and Four Corners region of the American Southwest. Important fossil-bearing regions discussed in chapters 12–14 are noted.

For this bounty of paleontological knowledge, we owe a great debt of thanks to the geologic forces that have shaped the land since dinosaurs lived here. Four Corners sits at the heart of an expansive, uplifted region known as the *Colorado Plateau*, a spectacular landscape encompassing 360,000 square kilometers (140,000 square miles) of deep canyons, multihued badlands, towering mesas, and sparsely vegetated plateaus. This vast area is dissected and drained by a single river system, the Colorado River and its tributaries. Although there remains some debate as to timing, it's generally thought that, beginning around 20 million years ago, plate tectonic forces associated with collision of the Pacific and North American plates shoved the Colorado Plateau skyward 1,200–1,800 meters (4,000–6,000 feet), such that the highest points today exceed the 2,100-meter (7,000-foot) elevation mark. After being lifted high above sea level, the gravity-driven waters of the Colorado River downcut aggressively, rapidly generating a series of deep canyons, including that tourist mecca appropriately dubbed the Grand Canyon.

Despite such a violent geologic history, the Colorado Plateau is surprisingly undeformed, composed of flat-lying sedimentary rocks deposited mostly during the Paleozoic and Mesozoic. In contrast, the regions surrounding the plateau have suffered tortuous deformation, the result of cracking, folding, thrusting, and stretching of the continent's crust, much of this activity associated with mountain building. Somehow, the Colorado Plateau—and thus the fossils entombed within its sediments—remained an island of order in a sea of geologic chaos. Today, the Rocky Mountains capture most of the moisture carried by eastward-blowing clouds. As a result, the plateau is largely high-desert terrain, dry and sparsely vegetated. It is this combination of geologic uplift, rapid erosion, flat-lying sediments, and paucity of vegetation that, together with an abundance of fossils, make the Four Corners and its four-state environs such an outstanding place to investigate the Mesozoic.

Over the past 500 million years, the Colorado Plateau has variously been inundated by saltwater seaways, Saharan-style deserts, giant lakes, dense forests, and open grasslands. During the Paleozoic, swarms of ornamented, ocean-dwelling trilobites grew, mated, and died here alongside the continent's edge. In the Mesozoic, Four Corners dinosaurs survived arid, dune-filled deserts at some times and walked the balmy beaches of an inland sea at others. Much more recently, during the Pleistocene ice ages, elephants and saber-toothed cats sought forest shelter here in an attempt to escape icy winds descending from nearby glaciers. Later they were joined by people, who became high-desert specialists alongside many other local organisms.

Keeping this environmental dynamism in mind, let's return now to the Mesozoic and the matter of dinosaurs as unexpected versus inevitable rulers of the land.

Petrified Forest. To the uninitiated, the name of this national park in northeastern Arizona conjures images of silent groves of majestic stone trees, the lithified equivalent of a living redwood forest. Concentrations of fossil wood in the park have magical names like Jasper Forest, Crystal Forest, and Rainbow Forest. One imagines some kind of prehistoric Pompeii, with animals as well as plants entombed by volcanic ash—a moment in time catastrophically, yet intimately, preserved for countless millennia. Perhaps such misguided expectations explain why a few visitors to Petrified Forest National Park are disappointed by the massive chunks of petrified logs chaotically strewn about the desolate landscape. Conversely, I—and I think the great majority of people who enter the park—have the opposite reaction. I am enchanted by the chunky crystalline trunks, some reaching 10 meters (33 feet) in length and 3 meters (10 feet) across—ghosts of a bygone world. Most of these trees are attributed to a single species of Triassic conifer with the daunting name *Araucarioxylon arizonicum*. The fossilized wood, made up principally of solid quartz, tends to bear muted shades of red, though handheld chunks held up to the sunlight can sparkle as if covered in glitter. Gaps in the logs are often filled with thick crystals of clear quartz, smoky quartz, yellow citrine, and purple amethyst. And the sporadic presence of chemical impurities has generated rainbow streaks of striking color. The massive trunks were fragmented long after they fossilized, resulting in flat cross-sectional breaks that resemble chainsaw cuts. Yet so hard are these fossilized logs that a diamond-bladed saw is required to slice through them.

The fossilized trees of Petrified Forest National Park come from a sequence of rocks called the Chinle Formation, dating to the earliest days of dinosaurs in the Late Triassic. In addition to *Araucarioxylon*, the Chinle has yielded many other types of fossils, including about two hundred additional kinds of plants represented by fossilized wood, leaves, pollen, and spores. These remains, and the many animal bones found with them, provide a crucial glimpse into Late Triassic western North America. Located just north of the equator at that time, the Four Corners region was humid and tropical early in the Late Triassic and then transitioned to a drier climate as Pangaea drifted northward. Braided streams and meandering rivers criss-crossed lowland basins, much of this water passing through lush swamps. Abundant vegetation lined these waterways, forming a messy patchwork of green belts teeming with life.

With the first flowering plants still millions of years off in the future, vegetation in Chinle times was dominated by mixture of water-dependent, spore bearing pteridophytes. The club mosses, or lycopods, were mostly low, herbaceous plants with trailing shoots, looking distinctly like seedlings of pine trees. In contrast, although some kinds of horsetails also took the form of low groundcover, others were treelike, reaching heights of 6 meters (20 feet). Most common were *Neocalamites* horsetails up to 2 meters (6 feet) tall. Ferns appear to have been plentiful, with several varieties attributable to one of several groups still living today. Yet, despite this variety of spore producers, the Chinle flora appears to have comprised mostly spore bearers: lycopods, horsetails, and ferns—and water-independent, seed-bearing gymnosperms—cycads, ginkgos, conifers, and bennettites (see chapter 5). Conifers like *Araucarioxylon* were likely the dominant aspect of the Chinle vegetation, but debate continues as to the extent of Triassic forests. Some investigators think that the trees were restricted to small pockets arrayed sporadically along the riverbanks, whereas others argue for large gallery forests that extended far from the water's edge.

Grasses are frequently the dominant groundcover in modern land habitats, so we naturally tend to picture ancient ecosystems as covered with grass, too. Yet grasses are angiosperms, members of the flowering plant family, which did not contribute to Mesozoic ecosystems until the end of the Cretaceous. What, then, covered the ground during the Triassic and the succeeding Jurassic? Lacking a definitive answer, artists and scientists often choose to reconstruct pre-Cretaceous Mesozoic landscapes as sparsely vegetated except for green corridors tracking the waterways. The result is that many Mesozoic scenes portray dinosaurs cavorting about in what looks like expansive dirt parking lots. This scenario is almost certainly incorrect. Today, land plants tend to fill all available space where productive soils are present, and there is no reason to suspect that the world of dinosaurs was any different. So it's likely that short herbaceous plants of some sort formed a frequent groundcover. Excellent candidates include horsetails, like *Neocalamites*, and ferns. Studies of modern horsetails show that these plants tend to be high in energy, which means that they could have been a dietary staple of dinosaurs.

Like their modern counterparts, the Chinle forests provided food, shelter, hiding places, and oxygen to a great diversity of animals, large and small. But what kinds of animals? The answer depends on exactly which strata of the Chinle Formation (i.e., which part of the Late Triassic) you look in, because these rocks span about 20 million years, from 225 million to 205 million years ago. Nevertheless, we can make some generalizations.

When the trees of Petrified Forest were green and vibrant, capturing sunlight and bending gently under the influence of Late Triassic winds, many of the waterways below were home to an array of amphibians. These were not frogs or salamanders, however; most were big-bodied beasts several meters in length, among them voracious predators. Unlike their distantly related modern cousins, some of these Triassic amphibians were fully adapted to life on land, with only tenuous, mostly reproductive, ties to water. More common, however, were the semiaquatic METOPOSAURS, sturdily built amphibians with broad, flat, tabletop-like heads bearing upward-facing eyes. Picture the union of an alligator and Kermit the frog, and you begin to get the idea. It's generally thought that metoposaurs, many with bodies more than 2 meters (6 feet) long, pursued a lifestyle akin to that of many crocodiles, congregating on banks around lake and river margins and entering the water to hunt for prey.

Another group of animals common throughout the Triassic (and much of the preceding Permian Period) were the THERAPSIDS, which ultimately gave rise to mammals. For a long time, therapsids were known as "mammal-like reptiles," a fair descriptor because many of these animals were reptilian in most respects, yet with a smattering of mammalian features. In recent years, however, this imprecise term has fallen out of scientific vogue. The therapsids were remarkably diverse and took their turn as the dominant terrestrial vertebrates during the Permian. As noted previously, the mammals first appeared around the same time as the dinosaurs and then remained small bodied for over 150 million years thereafter. In contrast, the Triassic therapsids evolved into a range of body sizes, including some of the largest animals of the time. One of the dominant herbivores in the Four Corners area during late Chinle times was a hefty, short-tailed therapsid known as *Placerias*. The skull of *Placerias* was blunt and toothless, save for a pair of tusks up front. A short distance southeast of Petrified Forest National Park, an extensive *Placerias* bonebed has yielded remains of dozens of individuals, prompting speculation that these cowlike herbivores moved in large, social groups.

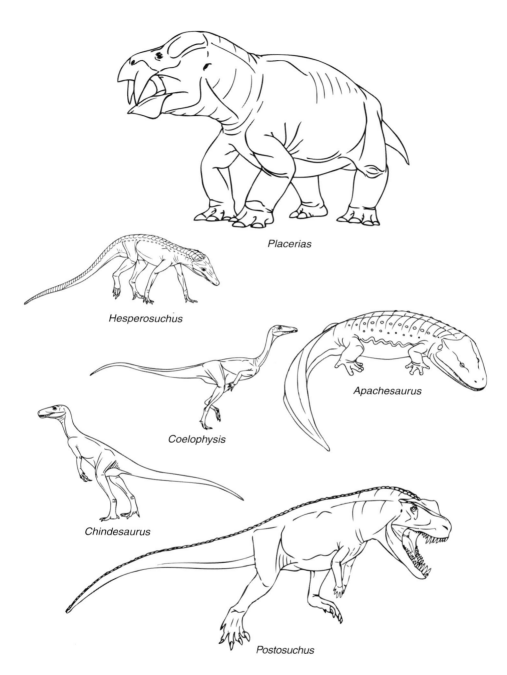

Placerias

Hesperosuchus

Coelophysis

Apachesaurus

Chindesaurus

Postosuchus

FIGURE 12.2

Representative animals from the Late Triassic Chinle Formation: the dicynodont therapsid *Placerias*; the crocodile relative *Hesperosuchus*; the theropod dinosaur *Coelophysis*, the temnospondyl amphibian *Apachesaurus*; the possible dinosaur *Chindesaurus*; the rauisuchian *Postosuchus*.

In terms of total species diversity, the amphibians and therapsids were exceeded during the Triassic by a bevy of reptiles. At the small end of the spectrum were predators such as the lizardlike sphenodontians. Also present were plant-eating procolophonids, the Triassic equivalent of modern-day horned toads. Most of the bigger reptiles emerged within the archosaurs, the group that gave rise to crocodiles, dinosaurs, and birds. Triassic archosaurs were a diverse bunch that included several unfamiliar, now-extinct groups. Early in the period, an evolutionary split resulted in two highly successful archosaur radiations. On one side were crocodiles and their immediate kin, and on the other were dinosaurs and their close relatives. Other than the dinosaurs, four archosaur groups stand out as most important for our Chinle saga, and all represent separate branches of the crocodile line.

First of these is a clan of semiaquatic predators known as PHYTOSAURS. Phytosaurs resemble modern crocodiles in having sprawled limbs, robust tails, armor plates embedded in the skin, and elongate skulls lined with sharp, cone-shaped teeth. Also like crocodiles, the nostrils faced upward from the top of the skull, a useful feature for air-breathing aquatic predators because they are able to move through the water with only their eyes and nostrils exposed above the surface. The nostrils of phytosaurs, rather than being positioned close to the tip of the snout as in crocodiles, were located far back on the skull near the eye sockets. Fossils of phytosaurs such as *Leptosuchus* are among the most common finds in Petrified Forest National Park. Members of the group are generally thought to have employed their sizeable, croc-like bodies to pursue a croc-like lifestyle, spending days lounging nearly motionless on the shores of rivers and lakes, and taking to the water to ambush their prey.

Second among the unfamiliar archosaur clans that prowled the Four Corners region during the Late Triassic is a group of land-dwelling carnivores called RAUISUCHIANS. The most prominent example from the Chinle Formation is *Postosuchus*, a 7-meter- (23-foot-) long beast that held the honor of top predator in this ecosystem. Show a picture of *Postosuchus* to most kids (indeed, nonpaleontologists generally) and they will typically call it a dinosaur. The large head is typical in many respects to those of theropod dinosaurs, being narrow, with long, serrated, and recurved teeth. Unlike theropods, most rauisuchians were quadrupedal, and it appears that most held their limbs in a semierect posture—not sprawled as in lizards, yet not fully upright as in most dinosaurs and mammals. These animals were true Triassic terrors, every bit as daunting as many large theropods. And some forms, including *Postosuchus*, may even have moved around bipedally like their distant carnivorous dinosaur cousins.

Third on our archosaur list are the armored AETOSAURS, the first major group of herbivores in the archosaur radiation. Within Petrified Forest National Park, aetosaur fossils are common finds, with a handful of types thus far recognized. *Stagonolepis* from the lower Chinle Formation and *Typothorax* from the upper Chinle are typical examples; both possess full-coverage body armor reminiscent of ankylosaur dinosaurs, which would evolve millions of years later. Aetosaur armor likely served an antipredator role, helping to fend off attacks from rauisuchians and other predators. However, it's also feasible that some of the elaborate armor, such as the shoulder spikes, were used to intimidate, or even do battle with, rivals of the same species. As we saw in an chapter 10, similar caveats apply to the ankylosaurs.

Finally, in addition to the phytosaurs, rauisuchians, and aetosaurs, another odd group of Triassic archosaurs is worthy of mention—the crocodilians. Although true crocodiles did not appear until the Jurassic, Late Triassic relatives in the crocodilian line were numerous and very different from today's semiaquatic, sprawled carnivores most easily envisioned lying motionless for hours on end by the water's edge. As described earlier, the phytosaurs apparently occupied this role for much of the Triassic. Instead, crocodilians of the time were mostly smallish, active, upright, land-dwelling predators fully capable of running down their prey. Although some had four feet, others appear to have been bipedal, looking more like *Allosaurus* than any alligator. Nevertheless, these Triassic archosaurs possessed several key features—particularly of the wrist, ankle, and skull—that conclusively reveal their close affinity with living crocodiles. A representative example from the Chinle Formation is *Hesperosuchus*, an agile little predator about 1 meter (3 feet) in length that may well have fed on a variety of still-smaller prey such as procolophonids and even insects.

By now you may be wondering, "OK, but what about the dinosaurs?" This turns out to be an interesting question and one for which the answer has changed in recent years.

Before tackling the issue of Triassic dinosaurs, let's step back for a moment and take a look at the big picture, because local diversity of plants and animals depends fundamentally on regional and global factors. In chapter 4, I addressed the docking of all the continents at the start of the Mesozoic to form a single giant supercontinent, Pangaea. Surrounding this supercontinent was a single giant ocean, PANTHALASSA. Were you to have been fortunate enough to sit in a spacecraft in low Earth orbit anytime during most of the Triassic, the view out your window would have consisted of prolonged, alternating vistas of sea and land. This grand merging of landmasses dramatically reduced the total amount of coastline. And, because the bulk of seagoing organisms tends to be concentrated along continental margins, the emergence of Pangaea severely shrunk the amount of habitat available for marine life.

Other influences of the supercontinent included strong climatic effects, with the predominantly temperate, stable climates of the preceding Permian Period replaced by increasingly hot and arid conditions, together with a surge in seasonality. Whereas the mean global temperature during the first half of the Permian was about 12°–15°C (54°–59°F), much like today, it was a balmy 22°C (72°F) for most of the Triassic. As far as we can tell, the poles remained ice-free during the Triassic and perhaps for the entire Mesozoic. In short, using the very broadest of brushstrokes, we can think of the Early and Middle Triassic as an arid time with relatively high temperatures. Keep in mind, however, that this hothouse world supported a diverse array of terrestrial ecosystems, including deserts, swamps, open coastal plains, and dense forests, among others.

For life on Earth, the Triassic was a time of evolutionary innovation predicated on disaster. The Mesozoic began with a bang, perhaps the single-greatest bang in the history of our planet. At the end of the Permian, 251 million years ago, a MASS EXTINCTION engulfed Earth's biosphere. Although the possible causes of this cataclysm are still hotly debated, its effects are not. Best we can tell, greater than 90 percent of all species alive at the time were exterminated in a paroxysm of death that has not been repeated before or since. Because the Permian was the final period of the Paleozoic and the Triassic kicked off the Mesozoic, this event is referred to as the end-Permian, or P-T, extinction (in contradistinction to the Cretaceous-Tertiary, or K-T, extinction, which closed the Mesozoic). It has also been dubbed, simply, "the mother of all extinctions." I will address this and other mass extinctions in a following chapter. For now, the salient point is that the Mesozoic started out with relatively empty, devastated ecosystems populated by few species. The lucky winners of this biotic lottery, less than 10 percent of the diversity that existed during the height of the Permian, included the ancestors of all lifeforms that have subsequently lived on Earth, among them dinosaurs and us. Needless to say, such a gargantuan reshuffling of the ecological deck, one in which most of the cards were destroyed, had profound and permanent effects on the planet's biosphere. More on that later.

During the Triassic, Paleozoic holdovers intermingled with evolutionary newcomers. The period opened with a radiation of therapsids, a group that had amply demonstrated its evolutionary mettle during the Permian. Also present in the depauperate Early Triassic faunas were primitive archosaurs, which would later spawn a great assortment of highly specialized reptiles, including crocodiles and dinosaurs. Many of the great archosaur clans that made their first appearance in the latter half of the Triassic—including the phytosaurs, aetosaurs, and rauisuchians—were extremely successful in terms of both species numbers and geographic distribution, yet they vanished by the close of the period. Much the same can be said for radiations of large amphibians like the metoposaurs. Over the course of the Triassic, therapsid fortunes waxed and waned; although a couple of groups were still radiating late in the period, only a handful of nonmammal therapsids survived the end-Triassic mass extinction. *Placerias*, one of the latest Triassic Chinle herbivores, is therefore a relic of sorts, one of the last representatives of a highly successful lineage. A central problem in Triassic paleontology is determining whether these various vertebrate groups went extinct gradually, fading away over many millions of years, or catastrophically, wiped from the planet in a geologic nanosecond (say, within a few million years versus tens of millions of years).

The Late Triassic was a time of great evolutionary novelty. And, whereas many Triassic newcomers did not persist into the Jurassic, others helped form the foundation of modern terrestrial ecosystems. First appearances in the Late Triassic include turtles, mammals, and three major clans of archosaurs—dinosaurs, pterosaurs, and crocodiles. Because birds have carried on the dinosaur legacy, pterosaurs are the only one of these groups that has not persisted to the present day.

As I write this paragraph, I can hear the songs of at least four different kinds of birds, their melodic calls wafting through my open window. So common and expected are birds that it's difficult to imagine most places without them. Other frequently heard animal calls include early evening insect and frog choruses. Were you to travel back in time to the early Triassic, however, one of the first things you would notice is the dearth of animal sounds. No birds. Few (if any) frogs. You might catch an occasional guttural grunt from a passing therapsid or metoposaur, but otherwise animal calls would largely be restricted to insects. Even the spectrum of bugs was limited, because all insects up to about 240 million years ago were apparently winged forms. Then, during the latter part of the Triassic, the terrestrial realm was rocked by an explosion in insect diversity. The vast majority of insect species living today belong to one of four major groups, or orders: Diptera (true flies), Hymenoptera (wasps and bees), Coleoptera (beetles), and Lepidoptera (butterflies and moths). Of these four, three of them—Diptera, Hymenoptera, and Coleoptera—first appeared during the Triassic. Given the profound importance of insects in terms of species numbers, biomass, and ecological impact, it's safe to say that terrestrial ecosystems underwent a profound shift toward the modern during the middle of the Triassic.

By the close of the Triassic, then, most of the essential building blocks of modern terrestrial ecosystems—including plants, insects, and vertebrates—were firmly established. It was into this world, in many ways similar to our own, that dinosaurs first entered.

Here our story turns abruptly to Georgia O'Keeffe, the famed southwestern painter. What does O'Keeffe have in common with dinosaurs? Is it the bones that were frequently her preferred subject matter? No. Perhaps it's the fact that O'Keeffe was recently honored by having a Triassic reptile named after her? Well, sort of.

If you depart Petrified Forest National Park and travel east on Interstate 40 for about 200 miles (320 kilometers), you arrive in Albuquerque, New Mexico. (Or, if you feel inclined to get your kicks, you can also get there a roundabout way via historic Route 66.) From Albuquerque, head north about 100 miles (160 kilometers), first on I-25 and then U.S. 84, until you come to the bustling metropolis of Abiquiu (population about 2,600). Finally, another 12-mile (19-kilometer) trip northwest in wide-open country gets you to Ghost Ranch, the location of Georgia O'Keeffe's longtime residence. Long before O'Keeffe ever put paint to canvas, Ghost Ranch was home to another famous inhabitant—a carnivorous dinosaur named *Coelophysis*.

The Ghost Ranch *Coelophysis* quarry is one of the most famous Triassic-aged dinosaur localities in the world. The site, which dates to about 205 million years ago (equivalent to the younger sediments preserved at Petrified Forest National Park), preserves an incredibly dense accumulation of these little meat eaters, on the order of a thousand articulated skeletons jumbled together as if dropped en masse from the sky. During the 1940s, the American Museum of Natural History conducted a series of expeditions to Ghost Ranch, led by museum paleontologist Ned Colbert. So tightly interlocked were the articulated skeletons that Colbert abandoned traditional methods and chose to excavate the fossils in large, arbitrarily defined blocks, each weighing on the order of 7,000 kilograms (15,500 pounds). Over the course of five field seasons, Colbert's crews removed a total of thirteen of these massive blocks. (O'Keeffe was apparently fond of visiting the dig site to check on progress.)

In 1981, a joint expedition involving three institutions—the Carnegie Museum, Museum of Northern Arizona, and Yale Peabody Museum—excavated an additional fifteen blocks so that today a total of twenty-eight bone-ridden chunks of Triassic sediment are being prepared and studied at various North American institutions. Late Triassic rocks in other parts of the Four Corners region, including Arizona's Petrified Forest, have yielded bits and pieces of *Coelophysis*, but all of these fossils combined represent but a tiny fraction of the bounty preserved in the Ghost Ranch quarry.

Coelophysis weighed about as much as a small adult human, tiny compared to later theropod giants like *T. rex*. It was also a gracile animal, with a long neck and tail, about 2.5 meters (7 feet) from stem to stern. As a small, upright, bipedal predator, *Coelophysis* approximates the ancestral animal that gave rise to all dinosaurs. Indeed, it is the namesake for a larger group (coelophysoids) that comprised the first major evolutionary radiation of theropods. Most carnivorous dinosaurs known from the Late Triassic and Early Jurassic are members of this group. How so many *Coelophysis* individuals came to be buried in one place is still debated, with flood, drought, and volcanism among the suggested causes of death.

Recently, two paleontologists at Columbia University and the American Museum of Natural History—Sterling Nesbitt and Mark Norell—opened one of the plaster jackets collected by Colbert's Ghost Ranch team in the 1940s. Inside they found not another *Coelophysis* but a strange, long-tailed, bipedal, toothless reptile that bore a distinct resemblance to ostrich dinosaurs (ornithomimosaurs). Further study revealed a range of bony features demonstrating that this animal was not a dinosaur at all but a 2-meter-(6-foot-) long herbivorous relative of crocodiles. Here, then, was another distinctly uncroc-like (at least to our eyes) crocodilian. Nesbitt and Norell named the animal *Effigia okeeffeae*. The first part of the name means "ghost," and the second part honors Georgia O'Keeffe, the more recent Ghost Ranch resident. *Effigia* is an example of the phenomenon known as convergent evolution, in which distantly related organisms (in this case, dinosaurs and crocodilians) independently evolve similar body plans, converging on parallel solutions. This find also underscores an important and very recent theme in our understanding of Triassic ecosystems—that is, many animals once regarded as dinosaurs have turned out to be members of the exceptionally diverse lineage of archosaurs that led to crocodiles.

Sterling Nesbitt, codiscoverer of *Effigia*, is one of three young investigators who have turned our understanding of Triassic dinosaurs on its head. The other two are Randall Irmis, now at the University of Utah, and William Parker, an employee of Petrified Forest National Park. Over the past several years, these three and their colleagues have initiated a systematic study asking, How many of the purported species of Triassic dinosaurs are actually dinosaurs? To date, their attention has been directed mostly at the plant-eating, bird-hipped ornithischians. A number of species of Triassic ornithischians have been named over the years, all based on fragmentary remains and most limited to isolated teeth. An example is *Revueltosaurus callenderi*, known from leaf-shaped teeth with several rounded cusps that bear a strong resemblance to the teeth of pachycephalosaurs and stegosaurs. In 2004, while looking for fossils in Petrified Forest, Parker came across some pieces of armor that looked crocodilian. These fragments led to a nearly complete skeleton, including a jaw with teeth that were identical to those previously ascribed to *Revueltosaurus*. So here again was a case of mistaken identity. *Revueltosaurus*, the putative herbivorous dinosaur, was instead a plant-eating croc relative, yet another member of the evolutionary radiation that brought *Effigia* into the world.

This finding disappointed a few rock hounds, who sold the relatively abundant teeth as those of early dinosaurs. More significantly, the discovery raised concerns that many other early dinosaur species known mostly from teeth may not be dinosaurs at all. It has long been thought that many kinds of both saurischian and ornithischian dinosaurs were running around in the Late Triassic. Yet Irmis, Nesbitt, and Parker see no convincing evidence of Triassic ornithischians outside South America and Africa; and even the South American record has been challenged. This conclusion raises some big questions: Where did the ornithischian dinosaurs come from? When did they originate? And when did they radiate around the world? For now, we have few answers.

Nevertheless, it's become apparent in recent years that dinosaurs were not nearly as diverse in the Late Triassic as we once thought. On the contrary, in many ecosystems, they appear to have been bit players, with most of the lead roles occupied by other kinds of back-boned animals. For example, in the Four Corners region, armored aetosaurs and tusked therapsids were the major vertebrate herbivores, whereas croc-like phytosaurs and theropod-like rauisuchians were the top carnivores. Certainly, dinosaurs such as *Coelophysis* were present, yet they were well outnumbered by other kinds of animals.

To be fair, the Chinle fauna is not entirely typical of Late Triassic ecosystems. In many other parts of the world, such as Europe, southern Africa, South America, and Asia, dinosaurs appear to have been considerably more common. Most successful by far were the plant-eating sauropodomorphs, particularly the prosauropods. Prosauropods were typically big bodied, including the largest land herbivores the biosphere had produced up to that point in Earth history. Quarries found in Argentina and Germany indicate that prosauropods were often the dominant herbivores, at least in terms of the total number of animals. The absence of Triassic prosauropods in the Four Corners region is puzzling, because all of the continents were then linked as Pangaea. As for Late Triassic theropods, for the most part they were relatively small bodied, and but one of several kinds of carnivores present in any given habitat.

So, with this great diversity of nondinosaur vertebrates in the Late Triassic, why were dinosaurs the selected ones, fated to become the principal large herbivores and carnivores on land for the remainder of the Mesozoic?

In an attempt to answer this question, let's return to the two main explanations for dinosaurian success: luck and superiority. In the first of these hypotheses, dinosaurs are a Cinderella story, unlikely successors whose destiny was driven by a series of external factors. Conversely, investigators supporting the superiority scenario have argued that dinosaurs possessed some internal feature, or suite of features, that enabled them to conquer all rivals and assume their rightful place as rulers of the terrestrial realm. Following this scenario, Attila the Hun would be a more apt analogue than Cinderella.

Alan Charig (Natural History Museum, London) was one of the first to articulate the competition model. In 1972, he suggested that upright posture was the key evolutionary advantage responsible for dinosaur success. He claimed that, over time, progressive changes to the hip, knee, and ankle enabled dinosaurs to achieve fully erect posture. With the legs acting as pillars to support the rest of the body, this new and improved stance resulted in longer strides and improved running capabilities. It may also have permitted more efficient breathing, because, in contrast to their sprawled brethren, the elbows no longer compressed the chest cavity with each step. Yet, although upright posture and gait were certainly important innovations for dinosaurs, it's difficult to support the view that this restructuring of the limbs somehow gave them a distinct advantage over their contemporaries. After all, various other kinds of archosaurs, including rauisuchians and crocodile relatives like *Effigia*, also exhibited erect or semierect posture.

Robert Bakker, the self-proclaimed heretic who helped kick off the dinosaur renaissance (see chapter 1), also argued strongly that dinosaur success resulted from some form of innate superiority, particularly given that mammals were already on the scene. For Bakker, superiority came in the guise of a hot-blooded metabolism, which enabled these early dinosaurs to outcompete both ectothermic reptiles and endothermic mammals. How else, Bakker queried, could dinosaurs have blossomed into the largest animals ever to walk the Earth while mammals scurried about in the shadows for so many millions of years, unknowingly awaiting their time in the sun?

Whatever the presumed advantage, the competition model was long held to be correct. In retrospect, this conviction appears grounded more in human psychology than in dinosaur biology. Ever since Darwin, the notion of progress as increasing superiority has been closely linked to the idea of evolution, with the history of life often characterized as the new and improved displacing the old and outdated. It was this same bias that caused people (predominantly those of European descent) to assume that Europeans must be somehow superior to peoples in other parts of the world. So ingrained was the assumption of progress that for a long time it was rarely even questioned, despite the fact that supporting evidence was weak or nonexistent.

Importantly, the competition hypothesis actually does make several testable predictions. If, for example, dinosaurs evolved one or more features that, over time, made them progressively more superior to therapsids and other archosaurs, we would expect to see a pattern of gradual replacement, with dinosaurs waxing while other groups waned. Viewed graphically, the image would be one of opposing wedges, with one wedge thinning out while the other thickened. Conversely, if some catastrophic event were responsible for the turnover, and dinosaurs emerged as victors by virtue of luck, we would expect a pattern of rapid replacement. (Note that "rapid" in this deep time context refers to time periods on the order of a million years, rather than the tens of millions of years required for gradual replacement.) Dinosaurs would then have to be regarded as opportunists rather than conquerors, and the graphic would look more like a series of abutting blocks, with older forms displaced abruptly by their evolutionary successors.

Michael Benton of the University of Bristol has been the strongest advocate of the rapid-turnover, opportunism hypothesis, basing his argument on a combination of logic and patterns observed in the fossil record. Benton makes the convincing case that, in order for the gradual, wedge hypothesis to apply, all of the competitors must be present to interact with one another. This requirement is difficult to conceive, given that dinosaurs (and other groups) include a diverse range of animals of different sizes and habits, including both herbivores and carnivores. All evolutionary competition happens in ecosystems, after all, which are strictly limited both in terms of geography and time. So it's difficult to envision how such local, short-term contests could be extrapolated to global dominance over deep time, generating the wedgelike pattern of progressive victory.

In addition, a major extinction occurred near the close of the Triassic. Although nowhere near the magnitude of the end-Permian catastrophe, the end-Triassic extinction is regarded as one of five global mass extinctions in the past 500 million years, a killing event that wiped out greater than 20 percent of all species in the oceans. It also hammered terrestrial ecosystems, emptying them of much of their biodiversity. Casualties included most (though not all) large amphibians, therapsids, and nondinosaurian archosaurs worldwide. The cause of this extinction is still widely debated, but no matter what the trigger, the event opened up many ecological niches, providing

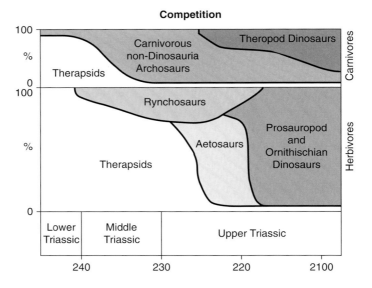

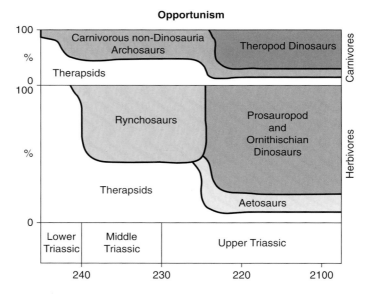

FIGURE 12.3

Two scenarios of evolutionary turnover during the Triassic, depicting alternative hypotheses (competition vs. opportunism) for major groups of vertebrates. A time scale is included at the bottom of both images, with numbers indicating millions of years ago. Modified after Benton (2005). Note the pattern of more gradual replacement in the top image (competition scenario) relative to more abrupt replacement in the bottom image (opportunism scenario). See text for discussion.

ample opportunities for a small cadre of dinosaurs to spin off numerous, often big-bodied descendants.

With this information in hand, can we now discard the notion of progress, insert an extinction, and conclude that dinosaurs were merely "Johnny-on-the-spot" opportunists? Not quite. The truth, it seems, is both more complicated and more interesting. The emerging picture is not a simple one of dinosaur survivors diversifying to fill empty eco-spaces. According to Benton, there were actually two mass extinctions, the first about 225 million years ago during the Late Triassic and the second at the Triassic-Jurassic boundary, about 200 million years ago. His contention is that the first event extinguished the dominant herbivores of the time—mostly therapsids and some near-archosaur relatives called rhynchosaurs—opening the way for prosauropods to become the main large herbivores in many parts of the world prior to the end of the Triassic. The second extinction, then, removed many large carnivores, such as the phytosaurs and rauisuchians, enabling theropods to radiate early in the Jurassic. Evidence for the earlier of the two extinctions remains controversial, and it is still unclear how other groups of animals were affected by these events.

Whatever the answer, opportunism looks like a much better explanation than competition to explain the rise of dinosaurs. Dinosaurs took advantage of emptied ecological niches brought about by the disappearance of other animal groups, becoming more of a rags-to-riches Cinderella story than a heroic epic in which all foes are vanquished.

But wait a minute. Even if there were one or two mass extinctions around the end of the Triassic, and even if a select few Early Jurassic dinosaur survivors were confronted by a sea of unexploited niches, they were not alone. In particular, the mammals were there as well. So we are still left with an unsolved problem. Why, if there was a relatively "clean ecological slate" (or at least a chaotic state of affairs) at the start of the Jurassic, did dinosaur diversity explode to include the largest land animals of all time, while mammals remained largely at rodent proportions for more than 150 million years thereafter? The standard view is that dinosaurs filled all the roles for large-bodied animals. On first blush, this idea seems reasonable. At the close of the Triassic, dinosaurs were already an ecologically diverse bunch, with plenty of large herbivores and carnivores. In contrast, virtually all mammals at this time appear to have been devoted insect eaters, hardly the foundation for an evolutionary radiation toward giant sizes!

Yet I wonder if other factors were involved in this key episode of the epic of evolution. Perhaps Mesozoic mammals were limited to small body sizes for reasons that still elude us. Or maybe dinosaurs evolved rapidly in the early Jurassic to fill the available niches. Might we rescue at least a portion of the competition hypothesis, perhaps even with Bakker's emphasis on metabolism? I wonder whether the intermediate metabolic rates described in the previous chapter are part of the answer. If Jurassic dinosaurs inherited intermediate physiologies from Triassic ancestors, they may have possessed the means (more energy available to be devoted to growth) to evolve rapidly to giant body sizes. If so, they may have occupied most of the large-bodied niches over a relatively brief deep time interval, thereby limiting opportunities for other animals to take on these roles. For now, such ideas are speculative, but I think that paleontologists will eventually make major inroads toward solving this problem.

The early evolution of dinosaurs, then, does not fit entirely well with either gradual or rapid scenarios of replacement. Although the three major lineages of dinosaurs (sauropodomorphs, theropods, and ornithischians) were present by the Late Triassic, they did not radiate in unison. Instead, the emerging picture is one of three waves of diversification. The first wave occurred in the Late Triassic, when sauropodomorphs, particularly the prosauropods, grew to large sizes and became the foremost herbivores in many places worldwide (although, curiously, apparently not in western North America). The second wave took place in the Early Jurassic with the diversification of theropods. In the Late Triassic, most theropods such as *Coelophysis* were small, sharing the meat supply with various, often larger, nondinosaur predators. The disappearance of other meat eaters at the end of the Triassic opened the door to a blossoming of coelophysoid theropod hunters in the earliest Jurassic (later followed by additional waves of predators generated from within the theropods). Finally, the plant-eating ornithischians make up the third wave of dinosaurs. Recent discoveries indicate that, contrary to previous paleontological wisdom, ornithischians were extremely limited in their geographic distribution (and thus ecological influence) during the Triassic, and it was not until well into the Jurassic that they diversified to challenge sauropodomorphs as the principal large herbivores on land. However, holes in the Jurassic fossil record may be skewing our view of these events, and it could turn out that the second and third waves occurred almost simultaneously beginning in the Early Jurassic.

By now it should be abundantly clear that we can't even begin to understand the evolution of dinosaurs without considering a host of nondinosaur factors. The Chinle Formation amply demonstrates this fact, offering a critical window into the Late Triassic world. Here, dinosaurs were merely one of many kinds of large vertebrates, with other archosaurs in particular taking on major roles. Certainly a visitor to the Four Corners region during Chinle times would hardly have suspected that a few millions of years hence, dinosaur relatives of the little predator *Coelophysis* would become the dominant large-bodied land animals worldwide.[1]

The staggering contingency inherent in the history of life is perhaps best understood by imagining alternative planetary histories. Rewind the tape of life to the end of the Permian, do away with the mother of all extinctions, and let the tape run again. Would dinosaurs have evolved? How about humans or even mammals? Granted, evolution might well have generated organisms that possessed key features of these groups—for example, upright posture, elevated metabolic rates, and/or big brains (interestingly, also features of humans). But, given the sheer complexity of the biosphere and its susceptibility to change in the face of perturbations, the odds are astronomically against those organisms being anything that we would today refer to as dinosaurs or mammals, let alone humans. And it wasn't just the big events, like mass extinctions, that altered the course of Earth history. Contingency is so deeply interwoven into the fabric of life that even small events can have profound, widespread consequences. (Recall the butterfly effect, in which the flapping of a butterfly's wings in the Amazon rain forest could result in a South Dakota tornado.) This means that other evolutionary transformations in the Triassic, such as the radiation of seed plants and turnovers in insect diversity, undoubtedly had cascading effects farther down the evolutionary path, with the reverberations felt all the way to the present day.

In short, dinosaurs appeared, and later became entrenched, into the Mesozoic world only through creative, mutually interdependent, ever-shifting interactions with other lifeforms and with the Earth itself. As we shall see, these very same sorts of dynamic interactions also dictated their subsequent Mesozoic history.

The Late Jurassic Morrison theropod *Ceratosaurus* prepares to defend a freshly dispatched meal—the armored ornithischian *Stegosaurus*—from another predator, *Allosaurus*. A group of mature sauropods, *Apatosaurus*, pass by in the distance.

13

JURASSIC PARK DREAMS

JURASSIC PARK, MICHAEL CRICHTON'S best-selling novel translated into Steven Spielberg's blockbuster movie, is a story of science gone wrong. Reckless humans embark on an entrepreneurial journey in which technology outstrips wisdom and the result is catastrophe. Despite this cautionary message, one of the questions now asked most frequently of paleontologists is whether such a re-created dino theme park might ever be realized. Will we one day possess the technological prowess (presumably still without the corresponding wisdom) to clone dinosaurs and set up a prehistoric zoo? For better or worse (I suspect the former), the short answer is no. However, such a succinct response is rarely satisfactory, so let me elaborate as a means of beginning our discussion of the Jurassic Period.

When scientists shoot down the Jurassic Park dream, they tend to focus almost exclusively on microscale obstacles associated with cloning. The central problem is recovery of a complete, intact sequence of genetic materials for a dinosaur. Crichton's clever fictional solution enlisted Mesozoic mosquitoes trapped in amber. As the story goes, these ancient mosquitoes bit dinosaurs, consuming small quantities of their blood. A small number of the pesky insects later became stuck in tree sap, which in turn became fossilized into amber, preserving both the mosquito and the dinosaur blood with its DNA. Find enough fossil mosquitoes, so the idea goes, and you can retrieve the DNA from a range of dinosaurs. Then it's a "simple" matter of cloning the dinosaurs and, voilà, Jurassic Park.

So far, dinosaur blood has not been found in any fossilized biting insects. However, in a stunning sequence of recent papers, Mary Schweitzer (North Carolina State University) and her colleagues have put forth evidence that dinosaur fossils occasionally preserve not only some of the original bone but also muscle fibers, blood vessels, and, yes, even genetic materials. Some of these discoveries are still questioned by the scientific community. More to the point of this discussion, however, a chasm separates identification of ancient genetic bits from the cloning of a *T. rex*. Because DNA deteriorates over time, any genetic materials preserved from the Mesozoic are almost certain to be grossly incomplete. If this information is not preserved somewhere in the fossil record—and to date we have no such indication—all the technology in the world will be unable to provide a solution. Even if you had the outstanding good fortune to recover all the genetic material from a dinosaur, the tiny fragments would have to be placed in the correct order, a gargantuan task equivalent to reconstructing a jigsaw puzzle of millions of pieces in which most of the pieces are shaped exactly alike. Imagine transforming a book like *War and Peace* into a basket full of words and then attempting to reconstruct the original text without any additional clues.

Some have suggested that we might not even need ancient DNA to make a dinosaur. Given that birds are the direct descendants of dinosaurs and carry the great bulk of this heritage in their genes, perhaps we could use our newfound understanding of genetics to "reverse-engineer" a dinosaur, figuring out which genes to modify during embryonic development so as to generate such features as teeth, bigger bodies, and a long, bony tail. With the present astounding rate of advancement in molecular biology, such genetic monstrosities may well be feasible by the end of this century. Yet it is difficult to imagine that these Frankenstein creatures would, in any real sense, be Mesozoic dinosaurs brought back to life.

Critics of the Jurassic Park scenario typically stop at this point. No clones, no dinosaurs. But our putative theme park creators would also have to overcome a host of (generally ignored) macroscale obstacles. First and foremost would be ecological interactions between and among species, with some of the biggest roadblocks pertaining to food. The literary and cinematic versions of Jurassic Park included not just dinosaurs, but ancient plants resurrected to feed the herbivores. Where did these plants come from? Surely their DNA was not extracted from plant-biting insects. We do not even know with any confidence what kinds of plants particular dinosaur species consumed. Little attention is given to this sticky point. You might object that surely these animals would be able to eat modern plants. Perhaps so, but no one can yet make this claim with confidence. So any would-be creators of a "Jurassic Park" or "Cretaceous Park" might also face the task of re-creating not just the plant-eating dinosaurs but their food supply as well.

Then there are the myriad nonfood life-forms necessary to support the dinosaurs. Every ecosystem is sustained by complex, finely tuned interactions between and among species. For example, many plants, most of them angiosperms, are closely tied to specific insect pollinators. Similarly, the guts of all animals contain specialized bacteria that aid digestion, a symbiosis that certainly applied to dinosaurs as well. In order to round out your prehistoric theme park ecosystem, you would need to borrow liberally from modern life-forms, a problematic proposition. To use a gross analogy, if an ecosystem is compared to a smooth-running engine, envision taking a handful of parts—say, from the engine of a 1957 Chevy—and trying to build a smooth-running motor by mixing them with parts from a 2001 Subaru. Not likely. And ecosystems are far more complex and interwoven than any machine. Given the incompleteness of the fossil record, re-creating an entire prehistoric ecosystem would be impossible. We are a long way from understanding modern ecosystems, let alone ancient examples. Consideration must also be given to the nonliving parts of the environment. Dinosaurs inhabited a hot-house world with higher concentrations of carbon dioxide and perhaps differing levels of oxygen as well, and these might have to be adjusted to keep the cloned prehistoric animals alive and well. In short, don't expect to visit a Mesozoic theme park any time soon.

Having trounced the idea of building a Jurassic Park, I would nevertheless like to use the concept to carry out a thought experiment, one that addresses key questions about the world of dinosaurs. Imagine that you could circumvent all of these obstacles and reconstruct a fully functioning Jurassic ecosystem with dinosaurs and all of the other animals, plants, fungi, protists, and bacteria. Imagine further that you have no interest in putting this Mesozoic menagerie behind glass or even in large enclosures. Rather, you decide to let these organisms roam and interact as they did 150 million years ago. As the (re-)creator of this Jurassic Park, it's your job to ensure the sustainable functioning of this ecosystem. By this I don't mean human interventions such as culling excess numbers of sauropods or launching a breeding program to save endangered stegosaurs. No, what I have in mind is more of a hands-off approach, allowing the ecosystem the space it needs to make the best use of its evolutionarily acquired "wisdom."

Perhaps the most fundamental question you would be faced with is, How much space is necessary to maintain this ecosystem? Not surprisingly, it turns out that answering this big question requires that we first tackle several smaller ones. How much food would different kinds of dinosaurs need to eat in any given day, week, or month? How large were dinosaur populations and exactly how many different kinds of coexisting dinosaur species were there in the ecosystem? At any given moment in Jurassic time, how much food (in both plants and meat) was available on the landscape? In other words, just how productive was this ecosystem? In particular, we need to know something of the quality, diversity, and amount of available vegetation, because plants largely determine the population sizes and numbers of herbivore species (and thus carnivore species) that an ecosystem can support.

Let's focus our thought experiment on a single Jurassic ecosystem. The best-known example is preserved in sediments of the Morrison Formation, broadly exposed on the COLORADO PLATEAU in the Four Corners region. The Morrison Formation, deposited about

150 million years ago, captures the culmination of the Jurassic dinosaur radiation. Today, the Morrison outcrops in broad, stacked bands of mud, silt, and sand that occur in pastel hues of beige, lavender, and burnt orange. So common are these rocks that it's hard to take a trip through the American Southwest without encountering them, even if you stick to the freeways. Entombed within these ancient sediments are remains of famous dinosaurs—plant eaters like *Apatosaurus* and *Stegosaurus* and meat eaters like *Allosaurus* and *Ceratosaurus*, among many others.

Starting with the big herbivores in the Morrison ecosystem, more than a dozen different kinds of long-necked sauropods have been recovered. The list of behemoths includes *Apatosaurus* (previously *Brontosaurus*), *Diplodocus, Brachiosaurus, Camarasaurus, Supersaurus, Haplocanthosaurus,* and *Barosaurus*. Apart from sauropods, the Morrison herbivore clan included several armored stegosaurs and ankylosaurs, as well as a range of small to medium-sized ornithopods. Yet these animals are spread over most of the 4- to 6-million-year range of the Morrison Formation, and it's likely that there was a significant amount of species turnover (species births and deaths) during this lengthy period. In order to be meaningful, our ecological reconstruction should include only those animals that overlapped in both time and space. We need to select an ecological snapshot in time.

Bonebeds—those fossil sites that preserve the remains of multiple animals—can be especially valuable for assessing species diversity at a particular time, because they often provide a direct record of animals that coexisted within a single ecosystem. The visitor center at Dinosaur National Monument, in eastern Utah, contains perhaps the best-known Jurassic bonebed, and one of the most famous dinosaur localities in the world.[1] The site, known as the Douglass Quarry (after Earl Douglass of the Carnegie Museum of Natural History), consists of a monolithic inclined wall of sandstone littered with hundreds of bones from a diverse range of Morrison dinosaurs. Many bones occur in isolation, but partial skeletons are also common within the assemblage. I've had the pleasure of climbing on this wall several times, moving from a jumbled string of *Stegosaurus* vertebra to an enormous *Camarasaurus* shoulder blade and then to a hip and leg of *Allosaurus*. The sheer volume of dinosaur bones (about 1,500) is staggering, and hundreds of additional specimens excavated here in the last century are housed in eastern institutions.

The herbivorous dinosaurs preserved in the Douglass Quarry include at least four different sauropods: *Apatosaurus, Diplodocus, Camarasaurus,* and *Barosaurus*. These giants came in a variety of sizes, ranging from about 7,000 kilograms (15,400 pounds, a little smaller than an adult bull African elephant) to a whopping 50,000 kilograms (110,000 pounds, equivalent to about six bull elephants)! Other herbivores include the lumbering, plated wonder *Stegosaurus*—still a heavyweight at about 5,000 kilograms (11,000 pounds), equivalent to a black rhinoceros—and two varieties of ornithopod dinosaurs: the bison-sized *Camptosaurus* and the mule deer-sized *Dryosaurus*. Together, these animals shared the bounty of plants that must have thrived here

FIGURE 13.1

A sampling of dinosaurs from the Late Jurassic Morrison Formation: the sauropod *Camarasaurus*; the theropod *Marshosaurus*; the stegosaur *Stegosaurus*; the theropod *Ceratosaurus*; the theropod *Allosaurus*. Not to scale.

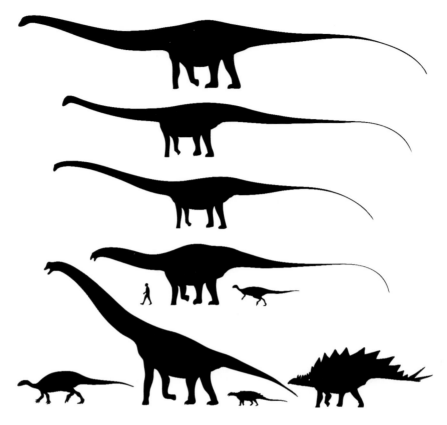

FIGURE 13.2

Variation in body size of the Morrison Formation herbivore dinosaur fauna. From top to bottom (and left to right where applicable): the sauropod *Supersaurus*; the sauropod *Diplodocus*; the sauropod *Barosaurus*; the sauropod *Apatosaurus*; the hypsilophodont ornithopod *Dryosaurus*; the iguanodont ornithopod *Camptosaurus*; the sauropod *Brachiosaurus*; the ankylosaur *Gargoylesaurus*; the stegosaur *Stegosaurus*. Human added for scale.

150 million years ago. Of course, insects and many other small herbivores accompanied them.

The sheer size of the biggest sauropods has long shrouded them in mystery. How could such gigantic animals with such small heads (at least compared to their body sizes) manage to consume enough food to sustain themselves? The mystery deepens when we look inside the mouths of sauropods. Cows are typical of big mammalian herbivores (e.g., elephants, deer, antelope, and the like) in possessing oversized teeth well suited for processing large amounts of fodder. Chewing serves not only to break food items into bite-sized chunks; it also grinds up the plant matter and begins the process of digestion by making the bits of vegetation more susceptible to chemical attack. By comparison, sauropods have puny teeth well suited to tearing or, at best, slicing, but not for grinding. In other words, the jaws and teeth of these long-necked dinosaurs appear better adapted for food procurement than for food processing, a bite-and-swallow approach instead of the chewing strategy used by mammals. An inability to chew would seem to be a serious detriment when you're one of the largest animals ever to walk the Earth.

Cows and other ruminant animals, in addition to fermenting plant food in the hindgut, possess a complex stomach with an additional fermentation chamber housing many millions of microbes. Foregut fermentation is complemented by food regurgitation, allowing animals to chew plant matter a second time. This process of "chewing the cud" means that ruminants do not have to chomp food thoroughly while foraging, because they have a second opportunity at a safe place of their choosing. There's no evidence that sauropods or any other dinosaurs had a foregut fermentation chamber, though it has long been argued that they may have had something analogous—a stone-filled gizzard. Birds, the direct descendants of dinosaurs, use stomach stones (or, more accurately, grit) in muscular gizzards to grind up plant matter and thereby circumvent their lack of teeth. For many years, polished and rounded stones ranging in size from pebbles to softballs have been found in the Morrison Formation, some in direct association with sauropod skeletons. These often colorful rocks have been interpreted as the gastroliths ("stomach stones") of Jurassic sauropods, and sold as such in rock shops throughout the Southwest. Unfortunately for rock shop owners, the hard evidence for sauropod gastroliths is minimal and disappearing fast. It's been argued that the total mass of stomach stones found in association with sauropods specimens is insufficient to support the idea of a gizzard. Even more damaging is a recent study indicating that the bulk of these supposed stomach stones, rather than being polished within the guts of ancient dragons, are merely water-rounded cobbles washed down into the Morrison Formation from overlying Cretaceous sediments. Nevertheless, occasional sauropod skeletons do show appropriately sized cobbles within the rib cage, suggesting that rock-filled gizzards may have been part of the digestive solution for at least some of these dinosaurian giants.

Either way, important answers to the problem of sauropod digestion are to be found farther along the gastrointestinal tract. The giant bodies of *Camarasaurus* and its kin, in addition to housing a massive heart, lungs, and liver, must have contained a truly impressive hindgut—the intestinal mass downstream of the stomach. Next time you go to a museum and see a skeletal reconstruction of a giant sauropod, imagine the cavernous abdominal cavity packed full with an enormous, glistening gut folded onto itself many times. It is here in the intestines that the real work of sauropod digestion took place. Within this thick, muscular, tubelike chamber was a fermentation vat driven by the metabolic activities of billions of symbiotic microbes. In short, an enormous gut, copious quantities of bacteria, and slow passage of food were likely the three major components in sauropod digestion. Many large mammals today also make use of this prolonged fermentation strategy to break down fibrous cell walls in order to access the nutrients locked within. Sauropods simply took the game to a new level.

How much food did giant dinosaurs eat? Bigger sizes translate into bigger appetites, whether you are a consumer of plants or meat or both. Elephants, the largest living land animals, consume such vast quantities of forage that they must spend about three-quarters of their time searching for food. So we can be certain that sauropods, some of them many times larger than the biggest elephant, ate an awful lot. Yet going beyond the gee-whiz level to put real numbers on estimates of food intake is a formidable task for paleontologists, because the animals in question haven't taken a breath, let alone a bite, for millions of years.[2]

Nevertheless, if you elected to venture down this paleontologic road less traveled, you would eventually run into a clean-shaven man with thick glasses and a wry sense of humor by the name of Jim Farlow, the same fellow I mentioned in the preceding chapter as my cocollaborator on the Goldilocks hypothesis. Farlow has a penchant for chasing thorny problems about dinosaur ecology and behavior that most other paleontologists eschew or actively disdain. While writing this book, I received a draft of a new paper from Farlow, who asked for my comments. The paper had a classic Farlow title: "A Speculative Look at the Paleoecology of Large Dinosaurs of the Morrison Formation, or, Life with *Camarasaurus* and *Allosaurus*." As soon as I had worked my way through this dense offering, backed by more than a dozen equations and myriad calculations, I knew that his general approach and conclusions would make a wonderful anchor for a discussion of Jurassic ecology. Thus, the following summary is based largely on his findings about Morrison dinosaurs, though (in case you were worried) without the math.

Farlow approached the problem of estimating dinosaur food consumption from a novel standpoint: ranching. He began by talking to some Wyoming ranchers about how much grazing land was required to maintain a 1,000-pound (455-kilogram) cow and her 250-pound (114-kilogram) calf for one month. The answer, it turns out, varies between about 10 and 20 acres, depending on such factors as rainfall, elevation, and soil quality. An acre is a little smaller than a football field, so cows need a lot of space. *Camarasaurus* is the most common Morrison herbivore, and an adult weighed on the order of 40,000 kilograms (88,000 pounds). Farlow did calculations to estimate the food intake of a *Camarasaurus*-sized cow and found that a single animal would require at least 18 times as much grazing area, somewhere between 180 and 360 acres (200–400 football fields)! Keep in mind that this is the estimated area necessary to support a single animal. And recall that warm-blooded (endothermic) animals like mammals require considerably more food than equivalent-sized cold-blooded (ectothermic) animals such as reptiles. It turns out that even a putative cold-blooded lizard of *Camarasaurus* proportions would need six times the acreage of a single cow-calf duo. If dinosaur metabolic rates were intermediate between those of ectotherms and endotherms, as suggested in chapter 11, the actual acreage necessary to support one *Camarasaurus* might lie between these values, closer to ten times that of a mammalian cow and calf, or more than a hundred football fields. If we assume that this food would not have been evenly abundant, but rather patchy on the landscape, the total required area grows rapidly.

In order to find enough to survive, then, an adult sauropod—whether cold-blooded, warm-blooded, or something in between—must have required a vast amount of food, and therefore a very large home range. If we now contemplate populations of sauropods with a mixture of adults, subadults, and juveniles, the necessary geographic range jumps to seemingly ridiculous proportions. As Farlow wryly notes, "No matter what their metabolism was like, dinosaur ranching would require a substantial investment in real estate."

Comparing cows and camarasaurs has its limitations. For one thing, wild ecosystems tend to be complex, with a diverse mixture of plant types instead of a monotonous sea of grass. For another, with the exception of an occasional deer or pronghorn, cows have little competition from other species. In the wild, plant offerings must be divided among not two or three but typically many herbivorous species. As we've seen, *Camarasaurus* shared the green gradient of the Morrison with a parade of giants. So we must assume that these plant eaters reduced competition by eating different plants. This conclusion is supported by differences in sauropod skulls (see chapter 7). *Apatosaurus* and *Diplodocus* have low, long, horselike skulls with slim, pencil-shaped teeth restricted to the front ends of the jaws. *Camarasaurus* and *Brachiosaurus*, in contrast, have shorter, boxier skulls with chunky, spoon-shaped teeth lining most of both the upper and lower jaws. Wear marks on teeth of adult camarasaurs indicate a diet of relatively coarse vegetation, perhaps cycads and/or conifers, whereas those on the slender teeth of *Diplodocus* are suggestive of soft, perhaps even aquatic plants. Interestingly, wear traces on the teeth of juvenile camarasaurs also point to a diet of soft plants, suggesting that this sauropod at least may have undergone a major shift in foodstuffs during growth.

No matter how these animals subdivided the plant spoils, we can be certain that they did so successfully because their bony remains are found in great abundance through millions of years of Mesozoic time. The Morrison fossil record in particular tells us that multiple species of sauropods shared the same ecosystems for about 5 million years. So all indicators suggest that these giants not only survived—they flourished. Evolution was clearly adept at weaving these sizeable strands into the Jurassic web.

A further point of discrepancy between cows and camarasaurs relates to diet. The great mammal radiation that ultimately spawned cows (and humans) was fueled largely by grasses. The abundant and diverse grasses, part of the radiation of flowering plants, provided the foundation for food webs over much of the planet (recall the Serengeti ecosystem described in chapter 7). But Jurassic dinosaurs could not have been grazers, because grasses did not show up until millions of years later, near the end of the Cretaceous. So what did *Camarasaurus* and other Jurassic herbivores eat? Was their plant food higher or lower quality than grass? Equally important is the amount of available vegetation. Were the plants locally abundant and distributed uniformly or clumped in well-separated patches? Answers to these questions require an understanding of Jurassic landscapes.

Pangaea, the early Mesozoic megacontinent, began to break apart in earnest during the Jurassic, mobilizing the dozen or so plates that make up Earth's crust. The source of all this geologic restlessness was extensive rifting at plate boundaries. Along rift margins, seawater rushed in to fill the growing gaps between continental plates, ultimately forming full-fledged oceans. Thanks to hothouse climates and a lack of ice tied up at the poles, many of these nascent oceans periodically flooded low-lying portions of landmasses, forming shallow inland seaways atop the continental crust. In the case of Europe, the land was so low and the flooding so extensive that the submerged continent was transformed into an island archipelago. Meanwhile, colliding plates caused extensive mountain building and explosive bursts of volcanism.

Overall, Jurassic climates were an extension of the hothouse conditions initiated in the Triassic. Yet, as Pangaea split apart into smaller landmasses surrounded by oceans, the close proximity of large bodies of water moderated the vast temperature swings typical of Triassic times. The moderating effect was heightened by the rise in global sea levels during the Late Jurassic and the Cretaceous. In general, the Middle and Late Jurassic appear to have been times of warm, equable climates, with much less seasonality than occurs today.

Not surprisingly, the biosphere also underwent major changes, as life-forms adapted to shifting geologic and climatic conditions. In the marine realm, thriving coral reefs took hold just offshore in the shallow and warm inland seas, supporting a tremendous array of marine life. These seas were also home to numerous bony fishes and nautilus-like ammonites, which in turn provided fodder for large marine predators like sharks, plesiosaurs, and ichthyosaurs. In the air, the great flying reptiles known as pterosaurs ruled; two major groups, short tailed and long tailed, spawned numerous specialists, from fish eaters and filter feeders to insectivores and scavengers. Birds also made their first appearance in the Late Jurassic, evolving from small raptorlike theropods. On land, new varieties of meat- and plant-eating dinosaurs filled the empty ecological spaces left by the mass extinction at the close of the Triassic. Although several successful groups of Triassic backboned animals—for example, aetosaurs, therapsids, and rauisuchian dragons—had fallen by the evolutionary wayside, terrestrial ecosystems of the Jurassic remained richly diverse with various other, more familiar vertebrates—fishes, frogs, salamanders, turtles, lizards, snakes, crocodiles, and small mammals.

Returning to the Morrison landscape in western North America, climate in the Four Corners region appears to have been dominated by warm temperatures and relatively low rainfall. Low precipitation levels were likely offset somewhat by rivers and groundwater flowing down from mountains off to the west, resulting in extensive, year-round wetlands. Plants in this preangiosperm world were dominated by seed bearers like conifers, ginkgos, cycads, and bennettites together with spore bearers like ferns and horsetails. Thick forests of evergreen conifers and tall shrubs likely tracked the meandering rivers. Farther afield from these watercourses, the vegetation was likely dominated by low herbaceous greenery.

As with the Chinle ecosystem example discussed in the preceding chapter, there is considerable disagreement about the distribution and amount of vegetation in the Morrison ecosystem, as well as the degree of aridity. Geologists tend to point to various sedimentologic indicators for dry conditions, whereas the paleobotanists are impressed by the abundance of fern spores, suggestive of a much wetter setting. Of one thing we can be certain, however: The vegetation was plentiful; it simply had to be to support such a diversity of giant animals. Overall, the scene may have resembled an East African savannah, with a combination of ferns, horsetails, and low-lying seed plants taking the place of grasses. Or the Morrison may have differed fundamentally from any modern ecosystem. At least for now we can get only the foggiest of views.

As discussed in chapter 7, giant herbivores forced to eat lower-quality plants must consume a greater volume in order to gain the same nutritional benefit. Given that sauropods carried out the bulk of their digestion in capacious hindguts, the passage rate of such low-quality food must have been very sluggish in order to release the stingy nutrients locked in plant cells. Slow passage through elongate intestines equates to huge volumes of digesting food. Farlow once estimated that the hindgut of a 35,000-kilogram (77,000-pound) sauropod might regularly have held in excess of 2,200 kilograms (5,000 pounds) of fermenting plant matter, equivalent to carrying around a Ford F150 extended cab pickup truck in your belly! But what plants did Morrison herbivores consume?

In an attempt to answer this question, Jürgen Hummel of the University of Bonn and a team of colleagues studied living relatives of the primary Morrison plants to assess their nutritive value. They exposed the fresh foliage of each to microbes obtained from a sheep's rumen, and they then measured the fermentation rate to get an estimate of the available energy in each plant type. Their experimental results indicate that cycads and tree ferns would have been poor choices for Mesozoic herbivores, but that ginkgos, conifers, horsetails, and other ferns were all high in energy and are thus likely candidates for sauropods and other Morrison plant eaters. The monkey puzzle conifer *Araucaria* turned out to be particularly energy-rich but slow fermenting. Given the widespread distribution of these majestic trees during the Jurassic, they were the likely fodder of choice for giant, high-browsing, slow-fermenting sauropods such as

Brachiosaurus and *Camarasaurus*. Conversely, the preferred food among lower-level browsers like *Stegosaurus* and *Diplodocus* may well have been horsetails like *Equisetum*, which showed the highest nutritive value of any of the plants examined, even exceeding that of grasses. Here, then, we are getting some of our first informed glimpses into the diets of dinosaurs.

Large quantities of poor- to medium-quality food also mean that much of the plant matter cannot be absorbed into the body as nutrients and therefore must be ejected as feces. So, although generally not included in television documentaries, it's likely that the terrain was strewn with megasized "dino-pies," transforming a simple walk across the Morrison landscape into a high-risk sport (as if the giant carnivores weren't enough). To make matters worse, the bacteria-ridden masses of half-digested plant matter in the hindgut probably would have generated vast amounts of rumbling flatulence. Thus, for any ranger walking through a Jurassic Park ecosystem, it would be important to stay downwind of the predators and upwind of the sauropods, if you get my drift.

Still unresolved is the question of total plant abundance. How much plant matter was generally available for Morrison herbivores? All land-living vertebrates depend on the energy provided by plants, so an estimate of the amount of vegetation would provide insights into the number of dinosaurs that may have lived there. Jurassic hothouse climates resulted in higher levels of the greenhouse gas carbon dioxide in the atmosphere. Because plants mix sunlight, water, and carbon dioxide to make themselves, it's been predicted (or, to be precise, "retrodicted," because we are referring to past events) that the increased carbon dioxide stimulated plant growth. Biologists use the term PRIMARY PRODUCTION to refer to the amount of organic material generated by photosynthesis, largely through plants in the terrestrial realm and algae in marine settings. It's possible to estimate primary production in distant time periods by running several pieces of information—for example, positions of continents and estimates of temperature, rainfall, and atmospheric gas composition—through mathematical models. This kind of analysis suggests that net primary production for the entire biosphere was indeed higher during the Jurassic than at present—about twice as high. If so, a given chunk of Jurassic landscape would, on average, have supported more biomass (i.e., more animals, plants, fungi, protists, and bacteria) than that same area today. Here, then, is a possible explanation to account for the humongous proportions achieved by many dinosaurs: perhaps there were more plants to feed greater numbers of larger herbivores, which in turn supported more and bigger carnivores.

But wait. The proviso "on average" turns out to be critical for our discussion. Plant matter is not distributed evenly over Earth's land surfaces. Compare the Amazon rain forest, for example, with Antarctica or the Sahara Desert. Biomass tends to diminish as rainfall declines, and the Morrison ecosystem apparently experienced low precipitation levels. The same study that calculated primary production worldwide for the Late Jurassic also found that available biomass in western North America would have been well below the average. Morrison dinosaurs were likely most abundant around permanent sources of water—including rivers, lakes, and wetlands—and fewer in number farther away from these wet areas. So it appears that the "more available-food" hypothesis fails, at least for this landscape, and we are left with some fundamental unsolved mysteries about Morrison herbivores.

Morrison Formation carnivores are also the source of a few mysteries. The largest obligatory meat eaters living today are big cats like lions and tigers, which weigh up to about 300 kilograms (660 pounds), and polar bears, around 600 kilograms (1,320 pounds). The only semiexceptions are Kodiak bears, which can grow to 800 kilograms (1,760 pounds), but these ursine giants augment their plant diet with meat. In addition to a single top predator, virtually all modern ecosystems include a diverse range of small to medium-sized carnivores as well. For example, where I live in northern California, the extremely rare mountain lion sits atop the food web, living alongside coyote, fox, bobcat, marten, raccoon, weasel, and skunk (among others), all of which tend to be more common.

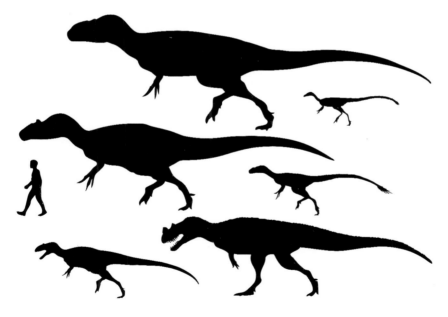

FIGURE 13.3
Variation in body size of the Morrison Formation theropod dinosaur fauna. From top to bottom (and left to right where applicable): *Torvosaurus, Coelurus, Allosaurus, Stokesosaurus, Marshosaurus,* and *Ceratosaurus.* Human added for scale.

Back in the Late Jurassic, the Morrison Formation was also home to a diverse bunch of carnivores, some of which grew surprisingly large. At the small end of the theropod spectrum is bobcat-sized *Koparion*, followed by coyote-sized *Tanycolagreus*, cougar-sized *Stokesosaurus*, and African lion-sized *Marshosaurus*. From here, the remaining theropods are all well beyond the size range of living carnivores. Next on our list is *Ceratosaurus*, with the characteristic bladelike horn atop its nose, weighing about twice as much as the largest cats alive today. Then comes *Allosaurus* at about 1,000 kilograms (2,200 pounds), more than three times the mass of the largest living cats. Finally, at an impressive 2,000 kilograms (4,400 pounds), there is *Torvosaurus*, a monster even compared with *Allosaurus*. (Just to keep things in perspective, the infamous Late Cretaceous tyrant *Tyrannosaurus* was about 5,000 kilograms [11,000 pounds], more than twice the size of *Torvosaurus* and about sixteen times the weight of the largest living predatory mammals!) For a number of reasons—not least of which is the impressive rows of serrated and recurved teeth lining the jaws—we can be quite certain that all of the Morrison theropods were strict meat eaters—no sissy plant supplements here. So, with at least three different kinds of carnivores (*Allosaurus*, *Ceratosaurus*, and *Torvosaurus*) far exceeding the biggest polar bears and cats, we are forced to conclude that, as with the herbivores already discussed, this was a carnivore community unlike anything we know of today.

Despite the presence of *Torvosaurus*, the dominant predator in the Morrison ecosystem appears to have been *Allosaurus*. Whereas *Torvosaurus* is known from a smattering of incomplete specimens, *Allosaurus* is represented by more than one hundred individuals. *Allosaurus* fossils run the gamut from isolated bones to virtually complete and exquisitely preserved articulated skeletons. However, our best glimpse of this Jurassic predator comes from a single bonebed in central-eastern Utah. The Cleveland-Lloyd Dinosaur Quarry has to date yielded the bony remains of about fifty allosaurs, from juveniles to adults, making this assemblage the largest known for any large theropod dinosaur. Also present at Cleveland-Lloyd, though in much smaller numbers, are *Stokesosaurus*, *Marshosaurus*, *Tanycolagreus*, *Ceratosaurus*, and *Torvosaurus*, evidence of a diverse guild of coexisting predators in the Morrison ecosystem. Remains of plant-eating dinosaurs like *Camarasaurus* and *Stegosaurus* are found at Cleveland-Lloyd as well, but in far fewer numbers. Over the years, much discussion has ensued about the formation of this site. Was it a predator trap, miring many carnivores attempting to make a meal of a single stuck herbivore? Or did drought events repeatedly congregate theropods around a waterhole where no herbivore would dare to tread? Disagreement persists among paleontologists about the genesis of this site. However, Cleveland-Lloyd provides another excellent snapshot of a Jurassic ecosystem, and it was deposited at about the same time as the Douglass Quarry at Dinosaur National Monument.

How much meat would an adult *Allosaurus* need to consume daily? Farlow made this calculation by inserting the mass of an adult allosaur (1,000 kilograms, or 2,200 pounds) into equations that have been used to estimate daily food intake for living endotherms and ectotherms. The results suggest that a mammal-like, endothermic allosaur would eat an average of 17 kilograms (37 pounds) a day, equivalent to consuming a border collie (or a 4-year-old child). At the other end of the metabolic spectrum, a lizardlike ectothermic allosaur would need only about 3.5 kilograms (8 pounds) a day, equal to eating a slightly overfed toy poodle. If allosaurs were mesotherms, as proposed earlier in this book, their caloric needs would have been intermediate between these extremes, equal to, say, one basset hound per day.

These food requirements may not seem like much for such a large animal. However, the ecological constraints of a predatory lifestyle become more apparent when translated into the amount of space required to support carnivorous giants. Farlow estimated the home range area of a single *Allosaurus* by making comparisons with living examples of ectothermic or endothermic carnivores scaled up to dinosaurian proportions. These results, which he readily admits are gross estimates, range from 85 square kilometers (33 square miles) for the ectotherm to about 18,000 square kilometers (6,950 square miles) for the endotherm! No matter what the true value was during the Jurassic, even the impressive minimum estimate confirms that allosaurs must have been on the move a great deal, either as individuals or as widely spaced packs. It looks like we'll need a vast allotment of land for our Jurassic Park.

The next logical question is, What kinds of animals were the favored prey of the various Morrison theropods? Once again we are restricted to speculative answers. One might suppose that, at least for the three largest predator species, sauropod steak (blood rare, of course) was the preferred menu item, because the long-necks were apparently abundant on the landscape. Sauropods would also seem particularly enticing because they were slow, likely incapable of outrunning any theropod, and contained a truly vast amount of meat.

Much has been written about the possible hunting tactics of theropod dinosaurs, and *Allosaurus* in particular, faced with the task of killing a sauropod. The skull of *Allosaurus* is relatively narrow up front, much taller than it is broad, and engineering studies confirm that this structure is much better suited to absorbing vertical forces than side-to-side forces. Some investigators have suggested that *Allosaurus* attacked its prey by opening the mouth wide and violently slamming the entire upper tooth row into the victim. Others propose that *Allosaurus* made use of its wide gape to take a sizeable chunk out of its prey using both upper and lower jaws. Perhaps the predators were adept at grabbing a quick bite and escaping before the wounded behemoth could turn on them—a hit-and-run approach to predation. Still others hypothesize that allosaurs resorted to pack hunting to take down adult sauropods.

But hold on a minute. Let's look at some basic numbers: *Allosaurus*, 1,000 kilograms (2,200 pounds); *Camarasaurus*, 40,000 kilograms (88,000 pounds). If ever there was an ultimate David-and-Goliath matchup, this is it (even if David in this instance is a 1-ton terror armed with intimidating recurved claws and teeth). As Farlow notes, from the perspective of an allosaur, *Camarasaurus* would have looked like a "moving mountain of meat." An allosaur with jaws agape running full speed into a full-sized sauropod is the equivalent of a Volkswagen bug broad-siding a fully loaded tractor-trailer truck. Natural selection would likely have regarded such "extreme" predation behavior most unkindly. Among modern megaherbivores, large size is perhaps the most effective predator deterrent, and it can even be a mighty weapon. Recall that large size renders adult African elephants virtually invulnerable to predation by lions. The same degree of immunity from predators likely applied to healthy adult sauropods, even if *Allosaurus* engaged in pack hunting.

So what's a poor allosaur to do? One option is to avoid the behemoths with the ridiculous necks and direct your hunting prowess at smaller targets such as *Camptosaurus*, *Dryosaurus*, and *Stegosaurus*. It's likely that these smaller herbivores, though by no means easy pickings (remember, *Stegosaurus* was up to five times the mass of *Allosaurus*), were frequently killed by larger theropods. Yet Farlow's calculations (based on rough estimates of the relative abundance of each species) suggest that the smaller plant eaters were not sufficiently abundant to support populations of big theropods. If so, sauropods must have been a necessary component of the diet of *Allosaurus* and other theropods. Assuming that healthy adult sauropods were off the menu, *Allosaurus* could have preyed on aged, sick, and wounded animals unable to defend themselves. Of course, there would also have been occasional windfalls of carrion as sauropods died from accident, disease, or old age.

Yet the most likely answer to this problem is that many, perhaps the majority of, sauropods passed down the gullets of theropods long before reaching adulthood. Despite rapid growth rates, sauropods had to survive about 10–15 years in the Morrison ecosystem before they achieved the stately status conferred by truly gigantic body sizes. As egg layers, dinosaurs were able to generate many more young than live-bearing mammals, which are limited by womb size and other factors. So the Jurassic landscape was likely full of hatchling, juvenile, and subadult dinosaurs, many of them tasty young sauropods all but incapable of running away.[3]

Before wrapping up our discussion of the Jurassic, let's return to the central question posed at the beginning of this chapter. As the Jurassic Park ranger in our thought experiment, you need to know how much space would be required to maintain an intact Morrison ecosystem. We can't yet be definitive, but, given the sheer size of the larger dinosaurs, we can say "one helluva lot." Jim Farlow's calculations suggest that, whether these dinosaurs were endotherms or ectotherms, the entire Four Corners region (including Utah, Colorado, Arizona, and New Mexico) was likely insufficient, with the answer lying somewhere between this size and that of the entire continent of North America. This is not to say that North America was home to a single ecosystem 150 million years ago. No, there would have been a great variety of ecosystems, as today. The point is that certain species, particularly the largest carnivores, were likely spread across several ecosystems.

As we've seen, a key constraint for the Morrison ecosystem was available energy. During the Late Jurassic, as today, solar energy kept the biosphere running. The energy captured by plants in the Four Corners region was transferred up an array of food chains within this complex web, first to herbivores, then to carnivores, and sometimes to secondary carnivores. In a very real sense, then, all carnivores can be thought of as plant consumers, at least indirectly, because the meat they ingest is made of energy derived from plants. It's often said that you are what you eat, but rarely do we think about this truism crossing links in the food chain, such that you are also made of what your dinner ate during its lifetime. Recall that energy flow is very inefficient; at each point of transfer—from plant to herbivore, herbivore to carnivore, and so on—about 90 percent of the energy is radiated out into the environment as heat, leaving just 10 percent for the next level up the chain. This constraint explains why the biomass of terrestrial ecosystems is made up mostly of plants, and why herbivores greatly outnumber carnivores. It also helps explain the lack of theropods in the gargantuan size ranges of sauropods—there was simply not enough food to sustain such Mesozoic godzillas.

Geography is another key constraint, particularly in a world of giants. In brief, the chain of causation goes like this: larger body size, more food; more food, more area (in order to find the food); more area, lower density of animals (because a given area of land can support fewer large animals); lower density of animals, larger species ranges. This last point requires clarification. Like virtually all complex systems, ecosystems are always in flux. Occasionally these fluctuations result in exceptionally bad times precipitated by such calamities as drought, flood, or fire. To reduce the odds of extinction, species must have the capacity to survive these bad times. The two most important factors for long-term survival are large numbers of individuals and an extensive range. A drought is far more likely to wipe out a species of two hundred individuals living in one valley than another of fifty thousand spread over many thousands of square kilometers. Standing counts of tens of thousands of individuals appear to be necessary for a species to have a reasonable chance of long-term success. Just as the stock market frequently penalizes those who invest too narrowly, the scythe of natural selection quickly weeds out species that drop in number much below this level.[4]

Once again we see the problems faced by giant animals, and particularly carnivores, because they must tap into the remnants of the energy budget that percolate to the top of the food web. In order to maintain enough animals to avoid extinction in the short term, any species of giant carnivore must occupy a vast range, because so few animals can live in a given area. Prior to the influence of humans, large cats, the largest strictly carnivorous mammals of recent times, occupied extensive, sometimes continent-sized ranges.

The difficulties faced by big meat eaters are multiplied if two or more species of top carnivores share the same ecosystem, as clearly occurred with our Morrison theropods.

Assuming some interspecies competition for meat, adding another predatory species to the ecological mix means that the total area necessary to support a single animal, and thus a single species, increases accordingly. With at least three truly giant carnivores living in the same Late Jurassic habitat, all of them much larger than any modern terrestrial carnivore, it's likely that the Morrison theropods required spacious ranges indeed, both as individual animals and, on a larger scale, as species.

Recall that Pangaea was still undergoing slow-motion fragmentation during the Jurassic, so plenty of land bridges still connected landmasses that we know today as distinct continents. These interconnections allowed for relative ease of movement, and dinosaur faunas appear to have been relatively homogenous during this period. For example, several Late Jurassic dinosaurs recovered from Europe and Africa are closely related to those found in the Morrison Formation. In a few instances, it's even been argued that the same species are present on two or more of these landmasses, although such claims seem to me to be suspect at best. Either way, this evidence points to a high degree of global cosmopolitanism in the Late Jurassic that rivals or exceeds that of any other time during the last 150 million years.

To bring our Jurassic Park dream into better focus, let's summarize some of what we know about the Morrison ecosystem. The emerging picture of this 150-million-year-old ecological web is at once familiar and fantastic. It includes broad rivers winding lazily through a landscape of lakes and wetlands. Particularly during the daylight hours, plant-eating dinosaurs large and small probably congregated amid buzzing insects and tall trees along the margins of these watery corridors. The thick vegetation offered not only sustenance but a measure of protection from both predators and the midday sun. Even in the shade, however, overheating was likely a major concern for dinosaurs. Like living megaherbivores, sauropods and stegosaurs may frequently have submerged themselves during the hottest portions of the day in order to cool down. Perhaps some species ventured into more open settings at night to feed. With the possible exception of adult sauropods, dinosaurian herbivores kept a watchful eye out for theropods, perhaps congregating into groups or even multispecies assemblages in order to reduce the threat of predation. Meanwhile, large and small theropods were ever-present, probably living singly and/or in smaller groups, some perhaps staking out territories on this mixed terrain. When one of the smaller carnivores—*Stokesosaurus* or even *Ceratosaurus*—made a kill, they likely wolfed down chunks of fresh meat as quickly as possible so as to get their fill before other predators took notice and challenged them for the carcass.

Together, this array of beasts, herbivore and carnivore alike, generated great heaps of dung that injected nitrogen and other key nutrients into the soil. The remains of all of these animals, as well as a great volume of unconsumed vegetation, were ultimately metabolized by a bevy of decomposing organisms—fungi, insects, protists, and bacteria—with the recycled nutrients used to construct future generations of plants and animals.

The final result was a magnificent, mind-bogglingly complex, ever-cycling whole. The fossil record tells us that this system had enough built-in resilience to withstand minor environmental hiccups and endure the comings and goings of species without jeopardizing the integrity of the whole. However, thanks to evolution, it also had the capacity to undergo major change, to literally transform itself in response to external pressures. Moving forward in time to the Cretaceous, we now turn to the topic of transformation, involving the interplay of ecology and evolution.

A scene from the Late Cretaceous Kaiparowits ecosystem, western North America. A herd of chasmosaurine horned dinosaurs moves across a floodplain, joined by a few *Gryposaurus* duck-billed dinosaurs. One of the *Gryposaurus* adults pauses to scan for predators.

14

WEST SIDE STORY

THE WALKIE-TALKIE CRACKLED with the disappointing news, realizing our collective fears. The helicopter had failed to lift the largest block containing the bulk of the duck-billed dinosaur skeleton. Our crews from the Utah Museum of Natural History (UMNH) had spent much of the last two field seasons working more than 3 kilometers (2 miles) from the nearest road, hiking daily through the hilly Utah badlands while carrying tools and supplies—rock saws, hammers, awls, brushes, plaster, burlap, and water—in order to unearth the 75-million-year-old bones. From stem to stern, this plant-eating dinosaur would have been about 11 meters (34 feet) long and weighed about as much as an African elephant. The skeleton's back end was found with the bones connected as in life. The front end, in contrast, was jumbled prior to burial, likely by flowing water. Although only part of the skull was present and little evidence was found of the legs, this specimen represented the most complete skeleton yet found in Grand Staircase–Escalante National Monument.[1] The bones were encased in brutally hard sandstone and spread over too great an area to be removed in a single chunk. So the crew had spent weeks dividing up the specimen into more manageable blocks, each consisting of multiple fossils together with the surrounding rocky matrix. Each received a mummy-like wrapping of plaster and burlap, a protective JACKET that would allow the bones to be safely moved back to the museum. Now the only problem was finding a ride for this dinosaur.

Because of the large size of the fossils, dinosaur workers regularly encounter logistical challenges not faced by most paleontologists. Smaller jackets (weighing up to about 25 kilograms, or 55 pounds) are carried by hand or placed into backpacks and walked to the nearest road. Larger specimens up to about 70 kilograms (154 pounds) can be secured to a stretcher and carried out by teams of four or more (an odd practice when one contemplates "rescuing" animals that have been dead for millions of years). Still heavier specimens (up to about 150 kilograms, or 330 pounds) can be roped down to a large platform—an old car hood used as a sled is often the tool of choice—and dragged over the ground. Occasionally, horsepower in the form of actual horses is brought in to replace humans for these drags. Yet the dragging technique is practical only for short distances. Finally, for remote specimens too large to be carried or dragged, paleontologists commonly resort to helicopter airlifts.

Dominated by rugged, sparsely vegetated, and spectacularly beautiful badlands, Grand Staircase–Escalante National Monument encompasses almost 2 million acres of south-central Utah and a vast portion of the greater Four Corners region. Because of its inhospitable terrain, the area now encompassed by the Monument was the last major region within the lower forty-eight United States to be formally mapped. The earliest white settlers (and almost everyone since) elected to circumnavigate this arid, desolate terrain rather than attempt a crossing. For the same reason, Grand Staircase now represents one of the last largely unexplored regions in the United States from a paleontological perspective. In 1996, Bill Clinton signed a Presidential Proclamation designating these fragile, high-desert wildlands a national monument to be administered by the Bureau of Land Management. The Monument was set aside in large part for the preservation and study of its diverse natural bounty, both living and fossil. It is surrounded by other public lands; some, like Zion, Bryce Canyon, and Capitol Reef national parks receive millions of visitors each year. In contrast, Grand Staircase retains a strong sense of remoteness and has far fewer visitors, largely because vehicle access to its interior is limited to a smattering of dirt roads.

The Monument's name refers to the staircase-like arrangement of rocks ascending from Grand Canyon National Park in the south to Bryce Canyon National Park in the north. In between is a series of gargantuan steps separated by spectacular, multihued cliffs of chocolate, vermillion, white, gray, and pink. This remarkable geologic sequence provides a localized glimpse of much of the Mesozoic, with Triassic rocks transitioning to Jurassic and then to Cretaceous, all capped by pink cliffs of the early Cenozoic Claron Formation. The three formations highlighted in this book—Chinle, Morrison, and Kaiparowits—occur in the Grand Staircase, although the Kaiparowits Formation only minimally so.

A few miles east, however, in the north-central Kaiparowits Plateau region of the Monument, the Kaiparowits Formation is abundantly exposed as gray, cliff-forming badlands. Together with the underlying Wahweap, Straight Cliffs, and Tropic Shale formations, these sediments preserve one of the most continuous records anywhere of the end of the dinosaur era. In 2000, together with Mike Getty of the UMNH,[2] I initiated a project to explore these badlands, and the Bureau of Land Management kindly agreed to fund the effort.[3] Prior to commencement of our work, most paleontological efforts within the Monument had concentrated on small Cretaceous denizens, particularly rodent-sized

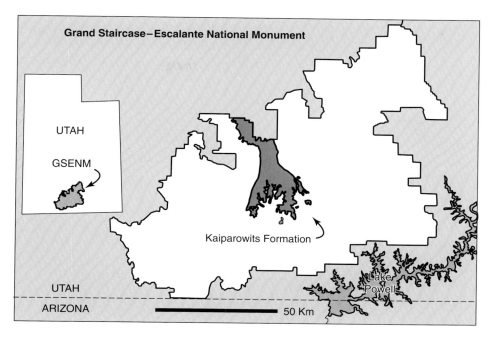

FIGURE 14.1

Map of Grand Staircase–Escalante National Monument, southern Utah. The dark shaded region depicts the area of greatest exposure of the Late Cretaceous Kaiparowits Formation. Inset depicts scale of the Monument relative to the state of Utah.

mammals. Dinosaurs were known, but limited to fragmentary remains, most simply picked from the surface. Were it not for the fact that this place is so inaccessible, the fossil riches of Grand Staircase would likely have been pillaged long ago, and I would have been forced to look elsewhere, likely in less productive sediments. I sometimes find this thought comforting, even inspiring, as I sit quietly and gaze out at the vast tracts of multihued, unexplored badlands. However, on that day, as we tried unsuccessfully to airlift a dinosaur, the thought offered minimal consolation.

Several of the hadrosaur blocks we were trying to get out of the Grand Staircase backcountry were well within the oversized, helicopter-only category, with the largest jacket roughly estimated at 500 kilograms (1,100 pounds). The National Park Service from nearby Zion National Park had kindly offered their firefighting helicopter and flight crew free of charge to perform the dinosaur extraction. We were extremely thankful and excited by the prospect of getting this beast on the move for the first time in millions of years. Up to this point, everything had gone off according to plan. Prior to airlift day, a mixed Park Service–UMNH crew had hiked out to the excavation site and manhandled the jacketed fossils into large, sturdy nets to which the helicopter lift cable would be attached. Several of our crew members had walked again to the site earlier that morning so that they would be present to attach the cable when the air support arrived. The helicopter arrived on

FIGURE 14.2

Top: Skeleton of the duck-billed (hadrosaurine) dinosaur *Gryposaurus monumentensis* in the ground in Grand Staircase–Escalante National Monument. Most of the visible bones are part of a tail in life position. Bottom: Fossilized skin impression found in association with this skeleton. This particular specimen is the "Dangling Dino" described in the chapter.

schedule, and the flight crew was briefed about the day's operation. An old flatbed truck would be waiting back at the nearest road so that the pilot could lower the fragile cargo directly into the vehicle. The weather was gorgeous, clear and crisp—typical for a fall morning in southern Utah.

Whereas it took us a good hour to hike out to the site from the road, the two-person crew flew the A-Star B2 helicopter out there in a few short minutes. Hovering above the hilltop, the ground crew attached the cable to the heaviest block, which contained most of the pelvis, hind limb, and tail of the duck-billed dinosaur. The pilot attempted to raise the oddly contoured, ivory white package several times before concluding that it was too heavy.

Upon hearing this news, my mind reeled at the thought of accessing (and paying for) a larger helicopter. I then recalled hearing of a similar airlift in Alberta that had gone much worse; the pilot had managed to get the dinosaur skeleton airborne only to find that he could no longer control the helicopter in the wind. Making a split-second decision, he elected to detach the cable. The jettisoned dinosaur experienced a brief freefall before colliding abruptly with the earth, transforming the precious cargo into dust. No, for the sake of humans and fossils alike, discretion is clearly the better part of valor in such endeavors.

Unable to handle the largest load, the crews began the process of airlifting the remaining skeletal blocks, all of which were considerably smaller. At the culmination of each trip, the pilot deftly maneuvered the cargo into the awaiting truck, where additional crew members were on hand to detach the lift cable. After the last of the smaller packages was secured, the pilot announced that he was going to give the largest block one last attempt. The ground crew explained to us that the fuel used in transporting the other blocks had reduced the weight of the helicopter enough that it might now be safe to handle the huge jacket that remained. Sure enough, the pilot's experience paid off. This time around, he was able to lift the massive block and carry it safely to join the others in the flatbed. Mission accomplished. A few weeks later I received a T-shirt from the Bureau of Land Management commemorating the event. It featured a cartoon of a smiling hadrosaur hanging from a helicopter, and beneath were the words "Operation Dangling Dino, September 29, 2004."

Months later, following hundreds of hours of meticulous preparation by a team of volunteers at the UMNH, we determined that this duck-billed dinosaur was new to science. The assessment was aided by discovery of an impressive, nearly complete skull of the same species found by the Alf Museum of Paleontology.[4] A characteristic crest atop the skull, resembling an exaggerated "Roman nose," revealed that this duck-billed dinosaur was a member of the genus *Gryposaurus*. Yet it clearly differed from all other gryposaur species. In particular, many bones are extremely thick and robust, giving the animal a "pumped-up" appearance—the Arnold Schwarzenegger of dinosaurs. One of my graduate students, Terry (Bucky) Gates, undertook the scientific description of this beast and elected to honor the national monument by naming the new species *Gryposaurus monumentensis*.

Gryposaurus monumentensis is one of many different kinds of dinosaurs and other animals found during 8 years of intensive work in Grand Staircase. The majority of our discoveries derive from the fossil-rich Kaiparowits Formation, which dates to between 76 and 74 million years ago. Thus far, our crews—including professional researchers, graduate students, and undergraduate students, but heavily dominated by volunteers—have found more than six hundred new fossil sites. Some consist of single, isolated bones; others are composed of handfuls of scattered elements; and still others contain partial to nearly complete skeletons like those of the dangling duck-bill. The Kaiparowits dinosaur menagerie includes two additional varieties of duck-billed dinosaurs: a tube-crested species that belongs to the genus *Parasaurolophus*, and yet another variety of *Gryposaurus*. Other herbivores include three kinds of large horned dinosaurs (ceratopsids); two kinds of armored ankylosaurs; one or two dome-headed dinosaurs (pachycephalosaurs); and at least one small, bipedal ornithopod (hypsilophodont). On the carnivore side of the ecological

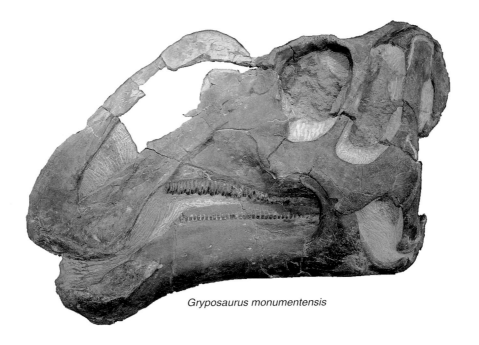

Gryposaurus monumentensis

Unnamed ceratopsian

FIGURE 14.3

Top: Actual skull of the duck-billed (hadrosaurine) dinosaur *Gryposaurus monumentensis*. Bottom: Reconstructed skull of an as yet unnamed chasmosaurine horned dinosaur. Both animals were recovered from the Late Cretaceous Kaiparowits Formation of southern Utah. Skulls are not to scale.

ledger is an array of large and small theropods. Small-bodied forms include at least three sickle-clawed predators (two dromaeosaurs and a troodont), an ornithomimosaur, and an oviraptorosaur that we named *Hagryphus*.[5] By far the largest theropod is a 1,000- to 2,000-kilogram (2,200- to 4,400-pound) tyrannosaur, much smaller than *T. rex* but typical of tyrannosaurs of that time. Also present was the monstrous 9-meter (30-foot) crocodile, *Deinosuchus*, which may have rivaled the tyrannosaur for the top carnivore spot.

The horned dinosaurs are particularly noteworthy. Of the three new species, one is a member of the short-frilled CENTROSAURINES. Most members of this group possess a small horn over each eye and a larger horn over the nose, but the new Grand Staircase centrosaur has a small, bladelike nose horn together with elongate eye horns curving straight forward instead of upward. The two remaining Kaiparowits ceratopsids are members of the long-frilled CHASMOSAURINES, the group that includes *Triceratops*. One of these is an enormous beast closely related to *Pentaceratops*, for which the skull alone exceeds 2.2 meters (7 feet) in length. The final member of the threesome is truly bizarre, adorned with a stout nose horn, elongate eye horns that arc sideways, and a short, broad frill rimmed at the back by ten curved hooks! With all of these bony bells and whistles, this beast ranks as one the most ornate dinosaurs ever found.

Preservation of the fossilized bones, most of which are a rich chocolate brown, is frequently exceptional. In contrast to all but a few dinosaur-bearing formations worldwide, many of the Kaiparowits dinosaur skeletons, particularly those of duck-bills, are found with abundant skin impressions. It's long been thought that these "dinosaur mummies," as they are called, formed under arid, desiccating conditions. Yet the mummies in Grand Staircase are buried in river channel sandstones, and a variety of geologic clues point to the Kaiparowits Formation being deposited in a very wet, perhaps even swamplike setting. So we clearly need to rethink the fossilization process, particularly as it relates to such instances of exceptional preservation. The many Kaiparowits dinosaurs are accompanied by a diverse variety of nondinosaurs—fishes, amphibians, lizards, turtles, crocodiles, and mammals, as well as abundant evidence of plants, insects, and clams—enabling us to get a meaningful glimpse of this ancient ecosystem.

Of the entire 160-million-year Mesozoic tenure of dinosaurs, perhaps the best-known interval worldwide spans the last 15 million years of the Cretaceous, from about 80–65 million years ago. Although spectacular finds have been made all over the globe—mostly recently in such places as China and Argentina—the most comprehensive record of the lives and times of dinosaurs occurs in western North America. This productive legacy is due in part to the many dinosaur hunters that have scoured the badlands of the American West for well over a century. Even more important is the fact that these badlands are some of the richest in the world for finding dinosaurs. In particular, rocks of Late Cretaceous age have yielded a great diversity of forms spanning millions of years of geologic time and a geographic range that stretches from Alaska to Mexico. Thousands of dinosaur specimens, including numerous skeletons and mass death assemblages, have been unearthed and studied. The rocks have been subject to intensive examination as

Unnamed tryannosaur

Hypsilophodont

Hagryphus

Parasaurolophus

Unnamed ceratopsian

FIGURE 14.4

Representative animals from the Late Cretaceous Kaiparowits Formation: an as yet unnamed tyrannosaur theropod; the oviraptorosaur theropod *Hagryphus*; an unnamed hypsilophodont ornithopod; the lambeosaurine hadrosaur *Parasaurolophus*; an unnamed chasmosaurine ceratopsid.

well, partially because of the remarkable fossils interred within but also because of the oil-, coal-, and gas-bearing potential of these sediments. Consequently, we know more about Late Cretaceous dinosaurs from western North America than from any other time-space slice of the Mesozoic. This bounty of information allows us to address a range of paleontological questions that cannot be broached elsewhere.

Latest Cretaceous dinosaur faunas are reasonably well known in the northern portion of the Western Interior—particularly in Alberta, Montana, and Wyoming—yet they remain poorly documented in the South. Certainly Late Cretaceous dinosaurs have been found in several regions in the Southwest, including Utah, Colorado, New Mexico, Texas, and various parts of Mexico, but the fossils are often fragmentary. Although these southern dinosaur finds have often been resolved to major group (tyrannosaurs, hadrosaurs ceratopsids, etc.), most are not sufficiently complete to make species identifications. The dearth of southern fossils heightens the importance of Grand Staircase–Escalante National Monument, which has now yielded the best documented Late Cretaceous dinosaur assemblage in the American Southwest.

In one sense, the Kaiparowits dinosaurs are unremarkable. The major groups described earlier are typical of dinosaur faunas of this age recovered from the North American Western Interior. In these other places, as well as in Grand Staircase, hadrosaurs and ceratopsids are invariably the dominant megaherbivores, with other ornithischians—pachycephalosaurs, ankylosaurids, and hypsilophodonts—present but considerably more rare. Among theropods, tyrannosaurs are invariably the sole large-bodied carnivores, although they are accompanied by a range of smaller forms, including ornithomimosaurs, dromaeosaurs, and troodonts. Yet in another sense the Kaiparowits fauna is stunning in its uniqueness, hinting at an amazing story that we have only begun to unravel.

During the 1960s, a couple of paleontologists noticed that dinosaurs and other Late Cretaceous animals found in the southern portion of the Western Interior represented distinct species relative to those found farther north. Then, in a series of papers beginning in 1987, Thomas Lehman of Texas Tech University documented this geographic variation in some detail, concluding that the evidence strongly supported the notion of distinctive plant and animal communities, or biomes, at different latitudes with the Western Interior. This finding came as a considerable surprise. Given their large body sizes, presumably with appetites to match, it had long been assumed that dinosaurs had extensive geographic ranges. Some paleontologists went so far as to suggest that, much like caribou today, herds of plant-eating dinosaurs undertook seasonal long-distance migrations, perhaps spending summers in the high latitudes of Alaska and overwintering much farther south. Lehman's results seemed to point in the opposite direction, toward much smaller, year-round species ranges.

The new Kaiparowits fauna being unearthed from southern Utah strongly supports Lehman's hypothesis. None of the sixteen dinosaur species found thus far have been conclusively documented in the northern region of western North America. Similarly, the even greater numbers of roughly coeval dinosaur species recovered up north in Alberta and Montana are absent from Utah. Not only is this finding inconsistent with migrations between the northern and southern regions of North America, but it also suggests that most kinds of dinosaur had remarkably diminutive species ranges, perhaps much smaller than those required by large-bodied mammals living today. To make matters worse (and more interesting), bonebed evidence elsewhere indicates that some dinosaur varieties, particularly among the horned and duck-billed dinosaurs, congregated at least occasionally in large groups of hundreds or even thousands of individuals, raising further questions as to how these animals found sufficient food to survive.

The mystery deepens when we understand the environmental setting in North America during Late Cretaceous times. For most of the Late Cretaceous, a shallow seaway extended from the Arctic Ocean in the north to the Gulf of Mexico in the south, splitting North America into two landmasses that we will refer to informally as "West" and "East" America.[6] WEST AMERICA, home to *Gryposaurus monumentensis* and dozens of other Late Cretaceous dinosaurs in the Western Interior, was effectively a narrow, elongate island continent. During Kaiparowits times, about 75 million years ago, the total combined area of the dinosaur-rich habitats in West America was about 4 million square kilometers (1.5 million square miles), less than 20 percent of the present-day size of North America. It remained mostly isolated for more than 30 million years. A north-south chain of mountains ran like a jagged spine down most of the length of the peninsular landmass, just as the Rocky Mountains do today.

The margins of the CRETACEOUS INTERIOR SEAWAY, far from being static, periodically underwent major expansions and contractions driven by the ebb and flow of global sea levels. Known formally as TRANSGRESSIONS and REGRESSIONS, respectively, these glacially paced seaway fluctuations produced a coincident shrinking and swelling of available habitat on land. The rising mountains and migrating seaways also resulted in a relatively complete geologic record, faithfully recording these marine incursions in the rocks. A paleontologist or geologist walking up a barren badlands slope in Utah, Montana, or Alberta is a time traveler of sorts, passing through millions of years of stacked, ancient environments—from river floodplain to nearshore coastal plain to seaside beaches to shallow marine settings, followed by the same sequence in reverse. Fortunately, this layer-cake deposition of sediments also preserved vast numbers of fossils.

Sandwiched between rising mountains to the west and a restless seaway to the east, West American dinosaurs had nowhere to run. So the fact that we find one set of giant dinosaurs living in Montana and an entirely different set of species living less than 1,000 miles south in Utah is, well, perplexing. How did these reptilian giants find sufficient food? Did they eat less than expected (suggestive of metabolisms at the cold-blooded end of the spectrum)? Was there an abundance of available plants, enabling the animals to occupy smaller ranges? Alternatively, is it possible that the dinosaurs we find in different regions lived at different times, giving the illusion of small ranges? Or is

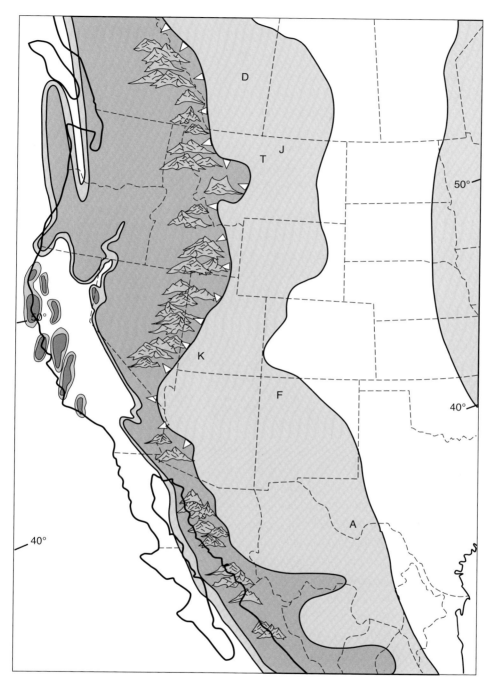

FIGURE 14.5

Late Cretaceous western North America (about 75 million years ago), with the Cretaceous Interior Seaway (white region on right side) covering the central portion of the continent. The positions of present-day states and provinces are noted, as are the locations of key geologic formations. Abbreviations: D, Dinosaur Park Formation, Alberta; J, Judith River Formation, Montana; T, Two Medicine Formation, Montana; K, Kaiparowits Formation, Utah; F, Fruitland Formation, New Mexico; A, Aguja Formation, Texas. Numbers along margin indicate the latitude during the Late Cretaceous.

there some other, as yet unconceived explanation? Before we can further plumb the depths of this mystery, we need to know more about the Late Cretaceous world.

Dinosaur evolution in Late Cretaceous North America occurred against a backdrop of global environmental changes. Much of this flux can be traced to the restless activities of plate tectonics. Beginning in the Late Jurassic, an increase in the flow of molten rock deep in Earth accelerated the movement of crust at the surface. The most obvious result was continental breakup—the final phases in the fragmentation of Pangaea. By the close of the Cretaceous, most of the major landmasses we know today were isolated or nearly so, separated from one another by the same oceans that continue to inundate most of the planet's surface.

The increasing isolation of Cretaceous continents segregated land-based floras and faunas. Separation of each "island" continent created new evolutionary opportunities, because this newfound isolation greatly limited the mixing of genes over large distances. Each continent became a semi-isolated evolutionary experiment with its own unique set of coevolving ecosystems. In the dinosaur realm, the great diversity of long-necked sauropods was largely replaced by representatives from a single group—titanosaurs. Late Cretaceous North America was an exception to this pattern, with sauropods largely displaced by smaller (though still giant) horned and duck-billed dinosaurs. A parallel pattern occurred among the top carnivores. Following the demise of the Jurassic allosaurs, torvosaurs, and their close kin, groups such as carcharodontosaurs and abelisaurs (distant relatives of the Late Jurassic *Ceratosaurus*), reigned south of the equator, whereas tyrannosaurs rose to unequivocal dominance in Asia and North America.

Occasionally, a drop in global sea level briefly reestablished land connections between the newly isolated continents, fostering a flurry of floral and faunal exchanges. North America, for example, shared geologically short-lived land connections with Europe and Asia in the Early Cretaceous and with Asia and South America during the Late Cretaceous, all of which added vital ingredients to the evolutionary mix and altered the fates of the constituent life-forms. Asia and North America, in particular, shared very similar faunas during parts of the Late Cretaceous, with dinosaurs and other life-forms crossing an ancient land bridge between eastern Asia and northwestern North America (present-day Alaska). (Tens of millions of years later, humans would use this same route to disperse from Asia to North America, and quickly thereafter to South America.) Nevertheless, for most of the Late Cretaceous, West America and Asia were fully isolated from each other.

Two further consequences of the Cretaceous pulse in tectonic activity were elevated sea levels and warm, equable climates. Recall that Mesozoic climates in general were much warmer than at present and that this trend applied throughout most of the Cretaceous. As with earlier periods of the Mesozoic, polar regions during the Cretaceous appear to have lacked ice caps. The Cretaceous biosphere was swept up in a major global warming event driven by a buildup of carbon dioxide in the atmosphere. Where did this extra carbon dioxide come from? Obviously not smokestacks, exhaust pipes, or jetliners. Increased seafloor spreading led to greater volumes of oceanic crust being subducted (driven down into Earth's mantle) along continental margins, a process that triggered widespread volcanism. Volcanic eruptions released massive quantities of carbon dioxide, which in turn increased atmospheric temperatures.

Yet the Late Cretaceous was no hothouse-hell-on-Earth. Instead, the climate was mild for most of the year, more like the Caribbean Islands than Death Valley in midsummer. Seventy-five million years ago, tropical and subtropical climates were widespread, extending at least to 70 degrees south and 45 degrees north latitude. Elevated sea levels were responsible for much of this moderating influence. Oceans expanded in part because of the higher atmospheric temperatures, which prevented water from being locked up in ice over polar landmasses. Higher rates of seafloor spreading also translated into raised midocean spreading ridges, which in turn displaced more water onto the continents, producing shallow seas on the low-lying regions of many continents. North America's Cretaceous Western Interior Seaway is a prime example of this phenomenon. The shallow sea had a strong moderating influence on climate, resulting in slower and smaller temperature swings seasonally (recall the "San Diego effect" discussed in chapter 4). This situation contrasts mightily with what we see today in the interior of North America—hot summers and cold, icy winters.

Early Cretaceous plants were similar to earlier Jurassic versions, with ferns, horsetails and other spore-bearing pteridophytes living alongside seed-bearing conifers, cycads, ginkgos, and bennettites. However, a plant revolution was kicked off in the Early Cretaceous with the appearance of another group of seed plants—angiosperms, or flowering plants. Although flowering plants apparently got off to a slow start, they had taken over many terrestrial habitats by the mid-Cretaceous. This domination has persisted to the present day, with living angiosperms numbering at least 275,000 species. As detailed in chapter 7, Late Cretaceous flowering plants were a radical departure from their gymnosperm forebears, offering an abundant, diverse, and readily renewable food source. Angiosperms also occupied a wider variety of habitats, recovered more quickly from damage, and possessed fewer chemical and mechanical impediments to digestion. Trees included such familiar names as beech, oak, maple, and magnolia (not the same species as today, but near relatives). Grasses served as the primary energy source of post-Mesozoic land-based ecosystems, but they were absent from West American habitats. What kinds of plants, then, made up the Late Cretaceous ground cover? Kirk Johnson's work suggests that, for most of this time, hadrosaurs, ceratopsians, and tyrannosaurs tromped through vast meadows of leafy, herbaceous flowering plants, including buttercups, hops, and nettles. Indeed, these low-lying plants may have been a mainstay in the diets of some herbivorous dinosaurs.

Let's return now to the problem of latitudinal variations in West American dinosaurs around 75 million years ago. Before we can be confident that different species of dinosaurs coexisted in the north and south of the West America, we must rule out the possibility of time-based differences. That is, while the known species of dinosaurs are different in Alberta and Montana than they are in Utah and New Mexico, it's conceivable that these animals did not actually overlap in time. If so, the presumed north-south provincialism would be illusory, with the actual story no more remarkable than geographically widespread dinosaur communities changing through time.

Eric Roberts is an exceptional young scientist at Southern Utah University and the geologist on the Grand Staircase project. While conducting a geologic survey of the Kaiparowits Formation, he found several ash layers deposited during powerful volcanic eruptions back in the Late Cretaceous. Each of these layers was dated using a common radiometric technique that compares relative amounts of two isotopes of the element argon (see chapter 2). The results indicate that the Kaiparowits Formation was laid down over a period of 2 million years, between about 76 and 74 million years ago. Roberts compared these numbers with dates obtained from other dinosaur-bearing formations farther to the north in the Late Cretaceous Western Interior—in particular, the Dinosaur Park Formation in Alberta and the Two Medicine and Judith River formations in Montana—and found that these geologic units overlapped in time. This finding suggests that many of the dinosaurs found within these formations co-occurred in time, if not in space, offering critical support for the provincialism hypothesis.

However, before we make the leap from rocks to bones and start comparing dinosaur faunas from north and south, we need to know how long, on average, a typical dinosaur species lived. Two million years is a heck of a long time, almost beyond human conception. If a typical dinosaur species persisted for, say, 100,000 years, then many different sets of species could have come and gone throughout the 2-million-year window of time, once again invalidating comparisons of northern and southern faunas.

Estimating the duration, or total lifetime, of a species turns out to be problematic. The best we can do is to provide a minimum estimate, equivalent to the time difference between the geologically oldest and youngest specimens. After all, someone could find an older or younger fossilized example tomorrow, extending that duration. The more complete the geologic and fossil records, the greater the confidence we can place in such estimates. For many kinds of hard-shelled marine organisms—for example, corals, clams, and ammonites—species durations can be determined with a relatively high degree of precision because the known fossil sample numbers in the hundreds or even thousands of specimens. Unfortunately, for the majority of dinosaur species (and vertebrates generally), species are known from only one or a few specimens, making it impossible to accurately assess their distribution in time. Nevertheless, enough dinosaurs are sufficiently represented in the fossil record to make some solid guesses. The average duration between first appearance (birth) and extinction (death) of dinosaur species (and species of other vertebrates) is on the order of 1 million years. So it appears we can be quite confident that we're comparing apples with apples when we compare, for example, dinosaurs from the Kaiparowits Formation of Utah and the Dinosaur Park Formation of Alberta, Canada.

The saga of West America becomes even more intriguing when we look closely at the temporal spans of dinosaur species. Rock formations bearing abundant remains of Late Cretaceous dinosaurs are arrayed along the Western Interior like jewels in a necklace. The crown jewel in this series is the Dinosaur Park Formation, abundantly exposed in a small (about 80 square kilometers, or 30 square miles), astoundingly fossil-rich pocket of badlands, Dinosaur Provincial Park, in southern Alberta. Arguably the greatest dinosaur boneyard on Earth, Dinosaur Park is a paleontologist's dream—spectacular badlands with minimal vegetation and fossils seemingly around every corner. After spending a while looking for fossils here, you get the feeling there's a skeleton buried in every other hillside. The first paleontologists to work this area over a century ago include legendary figures like Barnum Brown and Charles Sternberg. More recently, fieldwork conducted by crews from the Royal Tyrrell Museum of Paleontology, Alberta, has been led by Philip Currie, David Eberth, and Don Brinkman. In the intervening century, a little fewer than five hundred dinosaur skeletons have been found here, many superbly preserved. The bounty of dinosaur remains comprises more than thirty species, an astounding number for a single geologic formation.

Over the past two decades, a major effort was made to determine exactly when during the Late Cretaceous each of these dinosaur species existed. The results reveal some surprising patterns of origins and extinctions. For example, one kind of crested hadrosaur, *Corythosaurus casuarius*, occurs only in the lower portion of the formation, whereas a related form, *Lambeosaurus lambei*, is restricted to upper (and thus younger) sediments, suggesting that the latter replaced the former in time. Similarly, the short-frilled horned dinosaur *Centrosaurus apertus* is restricted to the lower portion of the Dinosaur Park Formation and is replaced by *Styracosaurus albertensis*, limited to the upper portion. Within the long-frilled horned dinosaurs (chasmosaurines), one species of *Chasmosaurus (C. russelli)* was replaced by another *(C. belli)*, and a third chasmosaurine species shows up at the very top of the formation. Similar patterns have been documented for some other dinosaur lineages.

Most interesting of all, preliminary evidence suggests that these turnover events within the Dinosaur Park Formation were not randomly distributed through time, instead occurring in concert across a number of groups. For example, *Centrosaurus apertus* and *Corythosaurus casuarius* are restricted to the lower part of the formation (before 75.5 million years ago), whereas *Styracosaurus albertensis* and *Lambeosaurus lambei* are found only in the upper portion of the formation (after 75.5 million years ago). This pattern suggests that some kind of external trigger may have caused multiple species to go extinct almost simultaneously, only to be replaced by close relatives. If so, we are getting strong hints of occasional, ecosystem-wide shakeups that resulted in rapid turnover of community membership.

Across the international border in Montana, Jack Horner has documented similar kinds of turnover patterns in the Two Medicine Formation. Horner and his colleagues linked these pulsed events to major incursions (transgressions) of the Cretaceous Interior Seaway, arguing that evolutionary turnover was driven by geographic bottlenecks, as the dinosaurs were jammed into a small space between the seaway to the east and mountains to the west. These workers noted further that, in most instances, the only bony features that exhibited significant change relate to bizarre structures interpreted as mating signals: ceratopsid horns and frills, hadrosaur crests, tyrannosaur horns, and pachycephalosaur skull caps.

We are now eager to find out if the dinosaur faunas in Utah and other southern regions of West America were also undergoing regular, perhaps even pulsed episodes of species turnover. If so, the next question to address is whether these turnover pulses occurred in unison in the northern and southern regions of the Western Interior. Thus far, we have been able to show that one species of *Gryposaurus monumentensis* occurs higher in the Kaiparowits Formation than another species of *Gryposaurus*, with the former presumably replacing the latter. It will be interesting to see whether other dinosaur lineages match this predicted pattern; either way, we should have a much better idea of the true story within the next 5 years or so.

Elisabeth Vrba, a renowned evolutionary biologist at Yale University, has documented similar pulses of evolutionary turnover in Pleistocene African mammals. In southern Africa, first appearances of new animal species—from antelope to birds to primates—tend to be almost simultaneous in the rock record, replacing other forms that apparently go extinct. Vrba has linked these TURNOVER PULSES to cyclical climatic changes that shifted the proportions of open versus forested habitats, driving the extinction of older species and the origin of newer forms. Vrba's interpretation (not without its detractors) is that environmental change fragments large populations into smaller subpopulations as animals attempt to maintain access to food resources by tracking their preferred habitats (e.g., grasslands or savannah). Once isolated, evolution modifies the subpopulations, adapting them to their new surroundings through the process of natural selection. At the cycle's conclusion, land habitats expand, reuniting the previously separated subpopulations, which may no longer recognize each other as members of the same species. As Jack Horner has suggested, it's feasible that Cretaceous seaway transgressions and regressions had a similar impact, driving turnover pulses in West American dinosaurs.

The origin of new animal species, particularly among large to giant forms, usually entails two fundamental steps (see chapter 6). First, populations of a given species become isolated from each other, often because of some environmental change such as seaway oscillations or climate change. Second, populations evolve independently to the point that they no longer recognize each other as members of the same species. Key factors in step 2 include Darwinian natural selection and sexual selection. In chapter 10,

I addressed the probable mating signal function of bizarre structures in dinosaurs—for example, the horns and frills of ceratopsids and the crests in hadrosaurs—and discussed their possible role in the origin of new species. If two populations of a given species become isolated from each other (step 1) and then diverge in their mating structures (step 2), they may no longer recognize each other as potential mates if subsequently reunited. In short, they will no longer interbreed, and new species will have been formed.

Diversifying into a bewildering array of shapes and sizes distinguished largely on the basis of bizarre structures, West American dinosaurs may turn out to be a prime example of this pattern. Fluctuations of the Western Interior Seaway undoubtedly transformed the world of these Cretaceous dinosaurs. During seaway expansions, the amount of habitat available to dinosaurs shrunk dramatically, and related changes may have occurred in regional climate and vegetation. So we have an environmental trigger with the potential to isolate populations (step 1) and a range of mating signal features targeted by evolution (step 2). An alternative model (to my mind, a less likely one; see chapter 6) is that the temporal sequence of, for example, horned dinosaurs within units like the Dinosaur Park of Alberta reflects speciation caught in the act within single, unbranching lineages. For example, *Centrosaurus brinkmani*, *Centrosaurus apertus*, and *Styracosaurus albertensis* may represent a sequence of ancestors and descendants that evolved in place. All of these intriguing ideas require further testing.

The emerging picture, then, is one of diverse communities of Late Cretaceous dinosaurs inhabiting a narrow strip of land, West America, where the available habitat fluctuated periodically in concert with the rise and fall of sea levels. Although the same major groupings of dinosaurs co-occurred in the north and south, the northerners and southerners appear to have belonged to distinct species. Despite body sizes that generally exceeded those of large-bodied mammals, West American dinosaurs apparently possessed small species ranges (although individual animals and herds still had plenty of room to roam). Given that most northern and southern dinosaur species within a given group appear closely similar, they may well have played similar ecological roles. For example, I would predict that long-frilled (chasmosaurine) horned dinosaurs in the north and south consumed very similar kinds of plants. If so, the ecological niches filled by dinosaurs may have changed very little for millions of years during the Late Cretaceous. Behind this apparent ecological stasis, however, a variety of factors—including seaway fluctuations and other environmental changes—resulted in relatively rapid species turnover. Just like those long-running Broadway shows, the players changed while the same story played out endlessly.

Such unexpected conclusions pose new questions. Particularly given their large size, what prevented the mixing of northern and southern faunas? Was there a physical barrier—perhaps a mountain range or a large river? Geologists have recovered no such evidence. As suggested by Thomas Lehman, it's more likely that the north and south had different climates (particularly differing amounts of rainfall) and, as a result, distinct plant communities. The fossil pollen evidence supports this view, with one major pollen type found in the south and another in the north. Inhabiting a narrow, north-south-oriented strip of land between mountains and sea, West American dinosaurs may have been sensitive to latitudinal variation in environments, despite the fact that temperature differences between the north and south were much less than those of the present day.

An even more perplexing question is how species of such big animals, many bigger than the largest living land mammals, existed on such diminutive chunks of real estate. Doesn't this contradict what we learned in the last chapter about the space needed by dinosaurs to sustain sufficient numbers? Maybe. Giant animals can make a living in small areas only if food concentrations are high, food requirements are low, or both. If most dinosaurs (and all giant ones) were indeed mesotherms, with intermediate grade metabolisms, as argued in chapter 11, much of the answer may relate to lower food requirements. Giant dinosaurs could well have had metabolic rates well below those of living mammals, enabling them to get by with much smaller individual ranges and species ranges. In addition, although we cannot confidently estimate plant biomass in Late Cretaceous West America, several indicators (soils, climate, and plant fossils) suggest a great diversity and abundance of vegetation. So the answer to this mystery may be that both slower metabolisms and greater food availability conspired to produce what was arguably the greatest florescence of Mesozoic dinosaurs.

Notwithstanding the evidence for small species ranges, a herd of hundreds (or even thousands) of horned or duck-billed dinosaurs simply could not have remained in the same place for long. They must have been moving somewhere over the course of the year; at present we just don't know where. Although current evidence does not substantiate dinosaur migrations between the northern and southern regions of West America, it is entirely possible that dinosaurs undertook long-distance migrations within the southern and northern regions, respectively. For example, animals living in the north may have migrated as far as 3,000 kilometers (1,900 miles), between present-day Montana and Alberta in the south and Alaska in the north. The longest annual migration undertaken by any land-living animal today, that of the caribou, is only about 600 kilometers (375 miles) as the crow flies. However, an individual caribou may walk about 4,800 kilometers (3,000 miles) during the course of a year. Within the northern and southern regions, then, there was still plenty of space for dinosaur migrations. The coming years will undoubtedly help us fill in critical pieces of this intriguing puzzle.

To conclude this discussion, I will consider perhaps the most fascinating case study in Cretaceous dinosaur evolution: *Tyrannosaurus rex*. *T. rex* lived in the latest part of the Late Cretaceous, the very end of the Mesozoic era, between about 68 and 65.5 million years ago. A close, though slightly smaller relative called *Tarbosaurus bataar* was the biggest predator in Asia at the same time. During the final 15 million years of the Cretaceous, the hothouse world that characterized most of the Mesozoic cooled somewhat, although the temperature difference between equator and poles remained much lower than the present day. At the same time, tectonic activity decreased, sea levels dropped, and shallow continental seaways retreated, taking with them the mild climates, which were replaced by more pronounced seasonality. Then, at the very end of this period, during the tenure of *T. rex*, climates warmed again, perhaps because of a burst of volcanic activity on the Indian subcontinent (see chapter 15).

In contrast to many dinosaurs that lived 7 million years earlier, *Tyrannosaurus rex* appears to have inhabited a broad region that minimally encompassed the entire Western Interior of North America. The rocks entombing *T. rex* fossils tell us that the giant carnivore inhabited at least three distinct environments. First were the humid, seasonally wet, coastal plain environments adjacent to the retreating Cretaceous Interior Seaway. Second were cooler, semiarid alluvial plain settings farther from the seaway yet still between the sea and the mountains. Finally, *Tyrannosaurus* remains are also found in semiarid, upland settings nestled within mountain ranges.

Importantly, each of these three environments appears to have been home to a distinct fauna of plant-eating dinosaurs. The coastal plain environment is dominated by the horned dinosaur *Triceratops*, with the giant hadrosaur *Edmontosaurus* present in small numbers. A little farther inland, alluvial plain sediments yield rare fossils of *Triceratops* together with those of the much smaller, distant relative *Leptoceratops*. Finally, the intermountain basins have produced abundant evidence of the titanosaur sauropod *Alamosaurus*, together with the giant ceratopsian *Torosaurus* and a hadrosaur of uncertain affinity. The simple discordance between the expansive geographic range of *T. rex* and the much more restricted ranges of coexisting herbivorous dinosaurs suggests that *Tyrannosaurus* was an ecological generalist that hunted an array of prey species. What did *T. rex* eat? Apparently, just about anything it wanted to.

The tyrant king represents the giant end member of an extremely successful lineage of theropods that were present in North America and Asia for most of the Cretaceous. During the Late Cretaceous, tyrannosaurs were the exclusive top theropods in ecosystems where they occurred. Tyrannosaurs existed in both eastern and western North America during at least the last 25 million years of the Late Cretaceous. Remember that, for all but the last few million years of this interval, North America was subdivided into eastern and western landmasses by the Cretaceous Interior Seaway. Then, about 69 million years ago, the seaway began to retreat, ultimately reconnecting eastern and western North America and more than doubling the available land area. Interestingly, tyrannosaurs before *T. rex* apparently never exceeded body masses in the range of about 1,000–2,000 kilograms (2,200–4,400 pounds), and there was typically more than one tyrannosaur species sharing the same habitats. Between about 69 and 68 million years ago, in the wake of the shrinking seaway, *Tyrannosaurus* evolved from a smaller-bodied cousin (perhaps *Daspletosaurus*) to emerge as the lone large theropod in its ecosystem and the largest terrestrial predator the planet has ever seen.

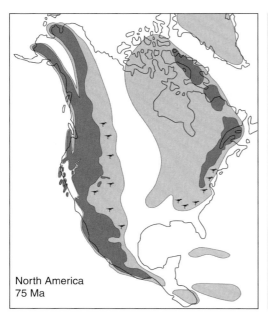

North America
75 Ma

North America
67 Ma

FIGURE 14.6

The evolution of tyrannosaur theropods on Late Cretaceous North America. Multiple species of giant (about 1,000- to 2,000-kilogram, 2,200- to 4,400-pound) tyrannosaurs were the top carnivores on "West America" 75 million years ago (left). Following retreat of the Cretaceous Interior Seaway and reconnection of eastern and western North America (about 67 million years ago), the "ultragiant" *Tyrannosaurus rex* (about 5,000 kilograms, or 11,000 pounds) appears to have been the sole tyrannosaur species in North America (right).

In brief, I and my colleagues (Mark Loewen, Jim Farlow, and Matt Carrano) think that three ecological factors were key in the evolution of *Tyrannosaurus rex*. First, an intermediate ("mesothermic") metabolism inherited from its tyrannosaur ancestors enabled *T. rex* to survive on substantially less food intake than would be required of a fully endothermic predator of similar size (see chapter 11). Second, the dramatic increase in landmass area following reconnection of eastern and western North America enabled *Tyrannosaurus* to spread out, permitting the lower population densities necessitated by extreme body masses. Third, whereas previously there had been two or more tyrannosaurs inhabiting western North America, for unknown reasons only the lineage leading to *T. rex* persisted. Without the competition for meat resources, there suddenly may have been sufficient ecological space for this survivor to expand to gigantic proportions. Interestingly, other groups of dinosaurs appear to have experienced coincident bumps in average body size. The heftiest ceratopsians *(Triceratops)*, ankylosaurs *(Ankylosaurus)*, and pachycephalosaurs *(Pachycephalosaurus)*, also appear in North America at this time. The recurrent evolution of these "giants among giants" in the latest Cretaceous may also be related to the dramatic increase in landmass size that followed retreat of the seaway. Or perhaps there is some other explanation that we simply have not yet identified.

North America is currently our best stage for watching Mesozoic ecosystems come and go, together with their dinosaurian inhabitants. It is a magnificent drama, though one we are just beginning to comprehend. Ultimately, of course, the curtain fell and dinosaurs departed stage left. The reasons behind this abrupt departure have been hotly debated, and it is to the matter of extinction, the large-scale comings and goings of life, that we now turn.

Extinction, the death of species, is a necessary part of evolution, clearing the way for new species and ecosystems. The transition from ground-dwelling theropod dinosaurs to flighted birds is an excellent example of evolution and extinction acting in tandem. From bottom to top: *Coelophysis* (Late Triassic), *Archaeopteryx* (Late Jurassic), and *Haliaeetus* (bald eagle; modern).

15

THE WAY OF ALL CREATURES

WHEN IT COMES TO the death of species, the Grim Reaper assumes several guises. The most common, referred to as BACKGROUND EXTINCTION, tends to be slow paced, ongoing, and small-scale, analogous to the daily sporadic deaths of people in every major city. The causes of background extinctions are typically small-scale as well—for example, flood, drought, or the arrival a new competitor species. The term *mass extinction*, in contrast, is reserved for extremely rare, large-scale events in which numerous species and even entire groups vanish over a relatively brief period. Whereas each event of background extinction tends to be localized, sometimes limited to single ecosystems, mass extinctions are global, impacting the full spectrum of marine and terrestrial habitats. To wreak such widespread havoc, the causes of mass extinctions must be of that rare sort capable of generating global effects. The list of suspects implicated in these horrific "biocidal" events encompasses both Earth-based and extraterrestrial perpetrators. In between the regular ticking of background extinction and the infrequent sledgehammer of mass extinctions are a variety of midscale events with intermediate effects—for example, climate change resulting in a killer long-term drought that causes extinctions across an entire continent.

Over the past half billion years or so, the time period for which we have a reasonable fossil record, there have been at least five significant mass extinctions. Our subdivision of Earth history into periods is based in part on the occurrence of these cataclysms, which occurred at the end of the Ordovician (450 million years ago), the Devonian (370 million years ago), the Permian (251 million years ago), the Triassic (213 million

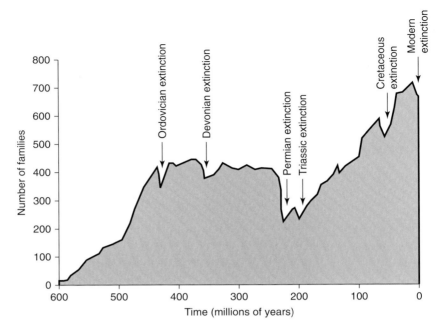

FIGURE 15.1

The five major mass extinction events on Earth during the past 500 million years. The shaded area represents the number of different families of marine invertebrates alive at any given time in the Phanerozoic Eon. Note the relative increase in families through time, with mass extinctions temporarily reducing diversity.

years ago), and the Cretaceous (65.5 million years ago).[1] Although greater than 98 percent of all species losses are likely attributable to the inexorable ticking of background extinction, this handful of mass extinctions has had profound and long-lasting effects on the biosphere. The unbalanced influence arises from the fact that mass death events simultaneously wipe out not only species but entire clans of species, opening the door to the evolution of new replacement forms from the handful of surviving lineages. And, of course, this effect is multiplied through time; every single organism alive today is part of an unbroken lineage that survived all of these mass extinction events. In effect, then, Mother Nature periodically undertakes a severe pruning of the tree of life, hacking away entire limbs. She then watches the tree regrow, budding off entirely new branches and twigs that ultimately form the foundations of new ecosystems worldwide.

Of the "big five," the Permian-Triassic (P-Tr) mass extinction—often called the "Great Dying" or sometimes the "mother of all extinctions"—was by far the largest, extinguishing about 70 percent of life on land and 90 percent in the seas, leaving in its wake a devastated biosphere of emptied ecosystems. The P-Tr extinction was perhaps the most dramatic punctuation mark in the history of life, bringing an end to the Paleozoic Era, the "Age of Ancient Life", and ushering in the Mesozoic Era, the "Age of Middle Life," sometimes known as the Age of Dinosaurs. Like the subsequent mass dying that wiped out the dinosaurs, the causes of the end-Permian event have long been debated. Possible culprits include asteroid impact, sea-level change, volcanism, and a sudden release of methane-rich water from the seafloor, causing a catastrophic drop in atmospheric oxygen. Of these alternatives, there appears to be growing consensus that the Great Dying was precipitated by large-scale volcanism in Siberia, which in turn led to profound atmospheric changes and ultimately the breakdown of ecosystems globally.

In a recent study, Sarda Sahney and Michael Benton (University of Bristol) suggest that the end-Permian extinction was actually the last (and biggest) of three major extinction pulses that took place in the Permian. These authors argue that Earth's ecosystems did not fully recover from this decimating trio of events for over 30 million years, until the Late Triassic, which happens to coincide with the appearance of dinosaurs, mammals, and several other groups (see chapter 12).

Viewed in this light, the origin and subsequent radiation of dinosaurs becomes part of the recovery of Earth's ecosystems following the mother of all extinctions. If one then considers the end-Permian and end-Cretaceous mass extinctions that bookend the Mesozoic, the dinosaur odyssey becomes inextricably interwoven with this pair of biological bottlenecks that literally transformed the face of the globe. Following the end-Permian event, vertebrate diversity on land was reduced to a smattering of amphibians, protomammals, and reptiles, paving the way for the eventual rise of dinosaurs and mammals. After the end-Cretaceous extinction—with the large dinosaurs extinguished and the amphibians, crocodiles, and turtles effectively tethered to freshwater settings—the world was primed for an explosion of mammals on land. Birds, which had begun diversifying prior to the close of the Cretaceous, also radiated following the extinction, perhaps because their major flying competitors, the leathery-winged pterosaurs, did not survive the event.

Presumably because mammals—and, in particular, our primate ancestors—diversified in the wake of the dinosaurs' disappearance, people are fascinated by the demise of dinosaurs. Indeed, it is one of the topics I and my paleontologist colleagues are asked about most frequently. So I would be remiss if I did not address the matter of their disappearance. Before delving into this story, however, it's important to emphasize three points.

First, although frequently touted as the ultimate losers and exemplars of backwardness, dinosaurs were anything but failures. As argued in chapter 1, dinosaurs are more aptly regarded as a one of life's great success stories. The misconception arises in large part from an unrealized truth: death is as inevitable for every species and group of species as it is for every organism. Although dinosaurs are the most famous members of life's prehistoric pageant, many other groups of organisms preceded them. In every instance, these ancient clans experienced a great flowering for a time, only to be unceremoniously pruned or whacked altogether from the tree of life by a major extinction event. Far greater than 99 percent of all species that have ever lived on Earth are now past tense. To make this point another way, the millions upon millions of species alive today are but a fraction of 1 percent of the total diversity that has inhabited this planet. Thus, the only thing inevitable about the reign of dinosaurs is that it would ultimately come to an end. Nothing to be ashamed of here—just the natural progression of things. Dinosaurs were the dominant large-bodied life-forms on land for over 150 million years. Primates have been around for less than half that duration, hominids have been walking upright less than 8 million years, and humans have been present for less than a half million years. So it takes a lot of gall (or at least a severe case of temporal myopia) to claim some sort of victory over dinosaurs. You may as well snub your nose at your dead relatives because, after all, you're alive and they're not.

The second point is that not all dinosaurs went extinct at the same time. As described in earlier chapters, the Triassic, Jurassic, and Cretaceous witnessed an ever-changing procession of dinosaur species and families of species. On the order of seven hundred dinosaur species are currently known; and, given all the fossil-free gaps around the globe and through time, we can be confident that this number represents but a small portion of the total. Extinction and evolution worked hand in hand, eliminating and originating species in such manner as to maintain the web of life. Sometimes a single dinosaur species in a single ecosystem was plucked from the web. At other times more sweeping extinction spasms devastated larger areas and many more species. So it's somewhat misleading to claim that the end-Cretaceous extinction wiped out the dinosaurs. Assuming that the seven hundred or so currently recognized dinosaur species are but a small portion of the total number that existed in the Mesozoic, far less than 1 percent of all dinosaur species were living at the end of the Cretaceous. Like water droplets passing through a swirling whirlpool, a long succession of species sustained the flow of dinosaurs for millions of years. In one sense, then, the surprising aspect of the end-Cretaceous extinction is not that the dinosaurs disappeared, because so many kinds of dinosaurs had vanished prior to that time. No, the remarkable thing is that there were no new dinosaurs to replace those that went extinct. Something happened that was incompatible with the perpetuation of dinosaurs, or, for that matter, any large-bodied animals.

The third and final point is one that I have made several times previously in this book. Dinosaurs actually did not go extinct 65.5 million years ago, at least not all of them. Birds, the direct descendants of dinosaurs, are dinosaurs themselves in a meaningful sense. With on the order of ten thousand living representative species (far outnumbering mammals), you could even make the argument that dinosaurs remain a thriving success story to the present day. And even if your bias is that birds are fundamentally distinct from most dinosaurs, it still must be conceded that avians are a robust legacy handed down by their Mesozoic forebears.

OK, keeping these points in mind, let's rephrase the question. Exactly what transpired 65.5 million years ago to take out *Tyrannosaurus rex* and most of its dinosaur contemporaries? Over the years, well over a hundred explanations have been proposed, running the gamut from the reasonable and testable to the inane and ridiculous. The list of proposed killing agents includes disease, slipped vertebral discs (because dinosaurs were so big), loss of interest in sex, poison plants (leading to diarrhea or constipation, depending on the hypothesis), fungal invasions, climatic change in the form of global cooling or warming (once again, depending on the hypothesis), cosmic radiation from a supernova, egg-eating mammals, sunspots, nasty aliens, and not enough room on Noah's Ark.

My personal favorite is the racial senility hypothesis. According to this long-outdated idea, dinosaurs as a group experienced a period of restless youth in the Late Triassic and Early Jurassic marked by great expansion in both form and diversity. This era of youthful exuberance was followed by a kind of adulthood during the latter Jurassic and early part of the Cretaceous, a "heyday" of sorts when dinosaurs supposedly reached their evolutionary peak. Finally, there was the inevitable period of senility toward the end of the Cretaceous. Evidence put forth in support of this notion included the marvelous variety of bony excrescences on Late Cretaceous dinosaurs—the frills and horns of ceratopsians, the crests of hadrosaurs, the boneheads of pachycephalosaurs, and the tail clubs of ankylosaurs. Adherents of the racial senility hypothesis perceived these unusual features to be unnatural and useless, perhaps the result of hormones gone wild and certainly symbolic of a group on its way out.

Whereas some of these extinction hypotheses are just plain wacky, most suffer from a common problem. They consider only the charismatic megafauna, dinosaurs, and ignore the many other groups that suffered major losses in the end-Cretaceous extinction. This event is usually referred to as the Cretaceous-Tertiary, or K-T MASS EXTINCTION. (Here *K* is used instead of *C* because the latter designates a much earlier period—the Carboniferous.) In the marine realm, the entire spectrum of top vertebrate predators, including the plesiosaurs and mosasaurs, disappeared together with several groups of fishes. The ichthyosaurs, those dolphin-look-a-like reptiles, disappeared about 30 million years before the end of the Cretaceous, so their demise must have had a different cause. Among nonverbebrates, several important, long-lasting groups were also wiped out in the K-T event; these include a variety of hard-shelled marine organisms, such as the nautilus-like ammonites and reef-making rudist clams. Also victimized were the chalk-making algae, famous for generating massive limestone deposits like the White Cliffs of Dover. Although some forms survived, decimation of the chalk producers likely had deep and cascading effects, because these microscopic organisms formed much of the base of the marine food web. Overall, it appears that the hardest hit in the marine realm were free-swimming or surface forms such as plankton and ammonites (although many varieties of fishes made out just fine). Among bottom-dwellers, several filter-feeding groups were decimated, including the colony-forming bryozoans, "sea-lily" crinoids, and corals, whereas scavengers took only minor hits. All of these losses inter-rupted the flow of energy and cycling of nutrients through the biosphere, in some cases shutting down nutrient cycles altogether.

Meanwhile, up on land, the pattern of extinction and survival has been more difficult to assess. The biggest problem is that few well-studied rock and fossil sequences span the time period immediately before and after the K-T extinction. Based on this very limited sample, it appears that lizards and mammals were severely decimated, and pterosaurs were extinguished altogether. Freshwater sharks, common predators in Mesozoic streams and rivers, also disappeared in the end-Cretaceous event. On the flip side, several groups of vertebrates weathered the storm with few to no species losses, including fresh-water fishes, amphibians, turtles, crocodiles, and a strange group of narrow-snouted, croc-odile-like reptiles called champsosaurs. At least two major conclusions emerge from this pattern of losers and winners. First, with few exceptions, all animals larger than about 10 kilograms (22 pounds) were killed off, so being big turned out to be a lethal liability. Sec-ond, freshwater vertebrates—including crocodiles, turtles, amphibians, champsosaurs, and fishes (but not sharks)—did much better than land dwellers. In the near future, it may be possible to compare these findings with similar inventories from other locales around the world. Nevertheless, the bottom line is that any reasonable hypothesis of the K-T ex-tinction must explain much more than the dinosaurian demise, accounting for patterns of extinction and survival across the full spectrum of marine and terrestrial life.

We are left with only two plausible hypotheses for explaining the K-T extinctions. The first of these I refer to as the SILVER BULLET HYPOTHESIS. This now-familiar idea posits that the extinction was brought about by a singular, devastating event—an asteroid impact. According to this view, first proposed in 1980 by the father-son team of Luis and Walter Alvarez and their colleagues, an asteroid about 10 kilometers (6 miles) in diam-eter collided with Earth one fateful day about 65.5 million years ago. Slamming into the planet at a velocity of about 100,000 kilometers per hour (62,000 miles per hour), the resulting explosive force of this extraterrestrial bullet exceeded that of all of the present stockpiles of nuclear weapons combined. The event triggered a massive earthquake (perhaps greater than 10 on the Richter scale), well beyond anything in recorded his-tory, causing tsunamis to radiate away from ground zero–like monstrous ripples on a planet-sized pond.

Upon impact, the asteroid disintegrated, vaporizing a chunk of Earth's crust with it. The huge volume of ash and pulverized rock ejected high into the atmosphere enveloped the world, blocked the sun, and transformed day to night. What followed was a pro-longed period of cold and dark that lasted about 4 months. Lacking a solar energy source, photosynthesis effectively ceased, shutting off the energy supply to most marine and terrestrial ecosystems worldwide. As if a giant asteroid plunging the biosphere into darkness were not sufficiently hellacious, additional lethal side effects may have included rampant wildfires and acid rain. Another possible killing agent was a massive pulse of infrared heat in the few hours following impact, as millions of bits of molten rock ejected into the atmosphere from the impact site rained down around the globe. In short, the Silver Bullet model claims that dinosaurs and their contemporaries were thriving during the latest Cretaceous until a giant rock collided with Earth, bringing their charmed lives to an abrupt halt.

The major competitor of the Silver Bullet scenario is an idea that I term here the BLITZKRIEG HYPOTHESIS. *Blitzkrieg*, a German compound word that means "lightning war," refers to an offensive military strategy developed in World War II by the German Wehrmacht. It entailed use of multiple weapons, including air bombardments and deployment of ground forces, designed to outmaneuver and overwhelm the enemy. With regard to the K-T event, the name fits well with the view that this mass extinction was precipitated not by a single cause but by multiple factors working in unison from the ground and air. Specifically, advocates of the Blitzkrieg view argue for three disruptive agents working in concert: receding sea levels, erupting volcanoes, and an asteroid impact. In other words, instead of a single strike from outer space, a barrage of attacks on the latest Cretaceous world reshuffled the biological deck and discarded the dinosaurs.

Unsurprisingly, the media immediately jumped on the Silver Bullet scenario. Here in one tidy package was an unimaginably violent solution to one of nature's greatest outstanding mysteries—the death of the dinosaurs. The idea had the added bonus of a possible repeat performance, with the potential to wipe out humankind next time round. News outlets have to wait a long time for science stories of this caliber. The impact hypothesis also sparked a worldwide research effort in a variety of scientific fields. Not long after the idea took hold, a pair of paleontologists reviewed the fossil record over the past 500 million years and reported that mass extinctions appeared to be cyclical, occurring about once every 26 million years. This announcement prompted astronomers, in turn, to search for an extraterrestrial cause to explain such metronome-like spasms of extinction. One innovative solution with a terrific name was the Deathstar hypothesis. Just outside the outer limits of our solar system orbits a thin band of comets known as the Oort cloud, named after its discoverer. Proponents argued that an unseen dwarf star—known as "Nemesis" or, simply, the Deathstar—might also exist not far beyond our solar system. According to this idea, the dwarf star perturbs the Oort cloud every 26 million years or so, unleashing a plague of comets on the solar system. Several sensationalist books were written on this topic before paleontologists determined that the original hypothesis of a 26-million-year extinction cycle was flawed, in part because of poor resolution of the timing of extinctions. Although debate about extinction cycles continues, the Deathstar hypothesis resides with cold fusion and other discarded ideas in the gutters of science.

Nevertheless, the Silver Bullet hypothesis is alive and well, with the asteroid impact scenario supported by multiple lines of evidence. The first (and in many ways still most pivotal) of these lines was identified by the Alvarezes in their 1980 article. These geoscientists found an anomalous abundance of the element iridium in a thin stratum of rock within the K-T boundary layer. Iridium, which comes from the same chemical group as platinum, is rare in Earth's crust but common in extraterrestrial bodies like asteroids. The "iridium spike," as it's called, was first identified in a clay layer near Gubbio, Italy, but has since been confirmed in K-T boundary sediments at many other locations around the world. Additional geologic evidence includes the presence of shocked quartz and glass spherules in the boundary layer. Shocked quartz is quartz modified by intense heat and pressure, leaving behind characteristic lines—exactly the kind of thing that might be produced by an asteroid impact. Similarly, glass spherules form during impact events as rock is melted, blasted into the air as a spray of droplets, frozen almost immediately, and then dropped back to Earth.

On the biological side, paleobotanists have documented a radical decimation of plant life at the end-Cretaceous boundary. Just below the iridium layer, fossil pollen and spores from many locations around the world indicate a diverse flora. Immediately above this layer, the pollen and spore evidence is heavily dominated by a single plant type—ferns. Ferns are one of the first plants to recover after major environmental catastrophes, and this "fern spike," as it's been called, is cited as evidence of a sudden cataclysm—that is, death by Silver Bullet.

The scientific world was stunned by the Alvarez impact hypothesis, with the iridium layer providing the geologic equivalent of a smoking gun. Almost equally surprising was discovery of the actual wound in Earth's surface—a crater about 200 kilometers (125 miles) across in the Gulf of Mexico, near the Yucatán Peninsula. The Chicxulub crater appears to be the right age and size to account for the impact of a 10-kilometer (6-mile) diameter asteroid. It's also associated with a number of physical effects consistent with a major impact, including shocked quartz, glassy spherules, and tsunamis or tidal waves. The sediments are even arranged in a layer-cake sequence consistent with the Silver Bullet model. First to be deposited were large particles like the glassy spherules, which would have rained down from the sky in the minutes following the impact. Immediately above this stratum are sediments indicative of tsunamis, or tidal waves, which in turn are topped by iridium and other fine sediments that would have remained airborne longer before falling back to the surface.

In short, although considerable debate persists as to the relative roles played by such factors as acid rain, wildfires, and the cessation of photosynthesis, there is strong consensus that an asteroid hit Earth about 65.5 million years ago and that its effects on the biosphere were catastrophic. But was the asteroid the overriding factor in the K-T extinctions (the Silver Bullet model), or were additional killing agents involved (the Blitzkrieg model)?

It turns out that strong support exists for protracted environmental changes in the latest Cretaceous, well prior to the asteriodal collision. As noted in the previous chapter,

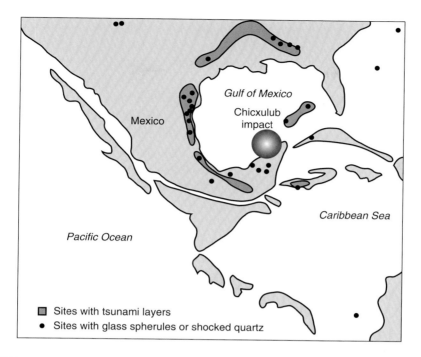

FIGURE 15.2

Map of Central America during the Late Cretaceous, showing the Chicxulub impact site and localities that have yielded additional evidence (glassy spherules and tsunami deposits) of an asteroid strike that may have caused the K-T extinction.

Earth's atmosphere experienced a gradual cooling between about 90 and 65 million years ago, culminating in an estimated 2°–3°C drop in temperature by the K-T boundary. Many investigators see evidence of a particularly intense pulse of global climate change in the last few million years of the Cretaceous, marked by a distinct warming trend. And there is now abundant evidence that much of the latest Cretaceous environmental shift resulted from a combination of volcanism and sea-level change, both cited as key factors of the Blitzkrieg model.

The tectonic pulse in the Late Cretaceous generated volcanism in various areas around the globe, especially in and around the Pacific Ocean. By far the greatest episode of volcanism, however, the largest since the end-Permian extinction, produced the DECCAN TRAPS, an incredible flood of lava that inundated much of present-day India and Pakistan. In all, the Deccan Traps released on the order of 1 million *cubic* kilometers (240 cubic miles) of basalt, equivalent to covering the entire continent of North America with lava about 20 kilometers (12 miles) deep! The duration of this event is still hotly debated. Whereas older estimates pointed to a lengthy duration, on the order of 10 million years, more recent estimates are consistent with a much briefer volcanic burst, with peak eruption levels occurring in the 1–2 million years immediately surrounding the K-T boundary. A bout of volcanism of such great magnitude undoubtedly had dramatic and devastating environmental effects, yet the exact nature of those effects remains unclear. If the Deccan Traps pumped vast amounts of carbon dioxide into the atmosphere, this event could have resulted in global warming. Conversely, ejection of tons upon tons of particulate matter into the atmosphere may have reduced available sunlight at the surface, resulting in a long-term cooling trend. What we do know is that the eruptions tended to be viscous rather than explosive—more akin to volcanoes on the Hawaiian Islands than to Mount St. Helens.

Meanwhile, the sea-level highs that resulted in shallow seas atop many of the world's continents finally subsided in the latest Cretaceous, leading to a global marine regression. Maximal retreat of the seaway occurred just prior to the K-T boundary, uncovering vast swaths of continental crust that had been underwater for millions of years. This transformative event greatly increased the amount of area available for land-dwelling and freshwater animals, while vastly reducing coastal and marine habitats. Additional effects included lengthening of stream systems (which had to flow much farther to reach the ocean) and emplacement of land bridges between previously separate continents. The land bridges resulted in exchanges of plants and animals between landmasses. One example mentioned in the previous chapter stretched from present-day Alaska to eastern Asia at least twice during the Late Cretaceous, forming a corridor for the movement of dinosaurs and other life-forms between North America and Asia. Viewed as a cause of the K-T event, however, regression is problematic, because even larger regressions occurred at other times in the Mesozoic without the consequent extinctions.

Nevertheless, with a wealth of evidence supporting the latest Cretaceous occurrence of an asteroid impact alongside dramatic episodes of volcanism and marine regression, it might seem that the Blitzkrieg hypothesis would emerge victorious over its Silver Bullet rival. Yet, before any such pronouncement can be made, we need more information about species extinctions during the K-T event. At the core of this debate is the pace of extinction. A rapid extinction—say, as brief as a week or a year, and no more than thousands of years—would support a single, devastating cause (Silver Bullet), whereas a more protracted period of dying—on the order of millions of years—might indicate multiple causes (Blitzkrieg). Many paleontologists claim that the fossil record shows numerous groups—for example, ammonites, plesiosaurs, pterosaurs—on the wane for millions of years prior to the K-T asteroid impact, with much lower diversities than earlier in the Cretaceous. Others argue that ecosystems remained relatively diverse until 65.5 million years ago, when a major event (presumably the asteroid impact) resulted in a virtually instantaneous killing strike. Unfortunately the rock record rarely provides the degree of resolution necessary to distinguish between such alternatives. An inch (2.5 centimeters) of sedimentary rock may record part of a day or many thousands of years. A study by Michael Benton found that the K-T extinction wiped out 64 out of 210 vertebrate families, with the greatest losses among the land-dwelling vertebrates, or tetrapods (43 percent of families lost). Yet these figures cover a period of about 5 million years. So we need more data on whether these losses were concentrated into a relatively brief instant of deep time or spread out over millions of years.

I think it's safe to say that, over the past couple of decades, most dinosaur paleontologists have tended to reside in the Blitzkrieg camp, arguing for a gradual extinction precipitated by multiple factors. However, despite all the fuss about their extinction, the fossil record of dinosaurs is remarkably sparse for the final stage of the Mesozoic. Indeed, to date, only one place on Earth—the Hell Creek Formation exposed most abundantly in eastern Montana and the Dakotas—has been investigated in detail. Several authors have argued that the Hell Creek fauna possessed far fewer species than earlier dinosaur faunas. In particular, comparisons have been made with the approximately 75-million-year-old Dinosaur Park Formation of southern Alberta, with its bewildering array of herbivorous and carnivorous dinosaurs (see the previous chapter).

For a long time, I regarded myself as a card-carrying member of the gradualist-Blitzkrieg camp. However, three lines of evidence have caused me to change my mind and join the ranks of Silver Bullet enthusiasts.[2] First is the documentation of many additional K-T boundary sites preserving that anomalous iridium layer; the presence of this asteroidal signature at far-flung locations around the globe confirms that the Chicxulub impact was truly a devastating, global event. Second is a growing fossil database indicating that the terminal Cretaceous world was not stressed to the breaking point, awaiting arrival of an extraterrestrial coup de grâce. With regard to dinosaurs in particular, recent work in the Hell Creek Formation (see chapter 5) and nearby Lance Formation have documented a much more diverse fauna than previously realized. Certainly some dinosaur lineages (e.g., short-frilled ceratopsids and crested duck-bills) went extinct prior to the end of the Cretaceous, but overall dinosaur diversity remained relatively high, with many groups represented by truly gigantic exemplars such as *Tyrannosaurus*, *Ankylosaurus*, and *Pachycephalosaurus*.

Third, new and improved estimates of the stratigraphic age and duration of dinosaur species from earlier in the Late Cretaceous suggest that these somewhat older ecosystems (dating to about 75 million years ago) were not nearly as diverse as previously supposed. For a long time, the Dinosaur Park Formation in Alberta, Canada, has been hailed as the exemplar of dinosaur diversity. Yet recall from the previous chapter that many dinosaur species recovered from these Alberta badlands do not co-occur in time. Instead, recent study of the rock record reveals that these dinosaurs had much lower ecological diversity than previously thought, a finding more consistent with rapid replacement of species (high rates of origins and extinctions) rather than numerous coexisting species at any one time. Considered in unison, the second and third lines of evidence suggest that the difference in co-occurring species between the two formations is not nearly as large as previously believed. So, even within the well-sampled Western Interior of North America (let alone the rest of the world, for which we currently have no hard data), I see no grounds for arguing that dinosaurs (or other groups of animals or plants) were undergoing a slow, attritional demise in the latest Cretaceous. We must keep in mind, however, that our knowledge of dinosaur diversity at the very end of the Cretaceous is effectively limited to one place in western North America. Fossil evidence from elsewhere in the world may one day tell a very different story.

Finally, it seems questionable to argue for a lethal cascade of agents when compelling evidence exists for a single agent capable of doing the job on its own. Would the dinosaur-laden biosphere have persisted long after 65.5 million years ago if an asteroid had not slammed into the Gulf of Mexico, bringing an abrupt close to the Mesozoic? We will never know for sure, but my hunch is that it would have, with evolution pumping out new and wondrous species. Why not, given that dinosaurs had already been around for 160 million years? Although the scythe of extinction was ultimately inevitable, it might have been millions of years before another such devastating event. Some authors, such as paleontologist Niles Eldredge of the American Museum of Natural History, now argue that not much happens in evolution until ecosystems are upturned by external forces like those that produce mass extinctions. Certainly without that asteroidal collision, the grand radiation of mammals would have been thwarted or at least significantly delayed, and there would be no humans today to contemplate this dramatic series of events.

I offer just one caveat. Research is growing that the Deccan Traps formed over a relatively brief period (less than a million years). One team of researchers recently argued that a 600-meter- (2,000-foot-) thick portion of the massive lava flows may have piled up in only 30,000 years. If further research supports these findings and points to a pulsed episode of volcanism in association with the K-T boundary, then we may have a problem. It would seem to be a gargantuan coincidence that a lethal asteroid strike would co-occur with one of the biggest bouts of volcanism known. Gerta Keller (Princeton University) and her colleagues have argued on geologic and biologic grounds that the Chicxulub impact evidence does not support a global killing event from space. Instead, they (and others) point to the Deccan Traps as the likely "Silver Bullet" that killed off the dinosaurs and other life-forms at the close of the Mesozoic. So the final chapter on this story may not yet be written.

Sometime during the mid-1980s, dinosaur paleontologist Jack Horner got fed up with being asked about the dinosaur extinction. One day, after facing that same question for the umpteenth time, Horner famously replied, "I don't give a damn how they died. I want to know how they lived." I confess to being with Horner on this matter. Indeed, this book serves as a testament to my passion for understanding dinosaur lives. To me, the most interesting aspect of mass extinctions is not the presumed causes of all the death and mayhem (OK, so volcanoes release lots of lava and Earth occasionally collides with big rocks). I am much more fascinated by life's response in the aftermath of such extinctions. How do the survivors get back on track? And what has prevented the biosphere from collapsing altogether when decimated by mass extinctions?

To provide at least a partial answer, I conclude this chapter with a brief discussion of a radical idea. Life is no mere innocent bystander constantly changing in response to environmental perturbations; instead, organisms actively generate and maintain the biosphere, in part by regulating the temperature and composition of the atmosphere. This dramatic, encompassing, and surprising idea, known as the GAIA HYPOTHESIS (named after the Earth goddess of Greek mythology), pertains as much to the present day as to the world of dinosaurs. Over the past few decades, this hypothesis has acquired increasing numbers of adherents, though it is still a long way from becoming mainstream, broadly accepted science. I position myself firmly within the ranks of its supporters.

Over the course of deep time, global climates on Earth have witnessed dramatic variations. This book has explored the hothouse conditions that characterized the Mesozoic, including degrees of global warming and sea-level rise well beyond the direst predictions of the present day. And there have been other, equally extreme hothouse intervals during the history of life. Yet Earth has been an icehouse world as well. You may be surprised to learn that the tenure of humanity has occurred largely within an icehouse interval (the so-called Ice Age), albeit one that may be coming to a rapid close thanks to human inputs into the atmosphere. Far more severe icehouse conditions than the Pleistocene ice ages occurred in the remote past. Between about 730 and 630 million years ago, during the Proterozoic eon, the planet may have been virtually enveloped in ice, a phenomenon aptly called "Snowball Earth."[3]

Underlying such dramatic extremes, however, is a story of astonishing climatic stability. Astronomers tell us that our star, the sun, has become markedly brighter over the past 4 billion years, increasing its radiation output by about 25 percent. Yet, rather than a progressive rise in temperature, as might be expected, the surface of Earth has tended to become cooler. Despite notable icehouse and hothouse fluctuations, Earth's surface temperature has remained relatively constant for most of its history, within the relatively narrow temperature range suited for life. What is the source of such enduring atmospheric stability and biological perseverance in the face of changing external conditions?

The apparent solution to this mystery came from a British atmospheric scientist by the name of James Lovelock. NASA hired Lovelock in the 1960s to design experiments for the Viking space probes going to Mars. His job was to determine how to use the probes to test for the presence or absence of life on the Red Planet. After contemplating the matter, Lovelock decided that we should be able to determine whether a planet has life simply by examining its atmospheric makeup. After all, he reasoned, life on the surface of any planet would have to use its atmosphere to cycle the elements necessary for existence. Nutrients must be supplied and wastes removed. Earth's atmosphere is composed of approximately two-thirds nitrogen and one-fifth oxygen, with several other elements such as carbon dioxide and methane occurring in trace amounts. Oxygen is highly reactive, particularly in the presence of methane. Therefore, in order to maintain constant levels, oxygen must be constantly replenished by the metabolic activities of organisms such as plants and certain bacteria, which continually produce oxygen as a waste product of photosynthesis. An increase in the amount of oxygen to a level of 25 or 30 percent could cause the atmosphere to burst into flames. A decrease to below 10 percent would be lethal to the vast majority of life-forms.

So it seems that something has kept the composition of the atmosphere, as well as its temperature, within narrow limits conducive to life for billions of years. In contrast, the atmosphere of Mars is effectively "dead"—or, more precisely, at equilibrium—composed almost entirely of carbon dioxide, with no free oxygen. Based on this observation, Lovelock concluded that Mars could not presently harbor any life. A similar claim could be made for the other nearby rocky world, Venus. The difference between the two is that carbon dioxide has turned the surface of Venus into an oven, whereas the atmosphere on Mars is so thin that it cannot retain heat. In contrast, here on our blue planet the oceans have not boiled away, and the atmosphere has maintained a temperature and composition suitable for life. Lovelock, sparked by a radical insight, argued that the planetary signature of life is an atmosphere far from equilibrium that includes highly reactive elements such as oxygen. Remove all traces of life on Earth, and our atmosphere would soon migrate to an equilibrium state dominated by carbon dioxide.

By all accounts, NASA was not overly enthusiastic about Lovelock's findings, because his "remote-sensing" argument could be construed as negating any need to travel to Mars in search of life. Lovelock's employment with the space agency was terminated soon thereafter. Of course, the presumed absence of life on Mars today does not mean that it has never existed there. On the contrary, the ancient presence of water, a key ingredient for life, was recently confirmed by geologic evidence recovered by Martian rovers. Remarkably, then, our best shot at discovering life elsewhere in the universe may rest with paleontologists exploring the rocks of Mars!

Unperturbed by his exit from NASA, Lovelock realized that he was on to something big. In contrast to the traditional view of environment as a purely physical phenomenon, his hypothesis implicated life in the generation of its own environment. Just as a thermostat keeps a room within a given temperature range, the gas-exchanging metabolic processes of life must interact with the physical processes of a planet in order to regulate the temperature and composition of the atmosphere. Some have misconstrued the Gaia hypothesis as claiming that Earth is "alive" or, worse, a single organism, pronouncements embraced by New Age spiritualists and scorned by hard-nosed scientists. However, at its core, this idea makes no such assertion. Rather, Lovelock and other Gaia supporters argue that the sum of planetary life has an integrated physiology that enables the biosphere to regulate its environment. One of Lovelock's closest colleagues, microbiologist Lynn Margulis, puts it this way: "The planet's surface is not just physical, geological, and chemical, or even just geochemical. Rather, it is geophysiological: it displays the attributes of a living body composed of the aggregate of Earth's incessantly interactive life."[4]

Yet how could such a regulatory process occur on a planetary scale? How does Gaia "know" when to raise or lower oxygen levels in the atmosphere, and what has prevented a catastrophic buildup of greenhouse gases like carbon dioxide? Whatever the mechanism involved, it must be driven by an unconscious, automatic process, akin to the way your body regulates its internal temperature by shivering or sweating, heals itself following an injury, and allows you to maintain balance while walking.

To answer this perplexing question, Lovelock and Margulis compiled a compelling argument. In short, they postulate that, merely by growing and reproducing, consuming nutrients and generating wastes, bacteria, fungi, plants, and animals participate in self-regulating feedback loops. Feedback loops, briefly introduced in chapter 5, are everywhere in nature. Instead of conscious monitoring, these loops employ circular control, gathering information about past performance and using this input to self-correct as necessary. Whereas negative feedback loops tend to bring a system back to a previous stable state, positive feedback loops drive a system to change until a new stable state in achieved. An example of negative feedback is the way in which the animals maintain relatively stable internal conditions (homeostasis) through adjustments to such regulatory factors as body temperature, blood pressure, and blood glucose levels. A biological example of positive feedback, or amplification, is the runaway evolution of bizarre features like peacock tails (chapter 10).

All this may sound a little mystical, so let's look at a concrete example—carbon. Over many millennia, volcanoes have spewed millions upon millions of tons of carbon dioxide into Earth's atmosphere. Yet, in contrast to Venus and Mars, it has remained a trace component of the atmospheric makeup. Therefore, there must be mechanisms that pump carbon dioxide out of Earth's atmosphere. One such mechanism comes in the form of plants, which consume huge amounts of carbon dioxide through photosynthesis, converting the carbon into a solid form that is then stored within their bodies. Animals balance this carbon sequestering at least in part by exhaling carbon dioxide with every breath. Another method of trapping carbon is rock weathering. As rocks break down in response to erosion, they combine with rainwater and carbon dioxide to form various carbon-based compounds called carbonates. The gaseous carbon dioxide is thus taken from the atmosphere and bound up into liquid form. Yet calculations show that the purely physical processes of rock weathering are insufficient to remove the necessary carbon dioxide from the atmosphere.

It turns out that many soil bacteria, through partnerships with fungi and plants, act as superb catalysts for rock weathering, increasing the rate of this process by many times. To give just one example, some bacteria secrete compounds that swell inside cracks and fissures within the rock, accelerating the pace of weathering about a thousand-fold. Most of this water-borne carbon ends up in the oceans, where a variety of life-forms tap into the resulting chemical compound, calcium carbonate, to make their shells. Shelled organisms include large-bodied animals such as clams and snails. Far more abundant are microscopic organisms called coccolithophores, the same life-forms responsible for the formation of vast reservoirs of Cretaceous chalk such as the White Cliffs of Dover. So the carbon dioxide that started out as a gas and was transformed to a liquid ultimately becomes locked up in a solid. When these animals die, their shells descend to the ocean bottom, piling up in massive layers that turn to limestone. After a lengthy residence on the sea floor, this limestone-entombed carbon is transported into Earth's interior via the process of subduction (see chapter 4). Ultimately, a portion of this carbon dioxide is released again as a gas through volcanic activity. The entire cycle, from volcanoes to rock weathering to soil bacteria to oceanic algae to limestone and back to volcanoes, forms a feedback loop that contributes to regulating Earth's temperature. Remarkably, it's estimated that this cyclical and ongoing process of rock weathering cools the planet by some 15°–45 °C (60°–110°F)!

As the theory goes, an increase in atmospheric carbon dioxide initiates a corresponding increase in photosynthetic and rock-weathering organisms, which in turn accelerate the rate of carbon capture from the atmosphere. Once carbon dioxide levels in the atmosphere drop sufficiently, the biomass of photosynthesizing plants and bacteria decreases, reducing the rate of carbon capture and allowing the system to reenter a state of equilibrium.[5] Other feedback loops involving biological and geologic processes have been identified, and many more likely await discovery. It is mind-boggling to think of life playing a pivotal role in its own fate, keeping the planet habitable for eon upon eon.

The Gaia hypothesis gives us a new perspective on the world of dinosaurs. The Mesozoic web of life, including the dinosaurs, was maintained by a complex, robust interweaving of life-forms, most of them microscopic. So fully integrated were these with Earth's physical processes that they formed a system of feedback loops regulating the Mesozoic atmosphere and perhaps the biosphere more generally. The end-Permian and end-Cretaceous extinctions—which precipitated the origin and demise of the dinosaurs, respectively—undoubtedly had major impacts on this regulatory process, as well as the composition of the web. Nevertheless, the biosphere, Gaia, has proved remarkably resilient, rebounding from such cataclysmic events multiple times over billions of years.

The Mesozoic odyssey of dinosaurs is a tale of fantastic creatures in distant lands. As with any such epic, the narrative includes lengthy periods of quietude punctuated by brief intervals of upheaval. We have seen that Mesozoic Earth bears a strong resemblance to the planet we live on today. Then, as now, ecology and evolution worked in tandem to sculpt resilient ecosystems bursting with splendor and diverse life-forms, many of which would look familiar to us. Yet this ancient Earth would seem alien as well, and not only because of the plethora of dinosaurian monsters. Dinosaurs inhabited a hothouse world lacking rain forests, icy polar caps, and grasslands. Pterosaurs graced the skies, while mosasaurs, ichthyosaurs, and plesiosaurs plied the oceans. Today, we still grope to understand even the basics of Mesozoic biomes—the large-scale communities of plants, animals, and soil organisms. Paleontology and its allied sciences will continue to refine our conceptions of these ancient worlds. An equivalent review of dinosaurs written 20 years from now will be very different indeed, achieving deep insights we cannot yet imagine. And future versions 40, 60, and 100 years hence will be equally distinctive. It's an exciting time to be a paleontologist, and the excitement shows no signs of abating anytime soon. For the few who elect to pursue paleontology, as either a professional or an amateur, rest assured that many, many lifetimes of groundbreaking work remain to be done. And for the vast majority who prefer a front-row seat on the sidelines, all I can say is, enjoy the show!

A rock wren alights on the horn of a *Triceratops* skull eroding out of the Hell Creek badlands, Montana, momentarily underscoring an evolutionary link between the present day and the deep past.

EPILOGUE
Whispers from the Grave

By now, it should be evident that these fascinating creatures called dinosaurs played a pivotal role in Earth's history. Perhaps your newfound knowledge of dinosaurs will enter conversation at parties or serve you in trivia contests with the youngsters in your life. Yet you might reasonably ask (as many have asked me), "What relevance do dinosaurs have today?" Do they serve any utility beyond their entertainment value? Let me try to persuade you that dinosaurs still have much to teach us, if we care to listen.

I'll begin my case with what may seem an outlandish claim:

Dinosaurs may well be crucial to the continued existence of humanity and much of the planet's biodiversity.

"Sure," I can hear you saying, "spoken like a true paleontologist." But consider this. The asteroid impact (Silver Bullet) hypothesis was put forth to account for the end-Cretaceous extinction of dinosaurs. Scientific study of this impact scenario led to a frightening realization: a future comet or asteroid strike could decimate human civilization and result in the extinction of millions of species, possibly even our own. Research has shown that such near-Earth objects (NEOs) pose a remote but very real threat; collisions with an asteroid greater than 1 kilometer (0.6 mile) in diameter (large enough to generate global havoc) occur every few hundred thousand years or so. In the wake of this insight, research programs led by NASA and other space agencies investigate technological avenues both for the detection and deflection of NEOs. In a related vein, study of the K-T "impact winter" scenario—that is, the global fallout resulting from a major asteroid strike—helped define the "nuclear winter" scenario—the global catastrophe that could follow an all-out exchange of thermonuclear weapons. The latter idea, presented to the U.S. Congress by such well-known scientists as Carl Sagan, has been a profound deterrent for those who might contemplate such military madness. Therefore, it's entirely feasible that dinosaurs, or at least their modern-day study by geologists and paleontologists, might one day be regarded as an important element in the persistence of humanity. This conclusion underscores the diverse interconnections binding the physical and biological realms, as well as the importance of pure research, which often yields unexpected outcomes.

There's another way in which dinosaurs just might help "save" our world. We live at arguably the most pivotal moment in human history. Overwhelming scientific consensus confirms that we've reached a tipping point in our relationship with the natural world, approaching (and, in some cases, exceeding) the limits of the biosphere to absorb the excesses of human existence.

To focus on a single aspect of this crisis—loss of biodiversity—Earth's biosphere is currently home to an estimated 13 million species. Biologists have formally named and described fewer than 2 million of these, with an overwhelming emphasis on larger, conspicuous critters. Between 10 and 40 percent of all species (including one in four varieties of mammals) are now threatened, with the numbers of imperiled life-forms rising rapidly. Over the past century, the pace of species extinction has skyrocketed, with recent estimates indicating a rate approaching a thousand times greater than prehuman levels! Paleontologists see evidence in the fossil record of five previous mass extinctions, and we are now on the brink of the sixth such cataclysm—the largest extinction event since the one that killed off the dinosaurs over 65 million years ago. The present panoply of biodiversity, the product of millions of years of evolution, is being wiped out in mere decades, an undetectable blip on the radar screen of deep time. If the current extinction trend continues, about half of all species alive today may be extinguished by the close of this century, with uncertain consequences for the survivors.

A unique aspect of this particular mass extinction is its precipitation by a single species—us. This time around, humans are the asteroid on a collision course with Earth. A recent assessment of global ecosystems—sponsored by the United Nations and carried out by 170 scientists from an array of disciplines—concluded overwhelmingly that this ecological crisis is being driven by human activities. The report's statistics are staggering: almost 60 percent of coral reefs threatened, more than half of the world's wetlands destroyed, 80 percent of grasslands imperiled, and 20 percent of drylands in danger of becoming desert. Causes of this dramatic, widespread environmental degradation are all too familiar to most readers: deforestation, human overpopulation, spread of toxic pollutants, overexploitation of animal species, and emission of greenhouse gases leading to global warming. Nature has simply been unable to absorb the punishment delivered by ballooning human populations and increasing consumption linked to industrialization.

Yet, despite the severity of this ecocrisis, we continue to spend our environmental "capital" like some crazed gambler unable to see beyond the next hand of Texas Hold'em. Without a rapid and radical shift in global priorities, we're headed for ecological bankruptcy, and the effects of this stunning environmental "downturn" will be felt much longer than any squandered pension, housing slump, or devalued currency. Study of previous mass extinctions makes it clear that it would take 5–10 million years to rebuild Earth's ecosystems to current levels of diversity.

You might now be thinking, "If all species eventually go extinct anyway, why should we care about curbing the current pace of losses?" At some level, the answer comes down to a moral decision. Do you feel that humans have the right to knowingly exterminate millions of species? Put another way, should the nonhuman members of the natural world have any intrinsic rights to survival? And what of future generations of humans that may never have the opportunity to see an elephant, polar bear, or humpback whale? Even if your scope of concern does not extend to other species, it's important to remember that human civilization, perhaps even the persistence of *Homo sapiens*, is also at risk.

In short, if we are to avoid a calamitous future, we need a dramatic course change in this century, and almost certainly within the next generation. Of the four thousand or so human generations that have lived and died, the present one may be unique in its capacity to determine the fate of all future generations. Yet the ecocrisis is greatly amplified by another problem—ignorance. Most people have not even begun to comprehend the depth and destructiveness of our present course, or the changes required to alter this path. Instead, as a society we tend to bury our collective heads in the sand while grasping steadfastly onto erroneous, outdated notions. Many presume, for example, that Earth is bountiful enough to supply our growing needs indefinitely and that technological innovations will bail us out of even our most egregious overindulgences. Others, overwhelmed by current environmental woes and the tempo of destruction, assume that there is little room for optimism. Yet awareness of this issue and possible solutions are accelerating rapidly, with more and more people coming to understand that their everyday decisions can have far-reaching, even global, effects. Surely there is hope and optimism to be found in recent efforts to reduce emissions of greenhouse gases and move in the direction of sustainability.

Nevertheless, any meaningful movement toward ecologically sustainable societies will require not only an outward change in technology but an inward shift in worldview. As many authors have argued, long-term sustainability will require that we expand our consciousness and redefine our relationship with nature. In essence, we must shift from a human-centered (anthropocentric) worldview to a life-centered (biocentric) one. As I see it, such a fundamental change will require no less than a transformation of our education system, from K–12 to higher education. We must equip parents and educators with the necessary tools both to communicate the science of natural systems and to foster a passion for nature. Numerous studies agree that this education must be place based and learner centered, and include a generous amount of intimate contact with the outdoors. Today, a variety of factors—from television and video games to fear of traffic and strangers—tend to keep children indoors and prevent them from forging bonds with nature. The affliction even has a name—nature deficit disorder. This indoor trend desperately needs to be reversed. Moreover, the educational curriculum must teach children—and adults—not merely how to further their careers but how to live well in the world and to embrace sustainability.

Recent interest in the importance of environmental education has spawned a movement in ecological literacy, or "ecoliteracy." Organizations such as the Center for Ecoliteracy in Berkeley, California, promote understanding of local ecological webs and human roles within these webs.[1] Proponents of ecoliteracy argue persuasively that designing entire curricula around basic ecological concepts and outdoor activities integrates children with the natural world and encourages growth of a more informed citizenry.

While I applaud the ecoliteracy approach, it seems to me that its advocates have neglected an essential element—transformation. As we've seen, everything changes, from the day-to-day composition of our bodies to the deep time parade of species and ecosystems. Life has undergone radical transformations, tracking unbroken lines of ancestors and descendants from humble, single-celled beginnings to astonishing explosions of plants and animals. We humans—upright apes and latecomers on the evolutionary scene—find relatives not only among primates and other mammals, but also among fishes, flowers, and fungi. If we are to survive as a species, it's imperative that we undergo an evolution of consciousness, adopting a worldview that recognizes this common journey and embraces other organisms as family members deserving of respect. So education must incorporate not only the myriad connections that bind us in this snapshot of ecological time but also the profound transformations that embed us in deep time. Ask any fiction writer or film director—transformation lies at the heart of all compelling narratives.

Cultural historian Thomas Berry goes so far as to claim that much of our present crisis with the environment comes down to a lack of story. We currently do not have a compelling narrative that places humanity into a larger context, so our lives tend to lack a sense of meaning or greater purpose. No longer do we feel the awe, wonder, or sense of sacredness about nature typical of many preindustrial (and present indigenous) cultures. But what should the new story be? Berry's solution, supported by growing numbers of scientists, theologians, and educators, is the Great Story—sometimes called the Universe Story, the New Story or the epic of evolution—that begins with the big bang and traces our sinuous path to the present day. I gave an abbreviated, dinocentric version of this epic tale in chapter 2, which shares a great deal in common with the human-centered version. Some people view this narrative as the greatest triumph of the scientific revolution, as well as an opportunity to help heal the rift between science and religion and provide that much-needed sense of meaning and purpose. I am one of those people.

Interwoven throughout the Great Story are the ideas of transformation and flow, highlighting the many ways in which we are not merely connected to but literally immersed in the current of life. Never before has it been so important to see ourselves as temporary, swirling concentrations of energy that arise from the background flow, persist briefly, and then return to that same flow. That flow is a single unfolding whole that stretches back through deep time to the origin of the universe. Humans are not separate from nature; we *are* nature. Yet science education today remains overwhelmingly stuck in the machine-based perspective of Descartes and Galileo (chapter 1), regarding the nonhuman world merely as a collection of objects rather than, to quote Berry, "a communion of subjects." In the end, if nature is regarded merely as natural resources, we deny ourselves a meaningful home.

The year 2009, in which I am now writing, is the 200th anniversary of Charles Darwin's birthday and the 150th anniversary of the publication of his famous tome, *On the Origin of Species*. Yet Darwinian evolution remains steeped in controversy within the American general public (and in a number of other countries) despite the fact that its core message—the common ancestry of all life on Earth—is accepted as the central tenet of biology by virtually the entire scientific community. I believe that this profound disconnect is due as much to a failure of education as to the antievolution efforts of religious fundamentalists. Universities have not made evolution relevant to their students, particularly to fledgling teachers. Elementary and high school educators not only lack confidence in communicating the nuts and bolts of evolution (mutation, natural selection, speciation, etc.); they are also missing the larger context. All but absent from evolution teaching are the numerous links with ecology and the moment-to-moment flow of matter and energy. Most critically, also missing is the Great Story, that grand epic that encompasses the unified history of life, the universe, and everything. As a result, even for those of us who accept its veracity, biological evolution is little more than an intellectual factoid, with virtually no meaning for our lives. To my mind, a bodily, internalized understanding of evolution in the grandest sense will be a key element in defining a sustainable future.

In short, we need more evolution in America's classrooms, not less. And it must be made explicitly relevant. Evolution isn't something limited to the distant past. It happens every day and will continue to happen on Earth as long as life persists. Recent research on human development shows that we are all "natural born creationists," spending most of our childhood with the firmly held belief that all varieties of plants and animals are distinct entities. In general, it is only from the age of 10 or 11 that children are capable of conceptualizing the kind of transformation inherent in evolutionary change. The Great Story, presented at various age-appropriate levels, has the potential to help us grasp the essence of both transformation and unity. If our education system is to move us toward sustainability, it must foster not only ecoliteracy but evolution literacy, or "evoliteracy." The combination of these two, perhaps best thought of as "nature literacy," should embrace both connections (ecology) and transformation (evolution). Bridging the current evoliteracy gap will require that we foster transformative thinking, which brings us back to those "long-dead" dinosaurs.

For several reasons, dinosaurs offer a superb vehicle for teaching science through a combined ecoevolutionary approach. First, unlike many topics in science, dinosaurs inspire rather than intimidate our imaginations. Second, as the primary exemplars of prehistoric Earth, dinosaurs serve as able, even ideal, guides to an exploration of the deep past. Third, the living descendants of dinosaurs, birds, are much beloved and keenly observed, forging a robust link between past and present. It's not as large a leap as you might think from your Thanksgiving turkey to an ancient tyrant named *Rex*. Fourth, as demonstrated in the preceding pages, the interdisciplinary nature of paleontology means that dinosaurs provide excellent access points to topics as diverse as genetic cloning and plate tectonics. To give two timely examples, the hothouse world of the Mesozoic is a surprising and informative way to begin discussing contemporary climate change, and the abrupt demise of dinosaurs is a natural starting place for addressing today's imminent extinctions. Finally, dinosaurs possess tremendous potential to help convey the transformational Great Story and, in doing so, foster evoliteracy.

Never before have humans needed to comprehend the vastness of time—until now. How will we ever promote ecologically sustainable societies unless we first reinsert ourselves into the fabric of time, deriving meaning from our past and contemplating our responsibility to future generations? In direct opposition to this need, our present experience of time tends to be ever accelerating. As adults, our lives revolve around endless series of tasks and pressing deadlines, completely obscuring the larger context of existence. Strangely enough, dinosaurs can offer an excellent means to help anchor us back into deep time. *Homo sapiens* has existed for a mere nanosecond on the scale of Earth history, and humans and dinosaurs (with the exception of birds) never lived together, separated in time by tens of millions of years. Yet, geologically speaking, dinosaurs disappeared quite recently. Gazing further back into deep time, Earth has persisted for billions of years, with dinosaurs inhabiting the planet for only a tiny (and relatively recent) fraction of that inconceivable duration. For us, a thousand years seems like an eternity, and it is virtually impossible to conceive the billions of years of Earth history. Placed within the context of the Great Story, the Mesozoic world of dinosaurs encourages us to imagine time differently, and thus to experience our present reality in new ways.

Another essential element of the Great Story is deep connections, including those forged through common ancestry (evolution) and the continuous flow of matter and energy (ecology). All life-forms at any given time are connected in vast webs of relationships that together encompass the biosphere; all of these webs are interwoven with those of earlier times and those of the future. Life in a contingent world means that whatever comes before affects everything that follows. Interconnections run so deep that change in one aspect can have cascading (and sometimes dramatic and unanticipated) impacts on many others. Dinosaurs can help us grasp this notion.

Teaching about dinosaurs is typically conducted as if they have no connection to our modern world, almost as if we are talking about strange creatures from another planet. At most, we inform students that the extinction of dinosaurs 65.5 million years ago cleared the way for mammals and, ultimately, the origin of humans. Had the asteroid missed and dinosaurs not gone extinct, so the argument goes, there would have been no diversification of large mammals—no whales, no elephants, no antelope, no primates, and no humans. I regard this contention as accurate but grossly insufficient, because it completely ignores life's interconnectedness. By implication, the paths of dinosaurs and mammals are unrelated; one group exits the stage and another enters the scene to continue the drama.

Yet dinosaurs persisted for 160 million years prior to the K-T extinction, coevolving in intricate organic webs with plants, bacteria, fungi, and algae, as well as other animals, including mammals. Together these Mesozoic life-forms influenced the origins and fates of one another and all species that followed. Flowering plants first evolved in the Early Cretaceous and underwent their initial florescence alongside a variety of dinosaurian herbivores. The key issue, then, is not simply the dinosaur extinction but the intertwining of global ecosystems through time. Had the disappearance of dinosaurs occurred earlier or later, or had dinosaurs never evolved, subsequent life-forms would have been wholly different, and we almost certainly wouldn't be here. In short, we owe as much to the dinosaurs' lives as we do to their deaths. Much the same could be said for almost any major group of life-forms.

How, then, might the dinosaur odyssey help us communicate the Great Story and achieve the lofty goals of sustainability education? There are many possibilities, and I will highlight just one here.

Birds are common, diverse, and a joy to watch. Their recently revealed status as feathered dinosaurs also prompts an entirely new perspective. I'm picturing an Audubon Society-sponsored "Christmas dinosaur count." Or better yet, imagine a nationwide "Backyard Dinosaurs" initiative, with the goal of getting kids outside to observe local birds. The tools necessary for this effort—for example, binoculars, bird feeders, and birdhouses—are relatively inexpensive and easily accessible. Kids (and parents) would learn to identify many birds that live around their schools and homes year-round, as well as migratory winged visitors. For youngsters, *Tyrannosaurus rex* and other ancient monsters would become intertwined with various modern "dinosaurs" like *Turdus migratorius* (American robin), *Corvus corax* (common raven), and *Buteo jamaicensis* (red-tailed hawk). Curriculum surrounding Backyard Dinosaurs might span a range of topics, but should be built on the duo of evolution and ecology. With regard to evolution, emphasis would be placed on the ancestor-descendant relationship with dinosaurs, including the place of birds within the larger epic of evolution. As for ecology, kids would learn about the roles birds play in local ecosystems—from plant eaters and pollinators to predators and prey—as well as their links to other animal and plant species.

In most areas, teachers and parents interested in backyard dinosaurs could find local support through national history museums, zoos, aviaries, and/or birdwatching groups. There might even be opportunities to team up with regional or national "citizen science" programs, designed to get laypeople participating in actual, hands-on scientific research. For example, parents and teachers could have children collect data on species identifications and counts of individual birds, which would then be added to a national, online database. The resulting information might then be used, for example, to track the effects of global warming on bird populations. Believe it or not, research programs like this are already well underway.[2] Birdwatching could also be a catalyst for service-learning efforts such as reclaiming watersheds, planting trees that attract birds and other animals, and raising awareness of endangered species and habitats. A vibrant Backyard Dinosaurs initiative, in addition to fostering nature literacy, would help build lasting human-nature connections and a stronger sense of place.

So how about it? Ready to start a backyard dinosaur revolution?

In wrapping up this epilogue, I return to the place where this journey began—the island of Madagascar. Here we will find an even more concrete example of how dinosaurs and paleontology can serve people living today.

Chapter 1 detailed some of the amazing discoveries of Late Cretaceous dinosaurs and other vertebrates made on the Red Island during the past 15 years—from predators like tiny-armed *Majungasaurus* and buck-toothed *Masiakasaurus* to herbivores such as the long-necked giant *Rapetosaurus* and the pug-faced crocodile *Simosuchus*. These finds and many more are the result of nine field expeditions (so far) led by David Krause of Stony Brook University to a tiny pocket of grassy badlands in northwestern Madagascar. I had the honor of participating in five of those expeditions. The project forever changed the course of my career and those of many others, students and professionals alike.

When working in Madagascar, we camp at the same site on the edge of tiny Berivotra village, a mere stone's throw from some of the one-room huts, with their mud or thatched-grass walls and palm-leaf roofs. We interact daily with the villagers, some of whom visit camp regularly and have become our friends. We've watched some children in this remote community grow from toddlers to adulthood, and we've been saddened by the premature deaths of others. We have shared in celebrations, deliberations, times of grief, and makeshift games of soccer (football). Berivotra villagers have guided us to promising fossil localities and helped carry the huge, fossil-filled plaster jackets across kilometers of rugged terrain to the nearest road. Always the community has welcomed us warmly, showering our mixed Malagasy–North American crew with broad smiles as we pass by each day on the way to and from our dig sites.

Back in 1995, while digging in some of our first dinosaur quarries, we noticed that village children were coming by regularly to watch us work, obviously curious about the strange *vazaha* (foreigners) who insanely chose to sit out in the midday sun picking at the chunks of rocklike bone strewn about the landscape and buried just beneath the surface. These bright-eyed, inquisitive youngsters were invariably rail-thin, although many had distended bellies from protein deficiencies, as well as liver- and spleen-enlarging parasites. We would later learn that diarrheal, dental, and respiratory infections are widespread in this and many surrounding communities. Lacking access to clean water, basic hygiene, and antibiotics, these diseases and others—with malaria topping the list—all too often prove fatal. Working with Malagasy interpreters from our crew, we spoke with village adults and found out that the children did not go to school because, well, there was no school.

The situation we found in Berivotra is mimicked over much of Madagascar, one of the poorest nations on Earth. About one-third of the country's population of 19 million cannot read or write, and literacy rates tend to be lowest in remote areas. Many rural Malagasy grow up without ever visiting a doctor or dentist, exacerbating abysmal health conditions and limiting both life expectancy and quality of life.

When we asked the Berivotra villagers how we might help and repay them for their many kindnesses, they were nearly unanimous in their request for a school. How much would that cost? They humbly requested U.S.$500, enough to cover the cost of a teacher for a full year! We readily agreed and took up a collection, making sure that sufficient funds were also available for books and supplies. It was decided that the nearby tin-walled church, the only sizeable structure in the community, would be used as the classroom. This simple act was the beginning of what would become the Madagascar Ankizy Fund, a nonprofit organization based out of Stony Brook University with the mission of providing education and health care to children (*ankizy* in the Malagasy language) in remote areas of the island.[3] The Ankizy Fund would eventually build a new school building in Berivotra, dig eight wells for fresh water, provide health care in the form of medical and dental teams, and initiate an arts and crafts "paleo-tourist" industry (with dinosaurs as a main theme). The latter was critical because many villagers were nearing starvation; the economics of the village revolved around the sale of charcoal, based on cutting down trees. But, though the village was literally inside the forest only 50 years ago, the trees have been cut back so far that it is now a 2-day walk to get there.

Today, all the children of Berivotra attend school, and, with assistance from the Ankizy Fund, many have gone on to a secondary school in the nearby city of Marovoay. It is a joy to see them learning and in better health, so proud to be able to read and write. The Ankizy Fund has also expanded its efforts to target other rural communities in Madagascar, thus far funding four new schools, renovating an orphanage, initiating large-scale parasite disinfection programs, and bringing dental and health care to a number of remote areas. Funds have also been acquired to build a medical clinic in a remote village. Community-wide education programs have addressed key issues surrounding nutrition, hygiene, and sexual health.

FIGURE EP.1

David Krause (left) stands with the first group of students to attend the newly built Sekoly Riambato (Stony Brook) school, in the village of Berivotra, northwestern Madagascar.

Numerous people in North America and Madagascar have been involved in these efforts—scientists, dentists, doctors, students, and volunteers among them. Funding has come from a variety of sources; most inspiring are the American grade school children who have conducted school fund-raisers, learning in the process about their counterparts on a Third World island and also that they can make a meaningful difference. As with many philanthropic efforts, however, one person has spearheaded the organization, serving as its heart, soul, and brains. In this case, that person is David Krause, the lead paleontologist on the Madagascar project. Krause has made Madagascar the core of his professional life, attempting to split his time between paleontological research and the Ankizy Fund (with the latter usually coming out ahead). As with any humanitarian effort in a developing country, the Ankizy Fund has encountered a number of unexpected pitfalls, but it has persevered through these challenges to make a tremendous difference in the lives of thousands of people in rural Madagascar.

Paleontologists regularly travel to underdeveloped countries like Madagascar in search of fossils. And many other academic disciplines include work in developing countries. Yet all too rarely do these international projects include a meaningful humanitarian component. Indeed, the Ankizy Fund is one of the only examples I am aware of—science with a social conscience. Here, then, is another way that paleontology has the potential to benefit people, in this case directly and immediately. I hope that many other scientists working internationally elect to make the same commitment that David Krause has in Madagascar.

NOTES AND REFERENCES

This section presents a chapter-by-chapter listing of notes and references. Throughout I have included both scientific references and more popular references that might be more approachable for a lay reader.

CHAPTER 1.   TREASURE ISLAND

NOTE

1.   Suzuki 1997:20.

REFERENCES

Bakker, R. T. 1968. The superiority of dinosaurs. *Discovery* 3:11–12.
———. 1971. The ecology of brontosaurs. *Nature* 229:172–174.
———. 1986. *The Dinosaur Heresies*. New York: Morrow.
Capra, F. 1996. *The Web of Life: A New Scientific Understanding of Living Systems*. New York: Anchor.
Chiappe, L. M., and Witmer, L. M. (eds.). 2002. *Mesozoic Birds: Above the Heads of Dinosaurs*. Berkeley: University of California Press.
Hay, W. W., DeConto, R. M., Wold, C. N., Willson, K. M., Voigt, S., Schulz, M., Wold-Rossby, A., Dullo, W.-C., Ronov, A. B., Balukhovsky, A. N., and Söding, E. 1999. An alternative global Cretaceous paleogeography. In E. Berrera and C. Johnson (eds.), *Evolution of Cretaceous Ocean/Climate Systems*. Geological Society of America Special Paper, 332:1–48.

Krause, D. W., O'Connor, P. M., Curry Rogers, K., Sampson, S. D., Buckley, G. A., and Rogers, R. R. 2006. Late Cretaceous terrestrial vertebrates from Madagascar: Implications for Latin American biogeography. *Annals of the Missouri Botanical Garden* 93:178–208.

Kuhn, T. S. 1970. *The Structure of Scientific Revolutions.* 2nd ed. Chicago: University of Chicago Press.

Ostrom, J. H. 1969. Osteology of *Deinonychus antirrhopus,* an unusual theropod from the Lower Cretaceous of Montana. *Bulletin of the Peabody Museum of Natural History* 30:1–165.

———. 1975. The origin of birds. *Annual Review of Earth and Planetary Science* 3:55–77.

Sampson, S. D., and Krause, D. W. (eds.). 2007. *Majungatholus atopus* (Theropoda: Abelisauridae) from the Late Cretaceous of Madagascar. *Society of Vertebrate Paleontology Memoir* 8.

Sampson, S. D., Witmer, L. M., Forster, C. A., Krause, D. W., O'Connor, P. M., Dodson, P., and Ravoavy, F. 1998. Predatory dinosaur remains from Madagascar: Implications for the Cretaceous biogeography of Gondwana. *Science* 280:1048–1051.

Suzuki, D. 1997. *The Sacred Balance: Rediscovering Our Place in Nature.* Vancouver: Greystone Books.

Waldrop, M. M. 1992. *Complexity: The Emerging Science at the Edge of Order and Chaos.* New York: Simon and Schuster.

CHAPTER 2. STARDUST SAURIANS

NOTES

1. Abram 1996:185.
2. Instead of the term *protist,* the more inclusive term *protoctists* is often applied to micro-sized protists and their macrosized cousins, including seaweed.
3. The Mesozoic is often called the "Age of Dinosaurs." The preceding Paleozoic era has been referred to as the "Age of Fishes," and the following era, the Cenozoic, has been dubbed the "Age of Mammals." Masked behind these designations are deep, long-held biases regarding the place of humans within nature. Specifically, the history of life on Earth has generally been viewed as a ladder of progression, or "scalae naturae," that begins with microbes and culminates in humankind. Yet, as argued by microbiologist Lynn Margulis and others, it is far more appropriate to regard the entire history of the Earth since the appearance of the first single-celled organisms as the "Age of Bacteria." Bacteria were the first forms of life and the only life-forms throughout most of geologic history. Today they remain dominant in terms of numbers of individuals and ecosystem importance. Moreover, as described in this chapter, bacteria gave rise to all succeeding kingdoms of life through key bacterial mergers, a symbiotic legacy that remains with us to the present day.
4. This quotation comes from Wilson 1999:107.

REFERENCES

Abram, D. 1996. *The Spell of the Sensuous.* New York: Vintage Books.

Darwin, C. 1859. *On the Origin of Species by Means of Natural Selection, or the Preservation of Favoured Races in the Struggle for Life.* London: Murray.

Gould, S. J. 1987. *Time's Arrow Time's Cycle: Myth and Metaphor in the Discovery of Geological Time*. Cambridge, MA: Harvard University Press.

Margulis, L., and Sagan, D. 1986. *Microcosmos: Four Billion Years of Microbial Evolution*. Berkeley: University of California Press.

Swimme, B., and Berry, T. 1992. *The Universe Story: From the Primordial Flaring Forth to the Ecozoic Era: A Celebration of the Unfolding of the Cosmos*. San Francisco: Harper.

Wilson, E. O. 1998. *Consilience: The Unity of Knowledge*. New York: Knopf.

CHAPTER 3.   DRAMATIS DINOSAURAE

NOTES

1. Several of the terms in the Linnaean hierarchy (e.g., *phylum*, *order*, and *family*) have recently fallen out of vogue. Biologists embracing the methodology known as phylogenetic systematics, or cladistics—used to reconstruct the evolutionary relationships of species and groups—have found it difficult to use such terms consistently across distantly related groups. As a result, an entirely new nomenclature is currently under discussion.

2. The advent of cladistics has led to another terminological revolution of sorts, at least as applied to categorizations of the natural world. Since birds are the direct descendants of dinosaurs, and dinosaurs are reptiles, then, in a very real sense birds are reptiles as well, a conclusion that seems counterintuitive. In an effort to reduce such ambiguities, biologists have invented a plethora of new terms relatively unknown outside the academic realm. Thus, the term *reptile* appears less and less in the scientific literature, replaced by *sauropsid*, regarded as the natural grouping, or *clade*, that includes lizards, turtles, crocodiles, and birds, as well as a number of extinct groups such as dinosaurs and pterosaurs. In this sense, birds are avian sauropsids descended from dinosaurs, which in turn belong to the larger sauropsid group known as archosaurs.

REFERENCES

Benton, M. J. 2005. *Vertebrate Paleontology*. 3rd ed. Oxford: Blackwell.

Raup, D. M., and Stanley, S. M. 1978. *Principles of Paleontology*. 2nd ed. San Francisco: Freeman.

Reisz, R. R., Scott, D., Sues, H.-D., Evans, D. C., and Raath M. A. 2005. Embryos of an Early Jurassic prosauropod dinosaur and their evolutionary significance. *Science* 309(5735):761–764.

Sampson, S. D., Carrano, M. T., and Forster, C. A. 2001. A bizarre predatory dinosaur from Madagascar: Implications for the evolution of Gondwanan theropods. *Nature* 409:504–505.

Wang, S. C., and Dodson, P. 2006. Estimating the diversity of dinosaurs. *Proceedings of the National Academy of Sciences, USA* 103(37):13601–13605.

Weishampel, D., Dodson, P., and Osmólska, H. (eds.). 2004. *The Dinosauria*. 2nd ed. Berkeley: University of California Press.

Zimmer, C. 1999. *At the Water's Edge: Fish with Fingers, Whales with Legs, and How Life Came Ashore but Then Went Back to Sea*. New York: Simon and Schuster.

CHAPTER 4. DRIFTING CONTINENTS AND GLOBE-TROTTING DINOSAURS

NOTES

1. For more on the topic of Alfred Wegener and his ideas on mobile continents, see Eldredge 1998.

2. The melting of polar ice and consequent elevation in sea levels is one of the primary concerns over the current global warming trend, since the majority of people on Earth live in low-lying coastal regions. Indeed, it is entirely possible that the current rise in sea level could lead to the flooding of entire countries such as the Netherlands.

REFERENCES

Eldredge, N. 1998. *The Pattern of Evolution*. New York: Freeman.

Krause, D. W., O'Connor, P. M., Curry Rogers, K., Sampson, S. D., Buckley, G. A., and Rogers, R. R. 2006. Late Cretaceous terrestrial vertebrates from Madagascar: Implications for Latin American biogeography. *Annals of the Missouri Botanical Garden* 93:178–208.

Kump, L. R., Kasting, J. F., and Crane, R. G. 2004. *The Earth System*. 2nd ed. Upper Saddle River, NJ: Prentice Hall.

Sereno, P. C., Wilson, J. A., and Conrad, J. L. 2004. New dinosaurs link southern landmasses in the mid-Cretaceous. *Proceedings of the Royal Society of London B, Biological Sciences* 271:1325–1330.

Suzuki, D., and McConnell, A. 1997. *The Sacred Balance*. Vancouver: Greystone Books.

Tatsumi, Y. 2005. The subduction factory: How it operates in the evolving Earth. *GSA Today* 15(7):4–10.

CHAPTER 5. SOLAR EATING

NOTES

1. A tiny fraction of life on Earth exists entirely off the solar grid, obtaining energy from deep sea hydrothermal vents, which channel energy from within the planet.

2. Humans have become extremely adept at exploiting the solar energy stored in these "fossil fuels." These ancient remains—in actuality, the energy of the sun stored in solid form—drive our economies and act as another pathway that returns carbon dioxide to the atmosphere. It is often stated incorrectly that fossil fuels are made from the remains of dinosaurs. Natural resources such as coal, oil, and natural gas are indeed based on the remains of dead organisms, some of which lived alongside dinosaurs. But the great bulk of these stored reserves are derived from plants and microbes, which together make up the bulk of the biomass in any land habitat.

REFERENCES

Callenbach, E. 1998. *Ecology: A Pocket Guide*. Berkeley: University of California Press.

Erickson, G. M. 1999. Breathing life into *Tyrannosaurus rex*. *Scientific American* 281:43–49.

Farlow, J. O., and Holtz Jr., T. R. 2002. The fossil record of predation in dinosaurs. Pp. 251–265 in M. Kowalewski and P. H. Kelley (eds.), *The Fossil Record of Predation*. Paleontological Society Paper 8. N.p.: Paleontological Society.

Hoagland, M., and Dodson, B. 1998. *The Way Life Works*. New York: Times Books.

Horner, J. R. 1994. Steak knives, beady eyes, and tiny little arms (a portrait of *T. rex* as a scavenger). Pp. 157–164 in G. D. Rosenberg and D. L. Wolberg (eds.), *Dino Fest*. Paleontological Society Special Publication. Knoxville, TN: Paleontological Society.

Johnson, K. R. 2002. The megaflora of the Hell Creek and lower Fort Union Formations in the western Dakotas: Vegetational response to climate change, the Cretaceous-Tertiary boundary event, and rapid marine transgression. Pp. 329–392 in J. Hartman, K. R. Johnson, and D. J. Nichols (eds.), *The Hell Creek Formation and the Cretaceous-Tertiary Boundary in the Northern Great Plains: An Integrated Continental Record of the End of the Cretaceous*. Geological Society of America Special Paper 361. Boulder, CO: Geological Society of America

Labandeira, C. C. 1997. Insect mouthparts: Ascertaining the paleobiology of insect feeding strategies. *Annual Review of Ecology and Systematics* 28:317–351.

Labandeira, C. C., Johnson, K. R., and Lang, P. 2002. Preliminary assessment of insect herbivory across the Cretaceous-Tertiary boundary: Major extinction and minimum rebound. Pp. 297–328 in J. H. Hartman, K. R. Johnson, and D. J. Nichols (eds.), *The Hell Creek Formation and the Cretaceous-Tertiary Boundary in the Northern Great Plains: An Integrated Continental Record of the End of the Cretaceous*. Geological Society of America Special Paper 361. Boulder, CO: Geological Society of America.

Russell, D. A., and Manabe, M. 2002. Synopsis of the Hell Creek (uppermost Cretaceous) dinosaur assemblage. Pp. 169–176 in J. H. Hartman, K. R. Johnson, and D. J. Nichols (eds.), *The Hell Creek Formation and the Cretaceous-Tertiary Boundary in the Northern Great Plains: An Integrated Continental Record of the End of the Cretaceous*. Geological Society of America Special Paper 361. Boulder, CO: Geological Society of America.

Ruxton, G. D., and Houston, D. C. 2003. Could *Tyrannosaurus rex* have been a scavenger rather than a predator? An energetics approach. *Proceedings of the Royal Society of London B, Biological Sciences* 270(1576):731–733.

———. 2004. Obligate vertebrate scavengers must be large soaring fliers. *Journal of Theoretical Biology* 228(3):431–436.

Wing, S. L., and Tiffney, B. H. 1987. The reciprocal interaction of angiosperm evolution and tetrapod herbivory. *Review of Paleobotany and Palynology* 50:179–210.

CHAPTER 6. THE RIVER OF LIFE

NOTES

1. In 2004, a remarkable Chinese theropod named *Mei long* was described by Xu Xing and Mark Norell. This duck-sized animal was found with its head tucked in birdlike fashion beneath its forelimb, apparently buried while sleeping.

2. The profound similarities in general structure among vertebrates explain why many paleontologists (including myself for several years) teach anatomy in medical and veterinary schools. Humans and all other vertebrates are sufficiently similar that, if you know

the anatomy of one animal in detail (e.g., bones, muscles, nerves, blood vessels, and organs), you know much of vertebrate anatomy in general.

3. The sieve metaphor, although certainly instructive, is misleading in that it gives the impression of evolution as a top-down phenomenon, with natural selection taking on the role of "Big Brother" (the sieve holder) to dispose of undesirables. The true process is much more collaborative and bottom-up, with species in a given ecosystem cocreating, or "inventing" each other through ongoing mutual adjustments. Whereas emphasis in the sieve perspective is biased heavily toward negative traits, the cocreation view directs attention toward positive traits that increase the success of a species within a given ecosystem.

4. In certain groups of organisms, particularly among plants, species tributaries can rejoin, and frequently do so, through interspecies crosses.

REFERENCES

Brett, C. E., and Baird, G. C. 1995. Coordinated stasis and evolutionary ecology of Silurian to Middle Devonian faunas in the Appalachian Basin. Pp. 285–315 in D. H. Erwin and R. L. Anstey (eds.), *New Approaches to Speciation in the Fossil Record*. New York: Columbia University Press.

Callenbach, E. 1998. *Ecology: A Pocket Guide*. Berkeley: University of California Press.

Chiappe, L. M. 2007. *Glorified Birds: The Origin and Early Evolution of Dinosaurs*. New York: Wiley-Liss.

Chimpanzee Sequencing and Analysis Consortium. 2005. Initial sequence of the chimpanzee genome and comparison with the human genome. *Nature* 437:69–87.

Currie, P. J., Koppelhus, E. B., and Shugar, M. A. (eds.). 2004. *Feathered Dragons: Studies on the Transition from Dinosaurs to Birds*. Bloomington: Indiana University Press.

Eldredge, N. 1995. *Reinventing Darwin: The Great Debate at the High Table of Evolutionary Theory*. New York: Wiley.

———. 1999. *The Pattern of Evolution*. New York: Freeman.

Eldredge, N., and Gould, S. J. 1972. Punctuated equilibria: An alternative to phyletic gradualism. Pp. 82–115 in T. J. M. Schopf (ed.), *Models in Paleobiology*. San Francisco: Freeman, Cooper.

Gould, S. J. 2002. *The Structure of Evolutionary Theory*. Cambridge, MA: Belknap.

Gould, S. J., and Eldredge, N. 1977. Punctuated equilibria: The tempo and mode of evolution reconsidered. *Paleobiology* 3:115–151.

Leakey, M. G., Spoor, F., Brown, F. H., Gathogo, P. N., Kiarie, C., Leakey, L. N., and McDougall, I. 2001. New hominin genus from eastern Africa shows diverse middle Pliocene lineages. *Nature* 410:419–420.

Sepkoski Jr., J. J. 1992. Phylogenetic and ecologic patterns in the Phanerozoic history of marine biodiversity. Pp. 77–100 in N. Eldredge (ed.), *Systematics, Ecology, and the Biodiversity Crisis*. New York: Columbia University Press.

Understanding evolution. 2008. http://evolution.berkeley.edu/. Berkeley: University of California Museum of Paleontology.

Vrba, E. 1985. Environment and evolution: Alternative causes of the temporal distribution of evolutionary events. *South African Journal of Science* 81:229–236.

Weiner, J. 1994. *The Beak of the Finch: A Story of Evolution in Our Time.* New York: Vintage Books.

Xu, X., and Norell, M. A. 2004. A new troodontid dinosaur from China with avian-like sleeping posture. *Nature* 431:838–841.

Xu, X., Zhonghe, Z., and Xiaolin, W. 2000. The smallest known non-avian theropod dinosaur. *Nature* 408:705–708.

Zimmer, C. 2001. *Evolution: The Triumph of an Idea.* New York: HarperCollins.

CHAPTER 7.   THE GREEN GRADIENT

NOTES

1. Caffeine in coffee plants and nicotine in tobacco plants are two familiar examples of plant-generated poisons that evolved to discourage herbivory. Of course, these toxic compounds have been utilized secondarily by humans as brain stimulants.

2. As demonstrated by human-introduced insecticides such as DDT, insect populations are often able to respond to chemical assaults through evolution, spreading successful mutations that enable subsequent generations of insects to metabolize the poison.

3. Another remarkable example of mutually beneficial symbionts, one discovered only recently, occurs in some species of conifers. Large evergreen conifers such as Douglas firs and coast redwoods often live for hundreds of years. In contrast, the insects feeding on their needles have brief life spans, generally less than 1 year. So, given that arms races require evolution, one might expect the insects to evolve faster than the trees and thereby overwhelm them. At least some trees, however, enlist symbiotic allies in the form of microscopic fungi that live inside their needles and derive sugars and starch from the tree. In return, the fungi produce chemicals that are toxic to local defoliating insects. Since the fungi are short-lived, they are better able to deal with the rapidly changing tactics of insect attackers, enabling their enormous, stately partners to survive for hundreds or even thousands of years.

REFERENCES

Bakker, R. T. 1978. Dinosaur feeding behaviour and the origin of flowering plants. *Nature* 274:661–663.

Barrett, P. M., and Willis, K. J. 2001. Did dinosaurs invent flowers? Dinosaur-angiosperm coevolution revisited. *Biological Reviews* 76:411–477.

Johnson, K. R. 2002. The megaflora of the Hell Creek and lower Fort Union Formations in the western Dakotas: vegetational response to climate change, the Cretaceous-Tertiary boundary event, and rapid marine transgression. Pp. 329–392 in J. Hartman, K. R. Johnson, and D.J. Nichols (eds.), *The Hell Creek Formation and the Cretaceous-Tertiary Boundary in the Northern Great Plains: An Integrated Continental Record of the End of the Cretaceous.* Geological Society of America Special Paper 361. Boulder, CO: Geological Society of America

Farlow, J. O. 1987. Speculations about the diet and digestive physiology of herbivorous dinosaurs. *Paleobiology* 13(1):60–72.

Farlow, J. O., Dodson, P., and Chinsamy, A. 1995. Dinosaur biology. *Annual Review of Ecology and Systematics* 26:445–471.

Fastovsky, D. E., and Smith, J. B. 2004. Dinosaur paleoecology. Pp. 614–626 in D. Weishampel, P. Dodson, and H. Osmólska (eds.), *The Dinosauria*. 2nd ed. Berkeley: University of California Press.

Tiffney, B.1997. Land plants as food and habitat in the Age of Dinosaurs. Pp. 352–370 in J. O. Farlow and M. K. Brett-Surman (eds.), *The Complete Dinosaur*. Bloomington: Indiana University Press.

Wing, S. L., and Tiffney, B. H.1987. The reciprocal interaction of angiosperm evolution and tetrapod herbivory. *Review of Paleobotany and Palynology*, 50:179–210.

CHAPTER 8.   PANOPLY OF PREDATORS

NOTES

1. Claims of the "biggest" theropod dinosaur are generally based on estimates of either body length or body mass. Whereas the South American *Giganotosaurus* may have equaled or slightly exceeded the total length of *Tyrannosaurus*, the largest thigh bones (femora) of each suggest that *T. rex* was the champion heavyweight of the pair. However, a hefty, isolated jawbone of *Giganotosaurus* may represent an even larger and heavier individual. Suffice it to say that both animals were truly enormous, and others of similar size (though probably not much larger) likely await discovery.

2. Pneumaticity is a remarkable phenomenon that is by no means unique to theropods, or even to dinosaurs. Air-filled pneumatic cavities are typical of many vertebrate groups, including mammals. As mammals, we humans have faces possessing several air pockets; these are the sinuses that tend to fill with thick fluid and become infected when we're sick.

3. One novel suggestion made by Nathan Myhrvold and Philip Currie is that the long tails of sauropods like *Diplodocus* were cracked like gigantic whips in order to generate sonic booms capable of deterring theropods. This hypothesis, though perhaps mechanically feasible, seems rather far-fetched and has received minimal support from other paleontologists.

4. In contrast to dinosaurs and reptiles generally, large mammals tend to have relatively long gestation times and give birth to fewer young, so a much smaller proportion of the animals alive at any given moment are nonadults.

5. Evidence that maniraptor sickle claws were used, at least on occasion, as weapons comes from one of the most spectacular fossils ever found—a *Velociraptor* intertwined in an apparent death pose with a ceratopsian, *Protoceratops*. In this specimen, found in Mongolia, one of the sickle claws of the predator is positioned in the neck region of the herbivore, and at the time of death it was likely embedded in the tissues surrounding the throat. One of the hands of the *Velociraptor* is holding or being held by the jaws of the *Protoceratops*. If interpreted correctly (and there are skeptics), it appears that the two animals became interlocked in a life-and-death struggle and were then quickly buried by the collapse of a sand dune.

6. Thanks to Lindsay Zanno for this provocative word image.

Barrett, P. M. 2005. The diet of ostrich dinosaurs (Theropoda: Ornithomimosauria). *Palaeontology* 48:347–358.

Barrett, P. M., and Rayfield, E. J. 2006. Ecological and evolutionary implications of dinosaur feeding behaviour. *Trends in Ecology & Evolution* 21(4):217–224.

Burness, G. P., Diamond, J., and Flannery, T. 2001. Dinosaurs, dragons, and dwarves: The evolution of maximal body size. *Proceedings of the National Academy of Sciences, USA* 98:14518–14523.

Carpenter, K. 2000. Evidence of predatory dinosaur behavior by carnivorous dinosaurs. *Gaia* 15:135–144.

Carrano, M. T. 1998. What, if anything, is a cursor? Categories versus continua for determining locomotor habit in mammals and dinosaurs. *Journal of Zoology (London)* 247:29–42.

Carrano, M. T., Blob, R. W., Gaudin, T., and Wible, J. (eds.). 2006. *Amniote Paleobiology: Perspectives on the Evolution of Mammals, Birds, and Reptiles.* Chicago: University of Chicago Press.

Chin, K., Tokaryk, T. T., Erickson, G. M., and Calk, L. C. 1998. A king-sized theropod coprolite. *Nature* 393:680–682.

Currie, P. J. 1998. Possible evidence of gregarious behavior in tyrannosaurids. *Gaia* 15:271–277.

Erickson, G. M., Makovicky, P. J., Currie, P. J., Norell, M. A., Yerby, S. A., and Brochu, C. A. 2004. Gigantism and comparative life-history parameters of tyrannosaurid dinosaurs. *Nature* 430:772–775.

Farlow, J. O., and Holtz Jr., T. R. 2002. The fossil record of predation in dinosaurs. Pp. 251–265 in M. Kowalewski and P. H. Kelley (eds.), *The Fossil Record of Predation.* Paleontological Society Paper 8. N.p.: Paleontological Society.

Farlow, J. O., and Pianka, E. R. 2003. Body size overlap, habitat partitioning, and living space requirements of terrestrial vertebrate predators: Implications for large-theropod body size. *Historical Biology* 17:21–40.

Henderson, D. M. 1998. Skull and tooth morphology as indicators of niche partitioning in sympatric Morrison Formation theropods. *Gaia* 15:219–226.

Hutchinson, J. R. and Garcia, M. 2002. *Tyrannosaurus* was not a fast runner. *Nature* 415:1018–1021.

Kirkland, J. I., Zanno, L. E., Sampson, S. D., Clark, J. C., and DeBlieux, D. 2005. A primitive therizinosauroid dinosaur from the Early Cretaceous of Utah. *Nature* 435:84–87.

Myhrvold, N. P., and Currie, P. J. 1997. Supersonic sauropods? Tail dynamics in diplodocids. *Paleobiology* 23:393–409.

Norell, M. A., Makovicky, P. J., and Currie, P. J. 2001. The beaks of ostrich dinosaurs. *Nature* 412: 873–874.

Rayfield, E. J. Aspects of comparative cranial mechanics in the theropod dinosaurs *Coelophysis, Allosaurus,* and *Tyrannosaurus. Zoological Journal of the Linnean Society* 144:309–316.

Rogers, R. R., Curry Rogers, K., and Krause, D. W. 2003. Cannibalism in the Madagascan dinosaur *Majungatholus atopus. Nature* 422:515–518.

Sereno, P. C., Larsson, H. C. E., Sidor, C. A., and Gado, B. 2001. The giant crocodyliform *Sarcosuchus* from the Cretaceous of Africa. *Science* 294:1516–1519.

Van Valkenburgh, B., and Molnar, R. E. 2002. Dinosaurian and mammalian predators compared. *Paleobiology* 28:527–543.

Witmer, L. M. 1997. The evolution of the antorbital cavity of archosaurs: A study in soft tissue reconstruction in the fossil record with an analysis of the function of pneumaticity. *Journal of Vertebrate Paleontology Memoir* 3:1–73.

Witmer, L. M., and Ridgley, R. C. 2008. The paranasal air sinuses of predatory and armored dinosaurs (Archosauria: Theropoda and Ankylosauria) and their contribution to cephalic architecture. *Anatomical Record* 291:1362–1388.

CHAPTER 9.   HIDDEN STRANDS

NOTES

1. Figures from A. Moldenke, cited in Luoma 1999:97.
2. Statistics derived from Wilson 2006.

REFERENCES

Carpenter, K., 2005, Experimental investigation of the role of bacteria in bone fossilization. *Neues Jahrbuch für Geologie und Paläontologie Monatshefte* 2005(2):83–94.

Chin, K. 2007. The paleobiological implications of herbivorous dinosaur coprolites from the Upper Cretaceous Two Medicine Formation of Montana: Why eat wood? *Palaios* 2007(22):554–566.

Chin, K., and Gill, B. D. 1996. Dinosaurs, dung beetles, and conifers: Participants in a Cretaceous food web. *Palaios* 11:280–285.

Chin, K., Hartman, J. H., and Roth, B. 2008. Opportunistic exploitation of dinosaur dung: Fossil snails in coprolites from the Upper Cretaceous Two Medicine Formation of Montana: *Lethaia*. doi: 10.1111/j.1502-3931.2008.00131.x (published online).

Grimaldi, D., and Engel, M. S. 2005. *Evolution of the Insects.* Cambridge: Cambridge University Press.

Hooper, L. C., Bry, L., Falk, P. G., and Gordon, J. I. 1998. Host-microbial symbiosis in the mammalian intestine: Exploring an internal ecosystem. *BioEssays* 20(4):336–343.

Luoma, J. R. 1999. *The Hidden Forest.* New York: Holt.

Labandeira, C. C., Dilcher, D. L., Davis, D. R., and Wagner, D. L. 1994. Ninety-seven million years of angiosperm-insect association: Paleobiological insights into the meaning of coevolution. *Proceedings of the National Academy of Sciences, USA* 91:12278–12282.

Sharma, N., Kar, R. K., Agarwal, A., and Kar, R. 2005. Fungi in dinosaurian *(Isisaurus)* coprolites from the Lameta Formation (Maastrichtian) and its reflection on food habit and environment. *Micropaleontology* 51(1):73–82.

Wilson, E. O. 2006. *The Creation: An Appeal to Save Life on Earth.* New York: Norton.

CHAPTER 10.   OF HORN-HEADS AND DUCK-BILLS

NOTES

1. Several groups of dinosaurs—in particular, hadrosaurs, ceratopsids, and sauropods—possess greatly enlarged noses, which may have had a thermoregulatory role, keeping the brain cool by shedding heat from warmer blood and returning cooler blood to the brain region.

2. One of the more famous examples is an innovative experiment conducted by biologist Malte Andersson on African widowbirds. Males of this species tend to have elongate tails, long thought to be useful in attracting mates. Andersson cut off the lengthy tail feathers of some of these males and glued them onto the already-long tails of other males. This left him with three groups to observe: one with regular adult-length tails, one with very short tails, and the other with very long tails. He watched their behavior and found that females most preferred males with artificially lengthened tails and least preferred those with missing tail feathers. So, despite the fact that elongate tails inhibit flight abilities and make these males more visible to predators, a strong selective advantage can arise that is passed on to subsequent generations.

3. The can-opener ceratopsid became *Einiosaurus procurvicornis*. *Eini* is the Blackfeet Indian word for "buffalo." Since the fossils were recovered from lands on the Blackfeet Reservation, and since these dinosaurs appear to have moved in large herds across the Western Interior plains just as buffalo did in more recent times, the name seemed apt. It also honored the Blackfeet Indians and an important animal in their history. The second name, *procurvicornis*, is Latin for "forward-curving horn." In full, then, the name translates as "buffalo-lizard with the forward-curving horn." The second ceratopsid, the one with low bony growths instead of horns, was dubbed *Achelousaurus horneri*. Achelous was a Greek river god that had the ability to change shape at will. To battle the famous Heracles (Hercules of Roman mythology) over a woman, Achelous turned himself into a bull. Heracles won the battle only after tearing a horn from the bull's head. This horned dinosaur lost its horns during the course of evolution. The second part of the name honors Jack Horner for all his contributions to paleontology, as well as for his generosity in allowing me to work on this remarkable fossil collection.

4. The reasons behind this female mimicry of males may relate to several factors, including predator defense—so that females don't stand out as targets for predators—and increased within-group competition—enabling females to fend off advances from younger males.

5. Recent innovations offer hope that we may be able to assess sexual dimorphism more accurately in the future. Over the past several years, it has become possible to accurately estimate the age at death of a given dinosaur, an important innovation that will help rule out the influence of growth in future analyses. As for decisive demonstration of sexual differences, a recently published study raises a glimmer of hope. This study, conducted by Mary Schweitzer (North Carolina State University) and colleagues, documents a specialized bone type in a specimen of *Tyrannosaurus* that appears very similar to the bone laid down by living female birds during egg-laying season. If confirmed by additional discoveries, this method could provide a means of attributing at least some specimens to females.

6. Among living animals, flank butting occurs in numerous groups whereas violent head ramming is restricted to a handful of specialized species. Some investigators have argued that, based on the microstructure of the dome, even flank butting would have been unlikely in pachycephalosaurs. Instead, they suggest a display function for pachycephalosaur domes. Still other investigators have countered this view by arguing that

the forces incurred in head-to-head butting would have been smaller than previously thought and could have been absorbed by the thickened domes. Finally, as if all this weren't enough, it has been proposed that the considerable species variation in the shape of pachycephalosaur skull domes is most consistent with a range of mate competition behaviors—with flat-domed forms engaging in flank attacks, more rounded domes used in head butting, and the tallest domes serving solely as display devices. Clearly, with ongoing disagreement and several outstanding hypotheses, paleontologists still have work to do on this matter.

7. Occasional exceptions are known; an exquisitely preserved skeleton of the hadrosaur *Maiasaura* in the collection of the Royal Ontario Museum includes the impression of what may be a dewlap, or a large flap of skin below the neck.

8. In rare instances, captive animals from two different species can produce offspring. For example, tigers and lions have been known to breed, with the product sometimes labeled a "tigron." However, most of these unusual unions do not result in viable offspring, and even those that do typically do not involve species that meet in the wild. So we still regard such species as distinct from each other.

9. The tenuous nature of some cichlid species barriers was recently revealed when nearby logging caused rapid erosion, followed by silt buildup in one of the lakes. Unable to see the coloration patterns used to identify members of their own kind, these fishes expanded their reproductive horizons, mating with fishes from closely related species. These unnatural unions resulted in offspring that in turn have also been able to reproduce, and the end result has been at least a temporary decrease in the diversity of these fishes.

REFERENCES

Andersson, M. B. 1982. Female choice selects for extreme tail length in a widowbird. *Nature* 299:818–820.

———. 1994. *Sexual Selection*. Princeton, NJ: Princeton University Press.

Chiappe, L. M., Coria, R. A., Dingus, L., Jackson, F., Chinsamy, A., and Fox, M. 1998. Sauropod dinosaur embryos from the Late Cretaceous of Patagonia. *Nature* 396:258–261.

Chiappe, L. M., and Dingus, L. 2001. Walking on Eggs: the Astonishing Discovery of thousands of Dinosaur Eggs in the Badlands of Pategonia. New York: Scribner.

Dodson, P. 1975. Taxonomic implications of relative growth in lambeosaurine hadrosaurs. *Systematic Zoology* 24(1):37–54.

———. 1976. Quantitative aspects of relative growth and sexual dimorphism in *Protoceratops*. *Journal of Paleontology* 50: 929–940.

———. 1990. On the status of the ceratopsids *Monoclonius* and *Centrosaurus*. Pp. 211–229 in K. Carpenter and P. J. Currie (eds.), *Dinosaur Systematics: Approaches and Perspectives*. New York: Cambridge University Press.

Emlen, D. J. 2008. The evolution of animal weapons. *Annual Review of Ecology and Systematics* 39:387–413.

Farke A. A, Wolff, E. D. S., and Tanke D. H. 2009. Evidence of Combat in *Triceratops*. *PLoS ONE* 4(1): e4252. doi:10.1371/journal.pone.0004252 (published online).

Farlow, J. O., and Dodson, P. 1975. The behavioral significance of frill and horn morphology in ceratopsian dinosaurs. *Evolution* 29:353–361.

Farlow, J. O., Thompson, C. V., and Rosner, D. E. 1976. Plates of *Stegosaurus*: Forced convection or heat loss fins? *Science* 192:1123–1125.

Geist, V. 1966. The evolution of horn-like organs. *Behaviour* 27:173–214.

Goodwin, M. B., and Horner, J. R. 2004. Cranial histology of pachycephalosaurs (Ornithischia: Marginocephalia) reveals transitory structures inconsistent with head-butting behavior. *Paleobiology* 30(2):253–267.

Hopson, J. A. 1975. The evolution of cranial display structures in hadrosaurian dinosaurs. *Paleobiology* 1:21–43.

Horner, J. R., and Gorman, J. 1988. *Digging Dinosaurs: The Search That Unraveled the Mystery of Baby Dinosaurs*. New York: Workman.

Horner, J. R., Weishampel, D. B., and Forster, C. A. 2004. Hadrosauridae. Pp. 438–463 in D. B. Weishampel, P. Dodson, and H. Osmólska (eds.), *The Dinosauria*. 2nd ed. Berkeley: University of California Press.

Jarman, P. 1983. Mating system and sexual dimorphism in large, terrestrial, mammalian herbivores. *Biological Reviews* 58:485–520.

Lehman, T. M. 1990. The ceratopsian subfamily Chasmosaurinae: sexual dimorphism and systematics. Pp. 211–229 in K. Carpenter and P. J. Currie (eds.), *Proceedings of the Dinosaur Systematics: Approaches and Perspectives*. New York: Cambridge University Press.

Maryanska, T., Chapman, R. E., and Weishampel, D. B. 2004. Pachycephalosauria. Pp. 464–477 in D. B. Weishampel, P. Dodson, and H. Osmólska (eds.), *The Dinosauria*. 2nd ed. Berkeley: University of California Press.

Padian, K. and Horner, J. R. 2004. Species recognition as the principal cause of bizarre structures in dinosaurs. *Journal of Vertebrate Paleontology* 24(3, suppl.):100A.

Sampson, S. D. 1995. Two new horned dinosaurs from the Upper Cretaceous Two Medicine Formation of Montana, USA, with a phylogenetic analysis of the Centrosaurinae (Ornithischia: Ceratopsidae). *Journal of Vertebrate Paleontology* 15(4):743–760.

———. 2001. Speculations on the socioecology of ceratopsid dinosaurs (Ornithischia: Neoceratopsia). Pp. 263–276 in D. Tanke and K. Carpenter (eds.), *Mesozoic Vertebrate Life*. Bloomington: Indiana University Press.

Sampson, S. D., Ryan, M. J., and Tanke, D. H. 1997. Craniofacial ontogeny in centrosaurine dinosaurs (Ornithischia: Ceratopsidae): Taxonomic and behavioral implications. *Zoological Journal of the Linnean Society* 221(2):293–337.

Schweitzer, M. H., Wittmeyer, J. L., and Horner, J. R. 2005. Gender-specific reproductive tissue in ratites and *Tyrannosaurus rex*. *Science* 308:1456–1460.

Varricchio, D. J., Martin, A. J., and Katsura, Y. 2007. First trace and body fossil evidence of a burrowing, denning dinosaur. *Proceedings of the Royal Society of London B, Biological Sciences* 274(1616):1361–1368.

Vrba, E. S. 1984. Evolutionary pattern and process in the sister-group Alcelaphini-Aepycerotini (Mammalia: Bovidae). Pp. 62–79 in D. Otte and J. A. Endler (eds.), *Living Fossils*. New York: Springer.

West-Eberhard, M. J. 1983. Sexual selection, social competition, and speciation. *Quarterly Review of Biology* 58(2):155–183.

Wheeler, P. E. 1978. Elaborate CNS cooling structures in large dinosaurs. *Nature* 275:441–443.

Witmer, L. M. 1995. The extant phylogenetic bracket and the importance of reconstructing soft tissues in fossils. Pp. 19–33 in J. J. Thomason (ed.), *Functional Morphology in Vertebrate Paleontology*. Cambridge: Cambridge University Press.

CHAPTER 11. THE GOLDILOCKS HYPOTHESIS

NOTES

1. A few lizards, particularly among the groups known as varanids and teids, are daytime hunters that have somehow circumvented the limitations of their physiology to be active for many hours.

2. Ever wonder why you get goose bumps when it's cold? It's because we evolved from much hairier primate ancestors. Goose bumps represent a futile attempt to retain heat by raising our body hair and thereby increasing the thickness of our (now nearly absent) insulation layer.

3. Prior to significant human impacts, what we today refer to as African lions spanned all of Africa plus some of Europe, the Middle East, and Asia, for a total area larger than North America. Thanks to human impacts, such vast ranges clearly do not apply today to large carnivores, yet considerable evidence suggests that continental-scale species ranges were typical in the not-too-distant past. Indeed, one of the major conservation challenges we currently face involves connecting chunks of wildlands with protected corridors in order to support sustainable populations of large predators such as bears, wolves, and lions. One of the most important conservation insights is that ecosystems cannot be preserved as tiny patches and sometimes even as large game parks. Huge expanses of land, or at least a series of interconnected patches, are needed to keep an ecosystem intact.

4. See Humphreys 1979 and McNab 2002.

5. See Farlow 1993.

REFERENCES

Barrick, R. E., and Showers, W. J. 1994. Thermophysiology of *Tyrannosaurus rex*: evidence from oxygen isotopes. *Science* 265:222–224.

Burness, G., Diamond, P. J., and Flannery, T. 2001. Dinosaurs, dragons, and dwarves: The evolution of maximal body size. *Proceedings of the National Academy of Sciences, USA* 98:14518–14523.

Chinsamy, A., and Hillenius, W. J. 2004. Physiology of nonavian dinosaurs. Pp. 643–659 in D. B. Weishampel, P. Dodson, and H. Osmólska (eds.), *The Dinosauria*. 2nd ed. Berkeley: University of California Press.

Erickson, G. M. 2005. Assessing dinosaur growth patterns: A microscopic revolution. *Trends in Ecology & Evolution* 20(17):677–684.

Erickson, G. M., Makovicky, P. J., Currie, P. J., Norell, M. A., Yerby, S. A., and Brochu, C. A. 2004. Gigantism and comparative life-history parameters of tyrannosaurid dinosaurs. *Nature* 430:772–775.

Erickson, G. M., Rogers, K. C., and Yerby, S. A. 2001. Dinosaurian growth patterns and rapid avian growth rates. *Nature* 412:429–433.

Farlow, J. O. 1990. Dinosaur energetics and thermal biology. Pp. 43–55 in D. B. Weishampel, P. Dodson, and H. Osmólska (eds.), *The Dinosauria*. Berkeley: University of California Press.

———. 1993. On the rareness of big, fierce animals: Speculations about the body sizes, population densities, and geographic ranges of predatory mammals and large carnivorous dinosaurs. *American Journal of Science* 293A:167–199.

Farlow, J. O., Dodson, P., and Chinsamy, A. 1995. Dinosaur biology. *Annual Review of Ecology and Systematics*, 26:445–471.

Farmer, C. G. 2002. Reproduction: The adaptive significance of endothermy. *American Naturalist* 162:826–840.

Humphreys, W. F. 1979. Production and respiration in animal populations. *Journal of Animal Ecology* 48:427–453.

McNab, B. K. 2002. *The Physiological Ecology of Vertebrates: A View from Energetics*. Ithaca, NY: Cornell University Press.

Padian, K., and Horner, J. R. 2004. Dinosaur physiology. Pp. 660–671 in D. B. Weishampel, P. Dodson, and H. Osmólska (eds.), *The Dinosauria*. 2nd ed. Berkeley: University of California Press.

Pough, F. H. 1980. The advantages of ectothermy for tetrapods. *American Naturalist* 115:92–112.

Reid, R. E. H. 1997. Dinosaurian physiology: The case for "intermediate" dinosaurs. Pp. 449–473 in J. O. Farlow and M. K. Brett-Surman, eds., *The Complete Dinosaur*. Bloomington: Indiana University Press.

Ruben, J. A., Hillenius, W. J., Geist, N. R., Leitch, A., Jones, T. D., Currie, P. J., Horner, J. R., and Espe, G. 1996. The metabolic status of some Late Cretaceous dinosaurs. *Science* 272:1204–1207.

Seebacher, F. 2003. Dinosaur body temperatures: The occurrence of endothermy and ectothermy. *Paleobiology* 29:105–122.

Xu, X., Norell, M. A., Kuang, X., Wang, X., Zhao, Q., and Jia, C. 2004. Basal tyrannosauroids from China and evidence for protofeathers in tyrannosauroids. *Nature* 431:680–684.

Xu, X., Tan, Q., Wang, J., Zhao, X., and Tan, L. 2007. A gigantic bird-like dinosaur from the Late Cretaceous of China. *Nature* 447(7146):844–847.

CHAPTER 12.   CINDERELLASAURUS

NOTE

1. Growing fossil evidence suggests that the Chinle will also be an important resource for unraveling the origin of dinosaurs from closely related, two-legged archosaurs sometimes called "dinosauromophs."

REFERENCES

Bakker, R. T. 1975. Dinosaur Renaissance. *Scientific American* 232:58–78.
———. 1986. *The Dinosaur Heresies*. New York: Morrow.

Benton, M. J. 1983. Dinosaur success in the Triassic: A noncompetitive ecological model. *Quarterly Review of Biology* 58:29–55.

Charig, A. J. 1972. Evolution of the archosaurs pelvis and hind limb: An explanation in functional terms. Pp. 121–155 in K. A. Joysey and T. S. Kemp (eds.), *Studies in Vertebrate Evolution*. New York: Winchester.

Diamond, J. 1997. *Guns, Germs, and Steel: The Fates of Human Societies*. New York: Norton.

Fraser, N. 2006. *Dawn of the Dinosaurs: Life in the Triassic*. Bloomington: Indiana University Press.

Galton, P. M., and Upchurch, P. 2004. Prosauropoda. Pp. 232–258 in D. Weishampel, P. Dodson, and H. Osmólska (eds.), *The Dinosauria*. 2nd ed. Berkeley: University of California Press.

Grimaldi, D., and Engel, M. S. 2005. *Evolution of the Insects*. Cambridge: Cambridge University Press.

Hummel, J., Gee, C. T., Südekum, K-H., Sander, P. M., Nogge, G., and Clauss, M. 2007. In vitro digestibility of fern and gymnosperm foliage: Implications for sauropod feeding ecology and diet selection. *Proceedings of the Royal Society of London B, Biological Sciences*, doi:10.1098/rspb.2007.1728 (published online).

Irmis, R. B., Nesbitt, S. J., Padian, K., Smith, N. D., Turner, A. H., Woody, D., and Downs, A. 2007. A Late Triassic dinosauromorph assemblage from New Mexico and the rise of dinosaurs. *Science* 317:358–361.

Irmis, R. B., Parker, W. G., Nesbitt, S. J., and Liu, J. 2007. Early ornithischian dinosaurs: The Triassic record. *Historical Biology* 19(1):3–22.

Nesbitt, S. J., and Norell, M. A. 2006. Extreme convergence in the body plans of an early suchian (Archosauria) and ornithomimid dinosaurs (Theropoda). *Proceedings of the Royal Society of London B, Biological Sciences*, doi: 10.1098/rsbp.2005.3426 (published online).

Parker, W. G., Irmis, R. B., and Nesbitt, S. J. 2006. Review of the Late Triassic dinosaur record from Petrified Forest National Park, Arizona. *Museum of Northern Arizona Bulletin* 62:160–161.

Parker, W. G., Irmis, R. B., Nesbitt, S. J., Martz, J. W. and Browne, L. S. 2005. The Late Triassic pseudosuchian *Revueltosaurus callenderi* and its implications for the diversity of early ornithischian dinosaurs. *Proceedings of the Royal Society of London, Biological Sciences* 272:963–969.

CHAPTER 13. JURASSIC PARK DREAMS

NOTES

1. As of the writing of this book, the future of the visitor center at Dinosaur National Monument is in question. The building has been deemed unsafe and closed until it can be repaired. It is my sincere hope that the reopening of the building will include a reinvigoration of the paleontology program at this famous national landmark.

2. Even for ecologists studying living plants and animals, unraveling diet and other aspects of life history is a remarkable challenge, generally resulting in partial answers acquired over many years of meticulous work in both the field and the laboratory.

3. Since theropods were also egg layers, the majority of carnivorous dinosaurs on the Morrison landscape were nonadults as well. Studies by Mark Loewen (University of Utah) and others indicate that juvenile allosaurs likely possessed greater speed and

agility than adults. Perhaps young allosaurs consumed a diet based on smaller, nimble prey such as lizards, amphibians, and mammals. Conversely, greater foot speed may have been necessary for juveniles to keep pace with adults. Still another (to my mind, unlikely) alternative is that allosaurs engaged in cooperative hunting, with the faster juveniles responsible for herding prey animals toward lurking adults. Finally, it's possible that the greater speed and agility of juveniles was not an adaptation at all but merely an evolutionary holdover. In other words, theropods ancestral to *Allosaurus* may have been smaller and more agile even as adults, and the loss of these qualities in adult allosaurs simply reflects an evolutionary shift toward bigger bodies.

4. This relationship between body size and range size explains why Loch Ness monsters, sasquatches, and yetis are so incredibly improbable. Obviously, a lone serpent ("Nessy") is effectively a dead serpent, since it cannot reproduce. Yet even a family or group of serpents is doomed by low numbers. Survival over deep time minimally requires thousands of animals alive at any given moment. So, while the idea of a small, relict population of prehistoric serpents persisting for countless millennia in a remote Scottish loch makes for wonderful stories (and attracts tourist dollars), it makes no sense from an eco-evolutionary perspective and underlines the myopic view of deep time still held by most of us. Die-hard believers might counter by invoking the lottery defense; that is, we just happen to be alive to witness (through occasional, fleeting glimpses caught on film only as dark, misshapen blurs) the very last descendant of a Mesozoic plesiosaur, today known as the Loch Ness monster. After all, some animal has to be last. Maybe, but the odds of this being the case, given the lack of representative fossils for the past 65 million years and the major faunal turnovers that have occurred in the marine realm since the dinosaur extinction, seem astronomically remote at best.

REFERENCES

Gates, T. A. 2005. The Cleveland-Lloyd Dinosaur Quarry as a drought-induced assemblage. *Palaios* 20(4):363–375

Farlow, J. O. 2007. A speculative look at the paleoecology of large dinosaurs of the Morrison Formation, or, life with *Camarasaurus* and *Allosaurus*. Pp. 98–151 in E. P. Kvale, M. K. Brett-Surman, and J. Farlow (eds.), *Dinosaur Paleoecology and Geology: The Life and Times of Wyoming's Jurassic Dinosaurs and Marine Reptiles*. Shell, WY: GeoScience Adventures Workshop.

Foster, J. R. 2003. Paleoecological analysis of the vertebrate fauna of the Morrison Formation (Upper Jurassic), Rocky Mountain region, U.S.A. In *New Mexico Museum of Natural History and Science Bulletin* 23. Albuquerque: New Mexico Museum of Natural History and Science.

————. 2007. *Jurassic West*. Bloomington: Indiana University Press.

Hummel, J., Gee, C. T., Südekum, K.-H., Sander, P. M., Nogge, G., and Clauss, M. 2007. In vitro digestibility of fern and gymnosperm foliage: Implications for sauropod feeding ecology and diet selection. *Proceedings of the Royal Society of London B, Biological Sciences,* doi:10.1098/rspb.2007.1728 (published online).

Rayfield, E. Norman, D. B., Horner, C. C., Horner, J. R., Smith, P. M., Thomason, J. J., and Upchurch, P. 2001. Cranial design and function in a large theropod dinosaur. *Nature* 409:1033–1037.

Schweitzer, M. H., Suo, Z., Avci, R., Asara, J. M., Allen, M. A., Teran Arce, F., and Horner, J. R. 2007. Analyses of soft tissue from *Tyrannosaurus rex* suggest the presence of protein. *Science* 316:277–280.

Schweitzer, M. H., Wittmeyer, J. L., and Horner, J. R. 2007. Soft tissue and cellular preservation in vertebrate skeletal elements from the Cretaceous to the present. *Proceedings of the Royal Society of London B, Biological Sciences* 274:183–187.

Stevens, K. A., and Parrish, J. M. 2004. Neck posture, dentition, and feeding strategies in Jurassic sauropod dinosaurs. Pp. 212–232 in V. Tidwell and K. Carpenter (eds.), Thunder-Lizards: *The Sauropodomorph Dinosaurs.* Bloomington: Indiana University Press.

Therrien, F., Henderson, D. M., and Ruff, C. B. 2005. Bite me: Biomechanical models of theropod mandibles and implications for feeding behavior. Pp. 179–237 in K. Carpenter (ed.), *The Carnivorous Dinosaurs.* Bloomington: Indiana University Press.

Upchurch, P., and Barrett, P. M. 2000. The evolution of sauropod feeding mechanisms. Pp. 79–122 in H.-D. Sues (ed.), *Evolution of Herbivory in Terrestrial Vertebrates: Perspectives from the Fossil Record.* Cambridge: Cambridge University Press.

Van Valkenburgh, B., and Molnar, R. E. 2002. Dinosaurian and mammalian predators compared. *Paleobiology* 28:527–543.

Zaleha, M. J., and Wiesmann, S. A. 2005. Hyperconcentrated flows and gastroliths: Sedimentology of diamictites and wackes of the Upper Cloverly Formation, Lower Cretaceous, Wyoming, U.S.A. *Journal of Sedimentary Research* 75(1):43–54.

CHAPTER 14.   WEST SIDE STORY

NOTES

1. We have since found and excavated more complete specimens from Grand Staircase–Escalante National Monument, including (in 2008) what appears to be an even larger duck-billed dinosaur.

2. Mike Getty is a truly remarkable fellow worthy of additional mention. In 1999, immediately after I began work in a joint position at the University of Utah and the Utah Museum of Natural History, I began a search for a paleontology collections manager. Mike, also a Canadian, was working for the Royal Tyrrell Museum of Palaeontology in Alberta and he came highly recommended for the position. In trying to describe Mike's qualities, one of my Alberta colleagues described him as follows: "Let's put it this way, Scott. If you dropped Mike naked into the middle of the badlands, many miles from the nearest road, not only would he make it out alive, he'd be carrying fossils with him." Hiring Mike Getty was one of the best decisions I've ever made. He has ably led both the field and laboratory programs at the museum, in addition to managing vertebrate fossil collections. In particular, the numerous successes of our field expeditions in Grand Staircase–Escalante National Monument—one of the most difficult places to do paleontology that I have ever seen or heard about—have occurred under Mike's capable direction.

3. Additional funding was subsequently secured from several sources, including the National Science Foundation, the Discovery Channel, and the University of Utah. However, the bulk of the funds have come from the Bureau of Land Management (BLM),

thanks in large part to the efforts of Alan Titus, Grand Staircase–Escalante National Monument, Laurie Bryant (BLM), and Scott Foss (BLM).

4. The Alf Museum of Paleontology, located in Claremont, California, is the only paleontology museum in the nation located on a high school campus. The complete *Gryposaurus* skull was found by a crew of students and educators from the school, and the leader of the expedition, Don Lofgren, kindly agreed to let us include the skull as part of our study.

5. Referring back to chapter 8, we know that it is possible, and perhaps likely, that some of these small coelurosaur theropods were omnivorous and/or, particularly in the case of the ornithomimosaurs, possibly even obligatory herbivores.

6. The formal scientific names for "West" and "East America" are Laramidia and Appalachia, respectively.

REFERENCES

Currie, P. J., and Russell, D. A. 2005. The geographic and stratigraphic distribution of articulated and associated dinosaur remains. Pp. 537–569 in P. J. Currie and E. B. Koppelhus (eds.), *Dinosaur Provincial Park: A Spectacular Ancient Ecosystem Revealed*. Bloomington: Indiana University Press.

Dodson, P. 1996. *The Horned Dinosaurs: A Natural History*. Princeton, NJ: Princeton University Press.

Gates, T., and Sampson, S. D. 2007. A new hadrosaur from the Upper Cretaceous (Campanian) Kaiparowits Formation of Utah. *Zoological Journal of the Linnean Society* 151:351–376.

Horner, J. R., Varricchio, D. J., and Goodwin, M. B. 1992. Marine transgressions and the evolution of Cretaceous dinosaurs. *Nature* 358:59–61.

Johnson, K., and Troll, K. 2007. *Cruisin' the Fossil Freeway*. Golden, CO: Fulcrum.

Lehman, T. M. 1987. Late Maastrichtian paleoenvironments and dinosaur biogeography in the western interior of North America, *Palaeogeography, Palaeoclimatology, Palaeoecology* 60:189–217.

———. 2001. Late Cretaceous dinosaur provinciality. Pp. 310–328 in D. H. Tanke and K. Carpenter (eds.), *Mesozoic Vertebrate Life*. Bloomington: Indiana University Press.

Roberts, E. M. 2007. Facies architecture and depositional environments of the Upper Cretaceous Kaiparowits Formation, southern Utah. *Sedimentary Geology* 197:207–233.

Russell, D. 1992. *An Odyssey in Time: The Dinosaurs of North America*. Toronto: University of Toronto Press.

Ryan, M. J., and Evans, D. C. 2005. Ornithischian dinosaurs. Pp. 312–348 in P. J. Currie and E. B. Koppelhus (eds.), *Dinosaur Provincial Park: A Spectacular Ancient Ecosystem Revealed*. Bloomington: Indiana University Press.

Sampson, S. D., Gates, T. A., Roberts, E. M., Getty, M. A., Zanno, L. E., Loewen, M. A., Smith, J. A., Lund, E. K., Sertich, J. and Titus, A. L. In press. Grand Staircase–Escalante National Monument: A new and critical window into the world of dinosaurs. *Learning from the Land, Vol. 2*. Washington, DC: U.S. Department of the Interior, Bureau of Land Management.

Vrba, E. S. 1987. Ecology in relation to speciation rates: Some case histories of Miocene-Recent mammal clades. *Evolutionary Ecology* 1:283–300.

NOTES

1. Most biologists are confident we are now in the midst of the sixth great extinction, this one precipitated by humans. See the epilogue for more on this topic.
2. I must credit two individuals, Kirk Johnson and David Fastovsky, who helped inspire my conversion to "the other side" of this debate.
3. Shortly after the melting of Snowball Earth, large multicellular life exploded in diversity, a coincidence that has led some workers to suggest that the two events were causally linked (Hoffman and Schrag 2000). That is, the blossoming of life, like that of dinosaurs, may have occurred in the aftermath of a global catastrophe.
4. The quotation comes from Margulis 1999:123.
5. Unfortunately, the potential for correcting this system has its limits, which currently seem to be exceeded by the rapid emission of vast quantities of greenhouse gases by human industrialization.

REFERENCES

Alvarez, W. 1998. T. rex *and the Crater of Doom*. New York: Vintage Books.

Alvarez, L. W., Alvarez, W., Asaro, F., and Michel, H. V. 1980. Extraterrestrial cause for the Cretaceous-Tertiary extinction. *Science* 208:1095–1108.

Archibald, J. D. 1996. *Dinosaur Extinction and the End of an Era: What the Fossils Say*. New York: Columbia University Press.

Archibald, J. D., and Fastovsky, D. E. 2004. Dinosaur extinction. Pp. 672–684 in D. Weishampel, P. Dodson, and H. Osmólska (eds.), *The Dinosauria*. 2nd ed. Berkeley: University of California Press.

Chenet, A.-L., Courtillot, V., Fluteau, F., Besse, J., Subbaro, K. V., Khadri, S.,Bajpai, S., and Jay, A. 2005. Magnetostratigraphy of the upper formations of the Deccan Traps: An attempt to constrain the timing of the eruptive sequence. *Geological Society of America Abstracts* paper 37–6.

Courtillot, V. E. 1990. A volcanic eruption. *Scientific American* 263(4):85–82.

Crespi, B. J. 2004. Vicious circles: Positive feedback in major evolutionary and ecological transitions. *Trends in Ecology & Evolution* 19(12):627–633.

Fastovsky, D. E., Huang, Y., Hsu, J., Martin-McNaughton, J., Sheehan, P. M., and Weishampel, D. B. 2004. The shape of Mesozoic dinosaur richness. *Geology* 32:877–880.

Fastovsky, D. E., and Sheehan, P. M. 2005. The extinction of the dinosaurs in North America. *GSA Today* 15(3):4–10.

Hoffman, P. F., and Schrag, D. P. 2000. Snowball Earth. *Scientific American* 282:68–75.

Johnson, K. R. 1992. Leaf-fossil evidence for extensive floral extinction at the Cretaceous-Tertiary boundary, North Dakota, USA. *Cretaceous Research* 13:91–117, doi:10.1016/0195–6671(92)90029-P (published online).

Keller, G. Abramovich, S., Berner, Z., Pardo, A., and Adatte, T. 2008. Did volcanism and climate change cause the K-T mass extinction? *Geophysical Research Abstracts* 10: EGU-2008-A-04804.

Lovelock, J. E. 1987. *Gaia: A New Look at Life on Earth*. Oxford: Oxford University Press.

Margulis, L. 1999. *Symbiotic Planet: A New Look at Evolution*. New York: Basic Books.

Raup, D., and Sepkoski, J. 1982. Mass extinctions in the marine fossil record. *Science* 215:1501–1503.

Robertson, D. S., McKenna, M., Toon, O. B., Hope, S., and Lillegraven, J. A. 2004. Survival in the first hours of the Cenozoic. *Geological Society of America Bulletin* 116:760–768, doi: 10.1130/B25402.1 (published online).

Russell, D. A., and Manabe, M. 2002. Synopsis of the Hell Creek (uppermost Cretaceous) dinosaur assemblage. Pp. 169–176 in J. H. Hartman, K. R. Johnson, and D. J. Nichols (eds.), *The Hell Creek Formation and the Cretaceous-Tertiary Boundary in the Northern Great Plains: An Integrated Continental Record of the End of the Cretaceous*. Geological Society of America Special Paper 361. Boulder, CO: Geological Society of America.

Ryan, M. J., and Evans, D. C. 2005. Ornithischian dinosaurs. Pp. 312–348 in P. J. Currie and E. B. Koppelhus (eds.), *Dinosaur Provincial Park: A Spectacular Ancient Ecosystem Revealed*. Bloomington: Indiana University Press.

Sahney, S., and Benton, M. J. 2008. Recovery from the most profound mass extinction of all time. *Proceedings of the Royal Society London, Biological Sciences*, doi: 10.1098/rspb. 2007.1370 (published online).

Schneider, S. H., Miller, J. R., Crist, E., and Boston, P. J. (eds.). 2004. *Scientists Debate Gaia: The Next Century*. Cambridge, MA: MIT Press.

Schultz, P. H., and d'Hondt, S. L. 1996. Cretaceous-Tertiary (Chicxulub) impact angle and its consequences. *Geology* 24:963–967, doi: 10.1130/0091–7613(1996) 0242.3.CO;2 (published online).

Siggurdson, H., d'Hondt, S. L., and Carey, S. 1992. The impact of the Cretaceous/Tertiary bolide on evaporite terrane and generation of a major sulfuric acid aerosol. *Earth and Planetary Science Letters* 109:543–559.

Wang, S. C., and Dodson, P. 2006. Estimating the diversity of dinosaurs. *Proceedings of the National Academy of Sciences, USA* 103(37):13601–13605.

EPILOGUE WHISPERS FROM THE GRAVE

NOTES

1. For more on the Center for Ecoliteracy, go to www.ecoliteracy.org.
2. An outstanding leader in this work is the Cornell Lab of Ornithology (www.birds.cornell.edu/).
3. The Madagascar Ankizy Fund is a 501(c)3 not-for-profit education corporation established through the Stony Brook Foundation. For those interested in learning more or getting involved, go to www.ankizy.org.

REFERENCES

Alters, B. J., and Nelson, C. E. 2002. Perspective: Teaching evolution in higher education. *Evolution* 56(10):1891–1901.

Berry, T. 1990. *The Dream of the Earth*. San Francisco: Sierra Club Books.

Capra, F. 1999. *Ecoliteracy: The Challenge for Education in the Next Century*. Liverpool Schumacher Lectures. Center for Ecoliteracy, www.ecoliteracy.org (published online).

Dowd, M. 2007. *Thank God for Evolution! How the Marriage of Science and Religion Will Transform Your Life and Our World*. San Francisco: Council Oak Books.

Evans, E. M. 2005. *Teaching and Learning about Evolution*. Chap. 3 in J. Diamond (ed.), *Virus and the Whale: Exploring Evolution in Creatures Small and Large*. Arlington, VA: National Science Teachers Association Press.

Fosnot, C. T. 1996. Constructivism: A psychological theory of learning. Pp. 8–33 in C. T. Fosnot (ed.), *Constructivism: Theory, Perspectives and Practice*. New York: Teachers College Press.

Louv, R. 2006. *Last Child in the Woods: Saving Our Children from Nature-Deficit Disorder*. Chapel Hill, NC: Algonquin Books.

Orr, D. W. 1992. *Ecological Literacy: Education and the Transition to a Postmodern World*. Stony Brook: State University of New York Press.

———. 1994. *Earth in Mind: On Education, Environment, and the Human Prospect*. Washington, DC: Island.

Pimm, S. L. 2001. *The World According to Pimm: A Scientist Audits the Earth*. New York: McGraw-Hill.

Raven, P. 2002. Science, sustainability, and the human prospect. *Science* 297:954–859.

Sampson, S. D. 2006. Evoliteracy. Pp. 216–231 in J. Brockman (ed.), *Intelligent Thought: Science versus the Intelligent Design Movement*. New York: Knopf.

Stone, M. K., and Barlow, Z. 2005. *Ecological Literacy: Educating Our Children for a Sustainable World*. Berkeley: University of California Press.

Suzuki, D. 1997. *The Sacred Balance: Rediscovering Our Place in Nature*. Vancouver: Greystone Books.

Swimme, B., and Berry, T. 1992. *The Universe Story: From the Primordial Flaring Forth to the Ecozoic Era: A Celebration of the Unfolding of the Cosmos*. San Francisco: Harper.

Wilson, E. O. 1984. *Biophilia: The Human Bond with Other Species*. Cambridge, MA: Harvard University Press,

———. 2002. *The Future of Life*. New York: Knopf.

GLOSSARY

ABELISAURS Carnivorous dinosaurs of the theropod dinosaur group Abelisauroidea, including small-bodied "noasaurids" and midsized "abelisaurids."

ADAPTATION A process of genetic change caused by natural selection operating within a population, resulting in one or more characters becoming better suited to function in a particular environment. (*Adaptation* is also used as a noun to refer to a particular feature that has become more widespread within a population because of some selective advantage.)

AETOSAURS Heavily armored, plant-eating archosaurs of the Triassic Period.

AFRICA-FIRST HYPOTHESIS The idea that Africa was the first of the Southern Hemisphere landmasses to become isolated in the fragmentation of Gondwana, allowing dinosaurs and other organisms to move between and among the remaining southern landmasses for millions of years.

AIR SAC An air-filled space that forms a connection between the respiratory system and the body, often invading bones. Air sacs, or pneumaticity, can cause the formation of sinuses in the skull or large spaces within the vertebrae, among other locations.

ALLOSAURS A major subgrouping of predatory theropod dinosaurs—formally known as "allosauroids"—from the Jurassic and Cretaceous, including allosaurids (e.g., *Allosaurus*), carcharodontosaurids (e.g., *Giganotosaurus*), and sinraptorids (e.g., *Sinraptor*). The name is sometimes also used informally to refer to members of the genus *Allosaurus*.

AMNIOTES Tetrapods that produce complex shelled eggs protected by multiple membranes. Amniotes include reptiles (including dinosaurs and birds) and mammals and their closest extinct relatives.

ANGIOSPERMS Flowering plants, a specialized group of seed-bearing gymnosperms that first appeared in the Early Cretaceous and diversified to become the dominant plants on Earth.

ANKYLOSAURS Plant-eating thyreophoran dinosaurs with heavily armored bodies and heads, including "nodosaurids"—with narrow heads, shoulder spikes, and no tail club—and "ankylosaurids"—with broader skulls, no shoulder spikes, and tail clubs.

ANTORBITAL OPENING An opening on the side of the skull of dinosaurs and other archosaurs, just in front of the eye, that was filled with an air sac during life. Known formally as the antorbital fenestra, this opening housed the antorbital cavity.

ARCHOSAURS Members of the group Archosauria, which includes birds, crocodiles, pterosaurs, and dinosaurs.

ATMOSPHERE A thin layer of gases surrounding planets and other celestial bodies of sufficient mass. Used in the text to refer to the air-based component of the Earth system, with the remaining spheres being the biosphere, geosphere, and hydrosphere.

ATP Adenosine triphosphate, the compound that cells use to store energy, regarded as the universal fuel of life.

BACKGROUND EXTINCTION Regularly occurring, isolated extinctions of individual species (in contrast to mass extinctions).

BACTERIA A very large group of single-celled microorganisms, usually only a few thousandths of a millimeter in length, that come in a variety of shapes, including spheres, rods, and spirals.

BEAK The sheath of keratin that covers the external portions of the upper and lower jaws in birds, and presumably covered the front end of the jaws in many dinosaurs. Also used to refer to the bony elements of the jaw covered in keratin.

BENNETTITES A group of seed plants bearing large palmlike leaves that was common in the Paleozoic and Mesozoic eras, but went extinct by the end of the Cretaceous Period.

BIOMASS The mass of living organisms in a given place (usually an ecosystem) at a given time.

BIOMINERALIZATION The process by which living organisms produce minerals, often to harden existing tissues. Biomineralization may play a key role in the process of fossilization.

BIOSPHERE The global sum of all ecosystems living on Earth at any given time. Used in the text to refer to the life-based component of the Earth system, with the remaining spheres being the atmosphere, geosphere, and hydrosphere.

BIPEDALISM The form of terrestrial locomotion that involves walking (and running) only on the hind legs.

BLITZKRIEG HYPOTHESIS A term coined in this book to refer to the hypothesis that the mass extinction 65.5 million years ago was caused by multiple factors: sea-level rise, volcanism, and an asteroid impact.

BONEBED A fossil locality that includes remains of at least two, and often tens or hundreds of, vertebrate individuals within a single geologic layer (stratum).

CAMBRIAN EXPLOSION The rapid appearance and diversification of complex visible life-forms at the beginning of the Cambrian Period, about 530 million years ago.

CARCHARODONTOSAURS A subgrouping of predatory theropod dinosaurs that includes some of the largest land-living carnivores known.

CARNIVORE A flesh-eating animal.

CENOZOIC ERA The "Age of New Life," often called the Age of Mammals, from 65.5 million years ago to the present.

CENTROSAURINES Short-frilled ceratopsid dinosaurs, or centrosaurs, typically with a deep snout, a large horn over the nose, and a highly ornate frill at the rear of the skull.

CERATOPSIANS The plant-eating horned dinosaurs, including primitive forms like psittacosaurs and intermediate forms like *Protoceratops*, as well as the larger and more specialized ceratopsids.

CERATOPSIDS The subgrouping of ceratopsians (horned dinosaurs) that includes centrosaurs and chasmosaurs.

CERATOSAURS Carnivorous dinosaurs of the theropod group Ceratosauria, including *Ceratosaurus* and the abelisaurs.

CHARACTER In biology, an isolated, abstracted feature of an organism. Specialized characters shared between two or more species are used as the basis of cladistic analysis.

CHASMOSAURINES The long-frilled ceratopsid dinosaurs, or chasmosaurs, typically with a small horn over the nose, large horns over the eyes, and an elongate frill with minimal ornamentation.

CHLOROPLASTS Organelles found in plant cells and some algae that capture solar energy through photosynthesis.

CHNOPS An acronym shorthand for remembering six elements essential to life: carbon, hydrogen, nitrogen, oxygen, phosphorus, and sulfur.

CLADE A group of organisms composed of an ancestor and all of its descendants.

CLADISTICS A quantitative method of assessing historical relationships of species based on evolutionary ancestry and the transmission of shared, specialized characters. Also known as phylogenetic systematics.

CLADOGRAM A hierarchical branching diagram connecting groups of species based on the shared possession of specialized features. Cladograms are used to graphically portray the results of cladistic analyses.

CLIMATE The long-term average of weather conditions over a large area.

COELOPHYSOIDS Carnivorous dinosaurs of the theropod group Coelophysoidea; the first major evolutionary radiation within theropods.

COELUROSAURS The advanced subgrouping of theropod dinosaurs that includes tyrannosaurs, ornithomimosaurs, and maniraptors.

COLORADO PLATEAU A large region of intermountain plateaus roughly centered on the Four Corners region of the southwestern United States. This sparsely vegetated area includes abundant dinosaur remains.

CONIFERS A group of woody seed plants with leaves that often take the form of needles.

CONTINENTAL CRUST The granite-rich rocks that form the continents and a portion of the lithosphere; continental crust is less dense than both oceanic crust and the underlying mantle.

CONVECTION The movement of molecules within fluids, often involving "convection cells," in which warmer, less dense fluid rises and cooler, denser fluid descends.

CONVERGENT EVOLUTION The acquisition of similar biological traits in distantly related lineages—for example, the wings of bats, birds, and butterflies.

COPROLITE Fossilized animal feces.

COSMOPOLITANISM In biogeography, the widespread distribution of an evolutionary lineage (or collection of lineages) over a relatively large area, sometimes encompassing two or more landmasses.

CRETACEOUS INTERIOR SEAWAY The shallow sea that inundated the central region of North America during the Late Cretaceous, extending from the Arctic Ocean in the north to the Gulf of Mexico in the south.

CRETACEOUS PERIOD The third and final period of the Mesozoic Era, from 144 to 65.5 million years ago. The Cretaceous is divided into the Early Cretaceous and Late Cretaceous epochs.

CRYPSIS In biology, the ability of an organism to avoid detection, often involving the use of camouflage among large-bodied animals.

CYANOBACTERIA Bacteria that obtain their energy through photosynthesis; also known as blue-green algae.

CYCADS A group of seed plants with short, stout trunks and long, pinnate leaves.

DECCAN TRAPS One of the largest volcanic features on Earth; formed on the Indian subcontinent at the end of the Cretaceous Period and implicated by some investigators as a key factor in the K-T mass extinction.

DEEP TIME Geologic time, measured in millions and billions of years rather than thousands.

DENTAL BATTERY Rows of closely packed teeth in the "cheek" region of the upper and lower jaws of certain dinosaurs (e.g., hadrosaurs, ceratopsids, some sauropods); used to slice or crush plant matter.

DIGITIGRADE Standing and walking on the digits, or toes, with the ball of the foot held off the ground. Present in dinosaurs, birds, and most mammals (but not humans).

DISPLAY In biology, a structure or behavior used by animals to "show off" to other animals. Displays may be used to attract mates, intimidate rivals, or deter predators, among other functions.

DIVERSITY In biology, the number and variety of species present in a given area.

DNA Deoxyribonucleic acid, which forms molecules that contain the genetic instructions used in the development and functioning of all life-forms on Earth.

DROMAEOSAURS Carnivorous theropod dinosaurs with well-developed forelimbs and a sickle-claw on the second toe of each foot; one of the major maniraptor lineages.

ECOLOGY The study of the interactions of organisms with their environments and with other organisms.

ECTOTHERMY Regulation of body temperature largely by exchanging heat with the surrounding environment. Associated with slower, "cold-blooded" metabolisms. The opposite is endothermy.

ELECTRON A negatively charged subatomic particle that resides in a cloud of equivalent particles around the nucleus of an atom.

ENDEMISM In biogeography, the narrow distribution of an evolutionary lineage (or collection of lineages) within a relatively small area.

ENDOTHERMY Regulation of body temperature by generation of internal heat. Associated with faster, "warm-blooded" metabolisms. The opposite is ectothermy.

EON The second-largest division of geologic time (e.g., Phanerozoic Eon).

EPOCH Subdivision of a period of geologic time, typically several million years in duration (e.g., Pleistocene Epoch).

ERA Subdivision of an eon of geologic time, measured in tens to hundreds of millions of years (e.g., Mesozoic Era) and composed of periods.

EUKARYOTES The category of life that includes animals, plants, fungi, and protists, whose cells contain a nucleus and other complex organelles.

EVOLUTION In biology, descent with modification; heritable changes in lineages of organisms through time. Also used to refer to the transformational history of the universe.

EVOLUTIONARY RADIATION The rapid origin and diversification of many species within a single evolutionary lineage.

EXTINCTION The death of one or more species, which occurs when the birthrate of organisms does not keep pace with the death rate.

FAUNA A group of animals that lives together within a region.

FEEDBACK LOOP Loops of information that make use of circular control, monitoring performance and making adjustments as necessary. Negative feedback loops tend to bring a system back to a previous stable state, whereas positive feedback loops tend to modify systems until a new stable state is reached.

FEMUR The upper leg (thigh) bone, common to most land-living vertebrates, including dinosaurs.

FERNS One of the major groups of spore-bearing pteridophyte plants.

FIBROLAMELLAR BONE Rapidly growing bone with a woven, fibrous appearance, often associated with endothermy in large vertebrates.

FOSSIL The remains of a living organism or its traces preserved in the rock record.

FRILL In horned dinosaurs, the often ornamented sheet of bone extending rearward from the skull and composed of the parietal and squamosal bones.

FUNGI A major group (often called a "kingdom") of eukaryote life-forms that reproduces mostly via spores and includes molds, yeasts, and mushrooms.

FURCULA The fused clavicles (collarbones) that comprise the "wishbone" of birds and some nonavian theropods.

GAIA HYPOTHESIS The idea that life (the biosphere) interacts with physical components of Earth to form feedback loops that, among other things, regulate the temperature and composition of the atmosphere, thereby making the planet habitable for life over deep time. Some argue that this idea has matured into "Gaia theory."

GASTROLITH Gizzard stone, or rock, swallowed by an animal to aid in digestion.

GENUS (GENERA) A low-level taxonomic rank used in the classification of living and extinct organisms. Every species is given a two-part name composed of genus and species (e.g., *Tyrannosaurus rex*).

GEOLOGY The study of Earth, its history, and its structure.

GEOSPHERE The land-based component of the Earth system, including the planet itself. The three remaining "spheres" are the atmosphere, biosphere, and hydrosphere.

GIGANTOTHERMY In ectothermic ("cold-blooded') animals, maintenance of relatively constant internal body temperatures by virtue of larger body size (as opposed to higher metabolic rates).

GINKGOS A group of seed plants with characteristic, fan-shaped leaves.

GOLDILOCKS HYPOTHESIS A term coined in this book to refer to the idea that most dinosaurs possessed intermediate metabolic rates (neither ectothermic, nor endothermic, but "mesothermic") that evolved in order to increase the amount of energy available for "building" activities such as growth, reproduction, and fat storage.

GONDWANA The southern supercontinent that formed after the initial breakup of the supercontinent Pangaea. Gondwana eventually fragmented into smaller landmasses that today include Africa, South America, Australia, Antarctica, Madagascar, and the Indian subcontinent.

GRADIENT In earth science and fluid dynamics, a difference (e.g., in density, pressure, or temperature) across a distance. The process of convection (whether in the oceans or Earth's mantle) is driven by temperature gradients, whereas the "green gradient" in the text refers to the abundant solar energy stored in plants that fuel terrestrial ecosystems.

GRADUALISM The hypothesis advocated by Darwin and many subsequent biologists that evolution occurs in slow, ongoing steps of small increments.

GYMNOSPERMS Seed plants, with the major Mesozoic groups consisting of cycads, ginkgos, conifers, bennettites, and angiosperms. All but the bennettites remain alive today.

HADROSAURINES The non-crested duck-billed dinosaurs, or hadrosaurs.

HADROSAURS Duck-billed dinosaurs, a group of plant-eating ornithopods from the Cretaceous.

HAVERSIAN CANALS A series of narrow tubes surrounded by dense, compact bone that contain blood vessels and nerves.

HERBIVORE A plant-eating animal.

HETERODONTOSAURS A group of primitive, small-bodied, plant-eating ornithopod dinosaurs.

HIP SOCKET A depressed region on the pelvis where the femur (thigh bone) articulates.

HOMEOTHERMY Thermoregulation in which a stable internal body temperature is maintained regardless of external conditions.

HOMOLOGY (HOMOLOGIES) In evolutionary biology, any similarity between two characteristics that is the result of shared ancestry. For example, the wings of bats and the arms of primates are homologous in that both are made of the same bones, although modified to serve very different functions.

HORSETAILS One of the major groups of spore-bearing pteridophyte plants.

HUMERUS The upper arm bone, common to most land-living vertebrates, including dinosaurs.

HYDROSPHERE The water-based component of the Earth system. The three remaining "spheres" discussed in the text are the geosphere, atmosphere, and biosphere.

HYPOTHESIS (HYPOTHESES) An idea that represents one possible answer to a question about a pattern observed in nature. Science largely revolves around the creation and testing of hypotheses.

HYPSILOPHODONTS A group of small-bodied, plant-eating ornithopods that existed from the Middle Jurassic through the Late Cretaceous.

IGUANODONTS A subgrouping of mid- to large-sized plant-eating ornithopod dinosaurs that lived from the Middle Jurassic through the Late Cretaceous.

ICHTHYOSAURS Dolphin-like marine reptiles of the Mesozoic Era.

INVERTEBRATES Animals lacking a vertebral column, including about 98 percent of present animal diversity.

ISOTOPE Any of two or more forms of a chemical element that have the same number of protons in the nucleus but differing numbers of neutrons.

JACKET In paleontology, a rigid, protective covering—usually made of burlap strips soaked in plaster—used to encase fossils for safe removal from a quarry site.

JURASSIC PERIOD The second period of the Mesozoic Era, from 206 to 144 million years ago. The Jurassic Period is divided into Early, Middle, and Late Jurassic epochs.

KINGDOM Traditionally the highest ranking category in biological taxonomy. Some biologists argue for the presence of five kingdoms (prokaryotes, protists, fungi, plants, and animals) whereas another recent scheme recognizes six.

K-T MASS EXTINCTION The mass extinction at the boundary of the Cretaceous (K) and the Tertiary (T) that extinguished the dinosaurs and many other lineages of animals and plants.

LAG An acronym for "line of arrested growth," in reference to growth lines present in the bones of many animals, including some dinosaurs.

LAMBEOSAURINES The crested duck-billed dinosaurs, or hadrosaurs.

LAMELLAR ZONAL BONE Slow-growing bone typically characterized by the presence of distinct growth rings, or lines of arrested growth (LAGs).

LAURASIA The northern supercontinent that formed after the initial breakup of the supercontinent Pangaea. Laurasia fragmented into smaller landmasses, including North America and Eurasia (Europe and Asia).

LITHOSPHERE The crust and uppermost mantle of Earth.

LYCOPODS A group of primitive, spore-bearing plants that tend to be evergreen and mosslike.

MAINTENANCE In animal physiology, energy devoted to activities that maintain the body, such as cell renewal, heat generation, and eating.

MANIRAPTORA The subgrouping of feathered, coelurosaur theropod dinosaurs that includes therizinosaurs, oviraptorosaurs, deinonychosaurs, and aves (birds).

MANTLE The highly viscous layer of Earth directly beneath the crust and above the outer core.

MARGINOCEPHALIA The "margin-headed" dinosaurs, including pachycephalosaurs and ceratopsians.

MASS EXTINCTION Rare, large-scale extinction events that extinguish numerous species and groups of species around the globe over a relatively brief span of time. Five mass extinctions are recognized for the last 500 million years, and humans may be precipitating the sixth such event of this magnitude.

MATRIX In paleontology, the rock that surrounds fossilized bones.

MESOTHERMY As used in this book, a hypothetical metabolic category applied to animals with physiologies intermediate between ectotherms and endotherms. Many dinosaurs may have been "mesotherms."

MESOZOIC ERA The "Age of Middle Life," often called the "Age of Dinosaurs," from 248 to 65.5 million years ago. Major subdivisions are the Triassic, Jurassic, and Cretaceous periods.

METABOLISM The chemical activity of cells necessary for maintaining life—in particular, the flow of energy and nutrients.

METOPOSAURS Small to midsized predatory amphibians that lived alongside dinosaurs.

MIDOCEAN RIDGE A deep-ocean mountain range, typically with a rift running along its axis, formed by plate tectonics.

MITOCHONDRIA Organelles, or cell parts, in eukaryotic cells that convert food into chemical energy using cellular respiration and ATP.

MOSASAURS Large marine lizards of the Cretaceous Period.

MYCORRHIZAE A symbiotic association of certain fungi with plant roots, enabling plants to increase their uptake of nutrients from the soil.

NATURAL SELECTION A natural process that increases the survival and reproductive success of some organisms over others within a species, based on the possession of inherited characteristics best suited to that particular environment.

NEUTRON An electrically neutral subatomic particle located in the nucleus of an atom.

NICHE In ecology, the role an organism (or population or species) occupies within its ecosystem.

NONAVIAN DINOSAURS All dinosaurs except birds.

OCEANIC CRUST The portion of Earth's crust, or lithosphere, that occurs beneath oceans and is denser than continental crust.

OMNIVORE An animal that eats both plants and other animals as primary food sources.

ORGANELLE A specialized structure within an eukaryotic cell that has a specific function. Examples include the mitochondria of almost all eukaryotes and the chloroplasts of plants and algae.

ORNITHISCHIA The bird-hipped dinosaurs and one of the two great branches within Dinosauria, including ornithopods, ankylosaurs, stegosaurs, ceratopsians, and pachycephalosaurs.

ORNITHOMIMOSAURS The ostrich-mimic dinosaurs, mostly small to midsized forms that form one major branch of coelurosaur theropods.

ORNITHOPODS The "bird-foot" dinosaurs, one of the major branches of the plant-eating Ornithischia, including heterodontosaurs, hypsilophodonts, iguanodonts, and hadrosaurs.

OSTEODERM Bones that form within the skin of many dinosaurs and other reptiles, taking such forms as scales, plates, and spikes.

OVERBURDEN The sediment above a fossil bone layer that must be removed prior to commencing excavation of the fossils.

OVIRAPTOROSAURS A major branch of the feathered maniraptor theropod dinosaurs, often referred to informally as "oviraptors."

PACHYCEPHALOSAURS The dome-headed dinosaurs, small- to large plant-eating, bipedal ornithischians.

PALEOBOTANY The study of prehistoric plants.

PALEONTOLOGY The study of all prehistoric life (as distinct from archaeology, the study of human cultures through analysis and interpretation of cultural remains).

PALEOZOIC ERA The "Age of Ancient Life," from 545 to 225 million years ago. Life actually preceded the Paleozoic by billions of years, but this era does mark the dramatic and diverse beginnings of visible life-forms.

PANGAEA The supercontinent that formed near the end of the Paleozoic Era and fragmented during the Mesozoic Era.

PAN-GONDWANA HYPOTHESIS The idea that major groups of dinosaurs and other vertebrates spread across Southern Hemisphere landmasses (Africa, South America, Antarctica, and Indo-Madagascar) either prior to the fragmentation of Gondwana or about 95 million years ago, the latter involving three temporary land bridges.

PANTHALASSA The "superocean" that surrounded the supercontinent Pangaea at the end of the Paleozoic Era and the beginning of the Mesozoic Era.

PARADIGM SHIFT A radical change in thinking within a scientific field from an older, widely accepted viewpoint to a new one, forcing reconsideration of long-held ideas.

PELVIS The hip bones of most vertebrates, including the ilium, ischium, and pubis.

PERIOD In geology, a subdivision of an era, consisting of tens of millions of years (e.g., Jurassic Period).

PHANEROZOIC EON The present eon of Earth history, composed of the Paleozoic, Mesozoic, and Cenozoic eras (about 540 million years).

PHOTOSYNTHESIS The process in green plants, as well as some bacteria and protists, by which sunlight, water, and carbon dioxide are combined to produce oxygen and sugar (energy).

PHYLOGENETIC SYSTEMATICS See *cladistics*.

PHYSIOLOGY The biological study of the functions of organisms and their parts.

PHYTOSAURS Long-snouted, aquatic, fish-eating archosaurs of the Late Triassic that closely resemble present-day crocodiles.

PLANTIGRADE Standing and walking on the foot bones as well as the toes. Present in humans and bears, and in contrast to the digitigrade stance and gait of dinosaurs, birds, and most mammals.

PLATE TECTONICS The theory that describes large-scale motions of Earth's lithospheric plates.

PLESIOSAURS Long-necked fish-eating reptiles with large flippers that lived in Mesozoic seas.

PNEUMATICITY In zoology, the presence of air spaces within bones caused by the bone-consuming activity of pneumatic epithelium. Examples include the facial sinuses of humans (and dinosaurs) and the hollowed-out vertebrae of sauropod dinosaurs and birds.

POIKILOTHERMY Thermoregulation in which an animal's internal body temperature fluctuates in response to changing external conditions. Present in most cold-blooded (ectothermic) animals.

PREDENTARY An accessory, beak-forming bone found at the front of the lower jaw in plant-eating ornithischian dinosaurs.

PRIMARY PRODUCTION The sum total of organic matter synthesized by organisms from sunlight and inorganic materials. All life-forms depend directly or indirectly on primary production.

PRODUCTION In animal physiology, energy devoted to "building" activities, particularly growth, fat storage, and reproduction.

PROKARYOTES Unicellular organisms that lack a nucleus or any other membrane-bound organelles. Traditionally, all prokaryotes were regarded as bacteria, but a recent scheme splits the group into bacteria and life-forms called Archaea.

PROSAUROPODS One of two major branches of sauropodomorph dinosaurs, including a variety of mid- to large-sized herbivores and the first major evolutionary radiation of plant-eating dinosaurs.

PROTISTS A highly diverse group of relatively simple unicellular and multicellular eukaryotes that gave rise to fungi, plants, and animals.

PROTON A subatomic particle with an electric charge (+1) located in the nucleus of an atom.

PSITTACOSAURS A primitive group of bipedal ceratopsian (horned) dinosaurs known only from Cretaceous-aged rocks in Asia.

PTERIDOPHYTES Spore-bearing plants, predominantly ferns, horsetails, and lycopods (club mosses).

PTEROSAURS A group of volant archosaurs closely related to dinosaurs (though not actually dinosaurs themselves), and the first vertebrates to evolve powered flight.

PUBIS One of three major bones in the vertebrate pelvis. In most saurischian (lizard-hipped) dinosaurs, the pubis points downward and forward, whereas in ornithischian (bird-hipped) dinosaurs, it is directed rearward.

PUNCTUATED EQUILIBRIUM The theory that sexually reproducing species change little during most of their durations, with evolutionary change generally concentrated in brief pulses around the formation of new species.

QUADRUPEDALISM The form of terrestrial locomotion in animals that involves four limbs or legs.

RADIATION See *evolutionary radiation*.

RAUSUICHIANS A group of mid- to large-sized predatory archosaurs that lived alongside dinosaurs during the Late Triassic.

REDUCTIONISM The idea that life is best understood through an examination of ever-smaller parts; the dominant theme of science for almost four centuries.

REGRESSION In geology, a period of falling sea level, resulting in an increase in terrestrial habitats.

SAURISCHIA The lizard-hipped dinosaurs, one of the two great branches within Dinosauria, including sauropodomorphs and theropods.

SAUROPODOMORPHS The group of plant-eating saurischian dinosaurs that includes prosauropods and sauropods.

SAUROPODS The giant, long-necked saurischian dinosaurs, including the largest animals to walk on land.

SEAFLOOR SPREADING The process by which oceanic crust is formed at midocean ridges and gradually moves away from the ridges.

SEXUAL DIMORPHISM Systematic differences in form between animals of different sex within the same species. Examples include sex-related variations in color and size, as well as the presence of elaborate structures such as horns, antlers, and tusks.

SEXUAL SELECTION A form of natural selection in which some organisms experience greater reproductive success than others of the same species because of one or more inherited characteristics.

SILVER BULLET HYPOTHESIS A term coined in this book to refer to the idea that the K-T mass extinction was caused by a single, catastrophic asteroid impact.

SPECIATION The evolutionary process by which new biological species are formed.

SPECIES In biology, a group of populations in which members can interbreed or potentially interbreed under natural conditions. In taxonomy, the smallest category of living things normally recognized.

SPINOSAURS A group of specialized giant predatory theropods characterized by elongate, crocodile-like skulls and conical teeth.

STEGOSAURS Plant-eating thyreophoran dinosaurs bearing distinctive combinations of bony armor plates and/or spikes arrayed along the midline of the back and tail.

SUBDUCTION The process that occurs at the junction of two colliding tectonic plates, in which one plate slides beneath the other and its material is shunted downward to the mantle.

SYMBIOSIS A prolonged association between two or more different organisms of different species, usually to the mutual benefit of both members.

SYSTEMATICS The branch of biology dealing with the study of life's diversity on Earth, in particular the identification, classification, and historical relationships of species and groups of species.

TETANURAE The group of advanced theropod dinosaurs that includes allosaurs, spinosaurs, and coelurosaurs.

TETHYS SEA The sea (or ocean) that existed between the supercontinents of Laurasia and Gondwana prior to the opening of the Indian Ocean.

THEORY In science, a well-substantiated explanation of some general aspect of the natural world that is based on an interconnected set of hypotheses and has withstood numerous attempts at falsification.

THERAPSIDS The group of synapsids that includes mammals and many of their extinct close relatives.

THERIZINOSAURS A lineage of highly specialized maniraptoran theropods that appears to have undergone an evolutionary transition from carnivory (or at least omnivory) to herbivory.

THERMOREGULATION The ability of an organism to keep its body temperature within certain boundaries, either through physiological or behavioral means.

THEROPODS One of the major branches of the lizard-hipped dinosaurs (Saurischia), including all carnivorous dinosaurs, as well as a range of likely omnivores and herbivores.

THYREOPHORANS The armored ornithischian dinosaurs, including stegosaurs, ankylosaurs, and several other primitive forms.

TITANOSAURS A diverse group of sauropod dinosaurs that achieved a nearly global distribution during the Cretaceous and included some of the largest land animals.

TRANSGRESSION In geology, a period of rising sea level, resulting in a decrease in terrestrial habitats.

TRIASSIC PERIOD The first period of the Mesozoic Era, lasting from about 248 to 206 million years ago. The Triassic Period is divided into Early, Middle, and Late Triassic epochs. Dinosaurs first appeared in the Late Triassic.

TROODONTS A group of sickle-clawed maniraptoran theropods that together with dromaeosaurs form the Deinonychosauria.

TROPHIC LEVEL The feeding level occupied by a group of organisms in a food chain.

TROPHIC PYRAMID The pyramid shape that results from having the bulk of an ecosystem's biomass at the lowest trophic level (plant producers) and successively lesser amounts at each higher (trophic) level (i.e., primary consumers, secondary consumers, tertiary consumers, etc.).

TURNOVER PULSE The idea that ecosystems remain relatively stable for long periods, yet periodically experience environmental disturbances that trigger widespread, nearly simultaneous extinctions and the origin of new species.

TYRANNOSAURS The group of coelurosaur theropods that includes *Tyrannosaurus* and other large to giant-sized carnivorous dinosaurs.

UNSTABLE ISOTOPE An isotope that decays spontaneously from an unstable form to a stable form. Unstable isotopes are critical to radioactive isotope age dating.

VERTEBRA (VERTEBRAE) A bone of the spinal, or vertebral, column of vertebrate animals. Among other things, vertebrae provide a bony canal for passage of the spinal cord as well as an area for attachment of many back muscles.

VERTEBRATES The group of animals possessing bony vertebrae along the back and tail, including fishes, amphibians, reptiles (including dinosaurs and birds) and mammals.

WEST AMERICA A term coined in this book to refer to the western North American landmass (otherwise known as Laramidia) formed by incursion of the Cretaceous Interior Seaway across the central portion of the continent.

ART AND PHOTOGRAPH CREDITS

American Geological Institute, © 2002 American Geological Institute and used with their permission, figure 1.1

David Backer, figure 9.1 (upper left and right image)

Fossil Museum, figure 9.1 (lower right image)

ReBecca Hunt-Foster, plate 14

Kirk Johnson, © Kirk Johnson, figure 9.2

Ashutush Kaushesh, Stony Brook University, figure EP.1

Marjorie Leggitt, figure 5.3

Mark Loewen, plates 3, 7, 8; figures 1.2, 2.1, 3.1, 3.4, 4.1, 4.2, 4.3, 4.4, 6.2, 8.2, 11.1, 11.2, 12.1, 12.3, 14.1, 14.2, 14.3, 14.5, 14.6, 15.1, 15.2

Lukas Panzarin, figures 6.1, 7.1, 7.2, 7.3, 7.4, 7.5, 8.1, 8.3, 8.4, 10.1, 10.2, 12.2, 13.1, 13.2, 13.3, 14.4 (Note: *silhouettes* of *Allosaurus, Brachiosaurus, Dryosaurus, Edmontosaurus, Torvosaurus, Triceratops, Tyrannosaurus,* and *Velociraptor* were modified after those of Gregory Paul; silhouettes of *Supersaurus, Apatosaurus,* and *Barosaurus* were modified after those of Scott Hartman.)

Santiago Ramirez, figure 9.1 (lower left image)

Scott Sampson, plates 1, 2

Michael Skrepnick, frontispieces for all chapters plus epilogue; plates 4, 5, 6, 9, 11, 12, 13; figure 10.3

Society of Vertebrate Paleontology, © copyright 2009 by the Society of Vertebrate Paleontology (reprinted and distributed with permission of the Society of Vertebrate Paleontology), figure 1.3

Jelle Wiersma, plate 10

Lindsay E. Zanno, figures 2.2, 2.3, 3.2, 3.3, 5.1, 5.2, 11.3

INDEX

Page numbers followed by *f* refer to figures, *n* refer to notes, and *t* refer to tables.

mixed-species groups, 134–135
Monolophosaurus, 161
Montana, 72, 157, 172
Morrison Formation, 41*f*, 215–218, 219, 223, 225–231
mosasaurs, 43, 44*f*, 259, 270
  defined, 312
Muir Woods National Monument, 143–144
multicellular life forms, 28–29
mushrooms, 154
mutation, 92, 98
mycelial mats, 154
mycorrhizae, 80, 154
  defined, 312
Myhrvold, Nathan, 290*n*3

NASA, 267–268, 273
National Park Service, 235
natural selection
  Darwin's theory of, 93–97
  defined, 312
  evidence of, 92
  evolutionary turnover, 248
  sexual selection, 96–97
Nature, 37
Nesbitt, Sterling, 204, 205
nesting, 89, 140, 170–173
neutrons, 25
  defined, 312
niche, 80
  defined, 312
nitrogen, 80, 154
noasaurids, 59–60
nodosaurids, 49
nomenclature, 36–39
nonavian dinosaurs, 40, 89
  defined, 312
Norell, Mark, 204
North Africa, 67
North America, 58, 62, 67–68, 228, 240–244, 251
north-south provincialism hypothesis, 246, 249–250
nuclear winter scenario, 273–274

ocean currents, 66
oceanic crust, 54–57, 68
  defined, 312
O'Keeffe, Georgia, 203, 204
omnivores, 74, 139
  defined, 312
On the Origin of Species (Darwin), 22, 276
opportunism model of dinosaur origins, 207–209
opposable thumb, 41, 136
Ordovician Period, 29
organelles, 27
  defined, 312
ornamentation
  of bird-hipped dinosaurs, 49
  of ceratopsids, 50
  crests, 49
  domes, 9–10, 49, 50, 133, 167–168
  hooks, 239
  hypotheses for, 159–163
  as mating signal, 163–170
  plates, 49

sexual dimorphism, 165–166
spikes, 49, 133, 158, 201
Ornithischia
  body size, 114
  cosmopolitanism, 58
  defined, 312
  diet and feeding strategies, 114–118
  emergence of, 32
  evolutionary relationships, 45*f*
  features of, 43, 48–50
  geographic distribution, 210
  groups, 48
  identification of, 205
ornithomimosaurs
  defined, 312
  ecosystem role, 74
  family tree, 129
  features of, 138–139
  foot speed, 135
  Kaiparowits Formation evidence, 239
  relation to other theropods, 128*f*
  skull, 139*f*
Ornithomimus, 122*f*
ornithopods
  defined, 312
  features of, 48–49, 50
  hatchlings, 172
  skull variation, 117*f*
Oryctodromeus, 172
osteoderms, 49
  defined, 312
Ostrom, John, 4, 5, 6, 88–89, 137, 179
overburden, 36
  defined, 312
Oviraptor, 139*f*, 174*f*
oviraptorosaurs
  defined, 312
  egg-laying behavior, 171, 189
  family tree, 128*f*
  features of, 47, 139–140
  Hell Creek Formation evidence, 74
  Kaiparowits Formation evidence, 239
  nesting and parental care, 172, 174*f*
oxygen, 267–268
Oxygen Catastrophe, 27

pachycephalosaurs
  defined, 312
  emergence of, 32
  evolutionary relationships, 45*f*
  features of, 48–49, 50
  flank-butting behavior, 168
  head-butting behavior, 167–168
  Kaiparowits Formation evidence, 237
  on Madagascar, 9, 10
Pachycephalosaurus, 50, 74, 111*f*, 116*f*, 252
Pachyrhinosaurus, 134, 156*f*
pack hunting, 137–138, 227
Padian, Kevin, 162
paleobotany, 13
  defined, 312
paleontology
  defined, 2
  paradigm shift, 4–6

COMPOSITION: Michael Bass Associates, Inc.

TEXT: 9.5/14 Scala

DISPLAY: Scala, Scala Italic, Scala Caps

PRINTER AND BINDER: Friesens Corporation